2012 INTERNATIONAL BUILDING CODE® STUDY COMPANION

2012 International Building Code
Study Companion

ISBN: 978-1-60983-155-4

Cover Design:	Ricky Razo
Publications Manager:	Mary Lou Luif
Project Editor:	Roger Mensink
Illustrator/Interior Design:	Mike Tamai
Manager of Development:	Doug Thornburg

COPYRIGHT 2011

ALL RIGHTS RESERVED. This publication is a copyrighted work owned by the International Code Council. Without advance written permission from the copyright owner, no part of this book may be reproduced, distributed or transmitted in any form or by any means, including, without limitation, electronic, optical or mechanical means (by way of example and not limitation, photocopying, or recording by or in an information storage and retrieval system). For information on permission to copy material exceeding fair use, please contact: ICC Publications, 4051 W. Flossmoor Rd, Country Club Hills, IL 60478, Phone 888-ICC-SAFE (422-7233).

The information contained in this document is believed to be accurate; however, it is being provided for informational purposes only and is intended for use only as a guide. Publication of this document by the ICC should not be construed as the ICC engaging in or rendering engineering, legal or other professional services. Use of the information contained in this workbook should not be considered by the user to be a substitute for the advice of a registered professional engineer, attorney or other professional. If such advice is required, it should be sought through the services of a registered professional engineer, licensed attorney or other professional.

Trademarks: "International Code Council" and the "ICC" logo are trademarks of International Code Council, Inc.

Errata on various ICC publications may be available at www.iccsafe.org/errata.

First Printing: September 2011

Printed in the United States of America

TABLE OF CONTENTS

Study Session 1:
    2012 IBC Chapters 1, 34 and 35—Administration 1
    Quiz 21

Study Session 2:
    2012 IBC Chapter 3 and Sections 508 and 509—Use and Occupancy Classification........ 29
    Quiz 54

Study Session 3:
    2012 IBC Chapter 6— Types of Construction...................................... 63
    Quiz 80

Study Session 4:
    2012 IBC Chapter 5—General Building Heights and Areas 89
    Quiz 112

Study Session 5:
    2012 IBC Sections 701 through 705—Fire and Smoke Protection Features I............. 121
    Quiz 142

Study Session 6:
    2012 IBC Sections 706 through 712— Fire and Smoke Protection Features II............ 151
    Quiz 168

Study Session 7:
    2012 IBC Sections 713 through 719—Fire and Smoke Protection Features III 177
    Quiz 198

Study Session 8:
    2012 IBC Chapter 9—Fire Protection Systems 207
    Quiz 226

Study Session 9:
    2012 IBC Sections 1001 through 1007, 1011 and 1013—Means of Egress I 235
    Quiz 256

Study Session 10:
    2012 IBC Sections 1008 through 1010, and 1012—Means of Egress II 265
    Quiz 285

Study Session 11:

    2012 IBC Sections 1014 through 1019—Means of Egress III 393

    Quiz 309

Study Session 12:

    2012 IBC Sections 1020 through 1029—Means of Egress IV 317

    Quiz 340

Study Session 13:

    2012 IBC Chapter 11—Accessibility 349

    Quiz 365

Study Session 14:

    2012 IBC Chapter 4—Detailed Occupancy Requirements 373

    Quiz 395

Study Session 15:

    2012 IBC Chapters 14, 15 and 18—Exterior Wall Coverings, Roofs and Foundations 403

    Quiz 422

Study Session 16:

    2012 IBC Chapters 16, 17, 19, 21, 22 and 23—Special Inspections, Concrete, Masonry and Wood 431

    Quiz 449

Study Session 17:

    2012 IBC Chapters 8, 12, 25, and 30—Interior Finishes, Interior Environments, Gypsum Board and Elevators 457

    Quiz 477

Study Session 18:

    2012 IBC Chapters 24 and 26—Glazing, Skylights and Plastics 485

    Quiz 502

Answer Keys 511

INTRODUCTION

This study companion provides practical learning assignments for independent study of the provisions of the 2012 *International Building Code*® (IBC®). The independent study format affords a method for the student to complete the study program in an unlimited amount of time. Progressing through the workbook, the learner can measure his or her level of knowledge by using the exercises and quizzes provided for each study session.

The workbook is also valuable for instructor-led programs. In jurisdictional training sessions, community college classes, vocational training programs and other structured educational offerings, the study guide and the IBC can be the basis for code instruction.

All study sessions begin with a general learning objective, the specific sections or chapters of the code under consideration and a list of questions summarizing the key points of study. Each session addresses selected topics from the IBC and includes code text, a commentary on the code provisions and illustrations representing the provisions under discussion. Quizzes are provided at the end of each study session. Before beginning the quizzes, the student should thoroughly review the referenced IBC provisions, particularly the key points.

The workbook is structured so that after every question the student has an opportunity to record his or her response and the corresponding code reference. The correct answers are indicated in the back of the workbook in the answer key.

This study companion was initially developed by Douglas W. Thornburg, AIA, C.B.O., for the 2000 *International Building Code*. In this publication, he has updated and revised the material based on the 2012 International Building Code. Doug is currently Technical Director of Product Development for the International Code Council®. In addition to authoring numerous educational texts and resource materials, he instructs seminars nationally on the IBC. Doug has over 30 years experience in the application and enforcement of building codes.

Questions or comments concerning this workbook are encouraged. Please direct your comments to ICC at *studycompanion@iccsafe.org*.

About the International Code Council

The International Code Council® (ICC®) is a member-focused association dedicated to helping the building safety community and construction industry provide safe, sustainable and affordable construction through the development of codes and standards used in the design, build and compliance processes. Most U.S. communities and many global markets choose the International Codes®. ICC Evaluation Service (ICC-ES), a subsidiary of the International Code Council, has been the industry leader in performing technical evaluations for code compliance fostering safe and sustainable design and construction.

Headquarters: 500 New Jersey Avenue, NW, 6th Floor, Washington, DC 20001-2070

District offices: Birmingham, AL; Chicago, IL; Los Angeles, CA

1-888-422-7233

www. iccsafe.org

Enhance Your Study Experience

ICC's bonus online quiz is a new practice tool just right for you.

Your 2012 *International Building Code Study Companion* includes a helpful online practice quiz developed to help you retain information and prepare for exams. The 60-question online practice quiz includes features such as:

- Many new questions to supplement those in the study companion;
- Format similar to computer based exams, with point and click convenience;
- Review of the quiz available upon completion, including code references;
- Questions covering topics from all study sessions of the study companion; and
- Ability to repeat the quiz multiple times.

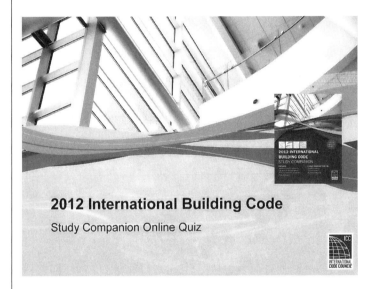

The study companion online quiz is beneficial when studying for the following 2012 Certification exams:

- Certified Building Official – Technology Module
- Commercial Building Inspector
- Building Plans Examiner
- Building Contractor

To take your online quiz, visit www.iccsafe.org/Quiz2012 and enter coupon code:

SM12002

2012 IBC Chapters 1, 34 and 35
Administration

OBJECTIVE: To obtain an understanding of the administrative provisions of the *International Building Code®* (IBC®), including the scope and purpose of the code, duties of the building official, issuance of permits, inspection procedures, special inspections, existing buildings and referenced standards.

REFERENCE: Chapters 1, 34 and 35, 2012 *International Building Code*

KEY POINTS:
- What is the purpose and scope of the *International Building Code*?
- What are the limitations for use of the *International Residential Code®* (IRC®)?
- When materials, methods of construction or other requirements are specified differently in separate provisions, which requirement governs?
- When there is a conflict between a general requirement and a specific requirement, which provision is applicable?
- When do the provisions of the appendices apply?
- What is the relationship between the I-Codes® and the referenced standards? How are any conflicts between the IBC and referenced standards to be addressed?
- What are the powers and duties of the building official in regard to the application and interpretation of the code? Right of entry?
- Under which conditions may the building official grant modifications to the code?
- How may alternative materials, designs and methods of construction be approved?
- What determinations must be made for the building official to grant modifications to provisions addressing flood hazard areas?
- When is a permit required? What types of work are exempted from permits?
- Is work exempted from a permit required to comply with the provisions of the code?
- What is the process outlined for obtaining a permit?
- Which documents must be submitted as part of the permit application process? What information is required on the construction documents?
- Which conditions or circumstances would bring the validity of a permit into question?
- When does a permit expire? What must occur when a permit expires prior to completion of a building?

KEY POINTS:
(Cont'd)

- For what is a registered design professional responsible? When is such an individual required?
- What submittal documents must be provided with each permit application? Under what conditions are submittal documents not required?
- What information is required on the construction documents? What special requirements apply to fire protection system shop drawings, means of egress layouts and exterior wall envelopes?
- What specific information must be included on a site plan?
- What is the process for the review and approval of construction documents? How are phased approvals to be addressed?
- What is the role of a design professional in responsible charge? When is such an individual to be utilized?
- What is a deferred submittal? What is the process for deferring the submittal of construction documents?
- What is the time limit of a permit for a temporary structure? To what criteria must a temporary structure comply?
- How are permit fees to be established? How is work started prior to permit issuance addressed?
- What types of inspections are specifically required by the code? When are inspections required?
- Are third-party inspection agencies permitted to perform the required inspections?
- Who is responsible for notifying the building official that work is ready for an inspection? Who must provide access for the inspection to take place?
- How shall an inspection be recorded? What if the inspection is failed?
- When is a certificate of occupancy required? For what reasons is revocation permitted?
- When may a temporary certificate of occupancy be issued?
- What is the purpose of a board of appeals? Who shall serve on the board?
- Which limitations are placed upon the authority of the board of appeals?
- When should a stop work order be issued?
- How must an existing building be viewed when additions or alterations are made?
- How does the *International Existing Building Code*® (IEBC®) relate to IBC Chapter 34?
- How are nonstructural repairs to an existing building to be handled? Glass replacement?
- What should happen when there is a change in use or occupancy of an existing building?
- How should moved structures be addressed?
- What accessibility issues are encountered when an existing building is altered or when its use changes?
- How can compliance alternatives be used to evaluate an existing building's degree of safety?
- What are referenced standards? How are they used in the IBC?

Topic: Scope
Reference: IBC 101.2
Category: Administration
Subject: General Requirements

Code Text: *The provisions of the* International Building Code *shall apply to the construction, alteration, relocation, enlargement, replacement, repair, equipment, use and occupancy, location, maintenance, removal and demolition of every building or structure or any appurtenances connected or attached to such buildings or structures.* See exception for dwellings and townhouses regulated by the *International Residential Code*.

Discussion and Commentary: The *International Building Code* is intended to regulate the broad spectrum of construction activities associated with buildings and structures. General provisions allow for a comprehensive overview of regulations; however, where more specific circumstances exist, any applicable specific requirements will take precedence.

Fundamental purposes of the provisions of the *International Building Code:*

- Safety of building occupants
- Stop panic
- Safety of fire fighters and emergency responders
- Safety and protection of others' property
- Safety and protection of own property

The appendices of the *International Building Code* address a diverse number of issues that may be of value to a jurisdiction in developing a set of construction regulations. It should be noted that the provisions contained in the appendices do not apply unless specifically adopted.

Topic: Dwelling Units
Reference: IBC 101.2, Exception

Category: Administration
Subject: General Requirements

Code Text: *Detached one- and two-family dwellings and multiple single-family dwellings (townhouses) not more than three stories above grade plane in height with a separate means of egress and their accessory structures shall comply with the* International Residential Code.

Discussion and Commentary: Many residential structures are exempt from the requirements of the *International Building Code* and are regulated instead by a separate and distinct document, the *International Residential Code*. The IRC® contains prescriptive requirements for the construction of detached single-family dwellings, detached duplexes, townhouses and all structures accessory to such buildings. Limited to three stories in height with individual egress facilities, these residential buildings are fully regulated by the IRC for building, energy, plumbing, mechanical and electrical provisions.

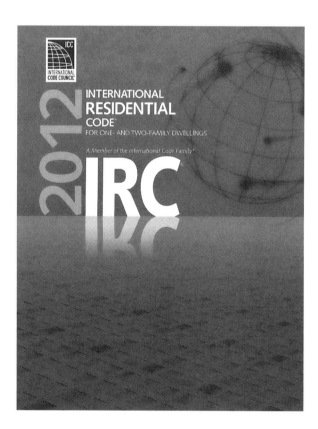

In addition to the general requirements, a townhouse is also limited by definition in the IRC. To fall under the scope of the IRC, each townhouse must be a single unit from the foundation to the roof, with at least two sides open to the exterior.

Topic: Appendices
Reference: IBC 101.2.1
Category: Administration
Subject: General Requirements

Code Text: *Provisions in the appendices shall not apply unless specifically adopted.*

Discussion and Commentary: The appendix chapters of the IBC address subjects that have been deemed inappropriate as mandatory portions of the code. Rather, the appendices are optional, with each jurisdiction adopting all, some or none of the appendix chapters, depending on its needs for enforcement in any given area. There are various reasons why certain issues are found in the appendices. Often, the provisions are limited in application or interest. Some appendix chapters are merely extensions of requirements set forth in the body of the code. Others address issues that are often thought of as outside of the scope of a traditional building code. Whatever the reason, the appendix chapters are not applicable unless specifically adopted.

IBC Appendix Chapters

Appendix A	Employee Qualifications
Appendix B	Board of Appeals
Appendix C	Group U — Agricultural Buildings
Appendix D	Fire Districts
Appendix E	Supplementary Accessibility Requirements
Appendix F	Rodentproofing
Appendix G	Flood-Resistant Construction
Appendix H	Signs
Appendix I	Patio Covers
Appendix J	Grading
Appendix K	Administrative Provisions
Appendix L	Earthquake Recording Instrumentation
Appendix M	Tsunami-Generated Flood Hazard

Appendix chapters, where not adopted as a portion of a jurisdiction's building code, may still be of value in application of the code. Provisions in the appendices might provide some degree of assistance in evaluating proposed alternative designs, methods or materials of construction.

Topic: Creation, Appointment and Deputies **Category:** Administration
Reference: IBC 103 **Subject:** Department of Building Safety

Code Text: *The Department of Building Safety is hereby created and the official in charge thereof shall be known as the building official. The building official shall be appointed by the chief appointing authority of the jurisdiction. In accordance with the prescribed procedures of this jurisdiction and with the concurrence of the appointing authority, the building official shall have the authority to appoint a deputy building official, the related technical officers, inspectors, plan examiners and other employees. Such employees shall have powers as delegated by the building official.*

Discussion and Commentary: The building official is an appointed officer of the jurisdiction and charged with the administrative responsibilities of the department of building safety. It is not uncommon for the jurisdiction to use a different position title to identify the building official, such as Chief Building Inspector, Superintendent of Central Inspection or Director of Code Enforcement. Regardless of the jurisdictional title, the code recognizes the individual in charge as the building official.

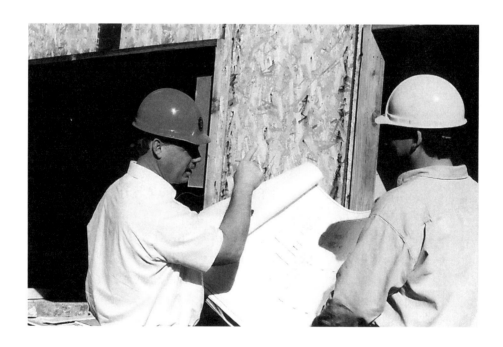

Inspectors, plan reviewers and other technical staff members are typically given some degree of authority to act for the building official in the decision-making process, including the making of appropriate interpretations of various provisions of the code.

Topic: General Requirements
Reference: IBC 104.1
Category: Administration
Subject: Duties and Powers of Building Official

Code Text: *The building official is hereby authorized and directed to enforce the provisions of the IBC. The building official shall have the authority to render interpretations of the IBC and to adopt policies and procedures in order to clarify the application of its provisions. Such interpretations, policies and procedures shall be in compliance with the intent and purpose of the IBC. Such policies and procedures shall not have the effect of waiving requirements specifically provided for in the IBC.*

Discussion and Commentary: It is important that the building official be knowledgeable to the point of being able to rule on those issues that are not directly addressed or that are unclear in the code. The basis for such a determination, which often takes some research to discover, is the intent and purpose of the *International Building Code*.

(Jurisdiction)
Department of Building Safety

[Photo]

_____
Name of individual

_____
Job function

The individual identified on the badge is a duly authorized employee of (the Jurisdiction) and is a designated representative of the Department of Building Safety.

Valid through _____   _____
　　　　　　　　Date　　　　Building Official

Sample of Required Identification
Section 104.5

Although the IBC gives broad authority to the building official in interpreting the code, this authority also comes with great responsibility. The building official must restrict all decisions to the intent and purpose of the code, with the waiving of any requirements being strictly prohibited.

Topic: Liability
Reference: IBC 104.8

Category: Administration
Subject: Duties and Powers of Building Official

Code Text: *The building official, member of the board of appeals or employee charged with the enforcement of* the IBC, *while acting for the jurisdiction in good faith and without malice in the discharge of the duties required by* the IBC *or other pertinent law or ordinance, shall not thereby be rendered liable personally and is hereby relieved from personal liability for any damage accruing to persons or property as a result of any act or by reason of an act or omission in the discharge of official duties.*

Discussion and Commentary: The protection afforded by the code regarding employee liability is limited to only those acts that occur in good faith. Absolute immunity from all tort liability is not provided where an employee acts maliciously. An employee is not relieved from personal liability where malice can be shown, and it is probable that the jurisdiction will not provide for the employee's defense.

Public officials should familiarize themselves with the laws of their respective states regarding their exposure to tort liability.

Topic: Alternate Materials
Reference: IBC 104.11

Category: Administration
Subject: Duties and Powers of Building Official

Code Text: *The provisions of the IBC are not intended to prevent the installation of any material or to prohibit any design or method of construction not specifically prescribed by the IBC, provided that any such alternative has been approved. An alternative material, design or method of construction shall be approved where the building official finds that the proposed design is satisfactory and complies with the intent of the provisions of the IBC.*

Discussion and Commentary: The building official is granted broad authority in the acceptance of alternative materials, designs and methods of construction. Arguably the most important provision of the IBC, the intent is to implement the adoption of new technologies. Furthermore, it gives the code even more of a performance character. The provisions encourage state-of-the-art concepts in construction, design and materials, as long as they meet the performance level intended by the IBC.

Building official may approve alternative materials, design and methods of construction, if:

- Proposed alternative is satisfactory
- Proposed alternative complies with intent of code
- Material, method or work is equivalent in:
  - (1) Quality
  - (2) Strength
  - (3) Effectiveness
  - (4) Fire Resistance
  - (5) Durability
  - (6) Safety

Advisable for building official to:
- Require sufficient evidence or proof
- Record any action granting approval
- Enter information into the files

The building official should ensure that any necessary substantiating data or other evidence that shows the alternative to be equivalent in performance is submitted. Moreover, where tests are performed, reports of such tests must be retained by the building official.

Topic: Research Reports
Category: Administration
Reference: IBC 104.11.1
Subject: Duties and Powers of Building Official

Code Text: *Supporting data, where necessary to assist in the approval of materials or assemblies not specifically provided for in* the IBC, *shall consist of valid research reports from approved sources.*

Discussion and Commentary: The most familiar research reports are probably the ICC Evaluation Service Reports maintained by ICC Evaluation Service, Inc. ICC Evaluation Service Reports are developed based upon acceptance criteria and provide evidence that products and systems meet code requirements. These reports are made available to building regulators, contractors, specifiers, architects, engineers and anyone else with an interest in the building industry or construction. The building official can also recognize the evaluation of products, assemblies and systems by other third-party agencies having the necessary credentials.

Although ICC Evaluation Service Reports are generally recognized nationally as valid reports developed by an approved source, the building official is the final authority on the acceptance of any research report for the purpose of accepting an alternate material, method or design.

Topic: Permits Required and Exempted **Category:** Administration
Reference: IBC 105.1, 105.2 **Subject:** Permits

Code Text: *Any owner or authorized agent who intends to construct, enlarge, alter, repair, move, demolish, or change the occupancy of a building or structure . . . shall first make application to the building official and obtain the required permit.* See 13 exemptions where a building permit is not required. *Exemptions from permit requirements of* the IBC *shall not be deemed to grant authorization for any work to be done in any manner in violation of the provisions of* the IBC *or any other laws or ordinances of this jurisdiction.*

Discussion and Commentary: Except in those few cases specifically listed, such as small accessory structures and finish work, all construction-related work requires a permit and is subject to subsequent inspections.

Work exempt from permit:

- One-story detached accessory buildings limited to 120 square feet
- Fences not over 7 feet in height
- Oil derricks
- Retaining walls limited to 4 feet in height, unless supporting a surcharge or impounding Class I, II or III-A liquids
- Water tanks supported directly on grade, limited to capacity of 5,000 gallons and a ratio of height to diameter not exceeding 2 to 1
- Sidewalks and driveways limited to 30 inches above grade, not over any basement or story below, and not part of an accessible route
- Painting, papering, carpeting, cabinets, counter tops and similar finish work
- Temporary motion picture, television and theater stage sets and scenery
- Prefabricated swimming pools accessory to a Group R-3 occupancy when capacity is limited to 5,000 gallons, depth limited to 24 inches and installed entirely above ground
- Shade cloth structures used for nursery or agricultural purposes
- Swings and other playground equipment accessory to detached one- and two-family dwellings
- Window awnings supported by an exterior wall in Groups R-3 and U, where the maximum projection is 54 inches
- Movable fixtures, racks, cases, counters and partitions limited to 5 feet 9 inches in height

---

Whether or not a building permit is required by the code, it is intended that all work be done in accordance with the code requirements. The owner is responsible for all construction being done properly and safely.

Topic: General Provisions
Reference: IBC 107.1

Category: Administration
Subject: Submittal Documents

Code Text: *Submittal documents consisting of construction documents, statement of special inspections, geotechnical report and other data shall be submitted in two or more sets with each permit application. The construction documents shall be prepared by a registered design professional where required by the statutes of the jurisdiction in which the project is to be constructed. See exception for projects where nature of work is such that review of construction documents is unnecessary.*

Discussion and Commentary: Submittal documents must be provided to the jurisdiction to allow for full review and approval prior to issuance of any required permits. In addition to plans and specifications, the submittal must include any applicable support documents such as a geotechnical report or a statement of special inspections. State law typically regulates the scope of buildings required to be prepared by a registered design professional; however, in special cases the building official can require preparation by such an individual where state law does not.

A minimum of two sets of submittal documents must be provided as part of the permit application. At least one set is to be retained by the building department, with the other set returned to the applicant. Many jurisdictions require that additional sets of such documents be submitted in order to provide sets to various departments that will be reviewing different portions of the submittal.

Topic: Payment, Schedule and Valuations **Category:** Administration
Reference: IBC 109 **Subject:** Fees

Code Text: *A permit shall not be valid until the fees prescribed by law have been paid. On buildings, structures, electrical, gas, mechanical and plumbing systems or alterations requiring a permit, a fee for each permit shall be paid as required, in accordance with the schedule as established by the applicable governing authority. The applicant for a permit shall provide an estimated permit value at the time of application. Permit valuations shall include total value of the work, including materials and labor, for which a permit is being issued, such as electrical, gas, mechanical, plumbing equipment and other permanent systems.*

Discussion and Commentary: Fees are typically established at a level that will provide enough funds to adequately pay for the costs of operating the various building department functions, including administration, plan review and inspection. It is common that the fee schedule be based on the projected construction cost (valuation) of the work. Although there are several different methods for determining the appropriate valuation, it is important that a realistic and consistent approach be taken so that permit fees are applied fairly and accurately.

Fee schedules are not established by the code. Rather, the local jurisdiction is required to establish the appropriate fees to fund the operation of the building safety department.

Study Session 1

Topic: Required Inspections
Reference: IBC 110.1

Category: Administration
Subject: Inspections

Code Text: *Construction or work for which a permit is required shall be subject to inspection by the building official and such construction or work shall remain accessible and exposed for inspection purposes until approved. Approval as a result of an inspection shall not be construed to be an approval of a violation of the provisions of the IBC or of other ordinances of the jurisdiction.*

Discussion and Commentary: The inspection function is possibly the most critical activity in the entire code enforcement process. At the varied stages of construction, an inspector often performs the final check of the building for safety-related compliance. If necessary, the building official may require a preliminary inspection of the site, building or structure to gain information that may be of assistance in the issuance of a permit.

Required inspections (where applicable):

- Footing and foundation
- Concrete slab or under-floor
- Lowest floor elevation
- Frame
- Lath or gypsum board
- Fire and smoke resistant penetrations
- Energy efficiency
- Others as required by the building official
- Special inspections
- Final

It may be necessary to have materials removed in order to provide inspection access or observation for a portion of the building. The responsibility rests with the permit applicant to make the work available for inspection, with no expenses to be borne by the jurisdiction.

Topic: Approval Required
Reference: IBC 110.6
Category: Administration
Subject: Inspections

Code Text: *Work shall not be done beyond the point indicated in each successive inspection without first obtaining the approval of the building official. The building official, upon notification, shall make the requested inspections and shall either indicate the portion of the construction that is satisfactory as completed, or notify the permit holder or his or her agent wherein the same fails to comply with the IBC. Any portions that do not comply shall be corrected and such portions shall not be covered or concealed until authorized by the building official.*

Discussion and Commentary: It is important that each successive inspection be approved prior to continuing further work. This practice helps to control the concealment of any work that must be inspected, resulting in the unnecessary removal of materials that might block access.

Any request for inspection is the responsibility of the building permit holders or their duly authorized agent. They must contact the building official when the work is ready for inspection, as well as provide access to that work.

Topic: Use and Occupancy
Reference: IBC 111.1
Category: Administration
Subject: Certificate of Occupancy

Code Text: *No building or structure shall be used or occupied, and no change in the existing occupancy classification of a building or structure or portion thereof shall be made until the building official has issued a certificate of occupancy therefor as provided herein. Issuance of a certificate of occupancy shall not be construed as an approval of a violation of the provisions of the IBC or of other ordinances of the jurisdiction.* See exception for work exempt from permits.

Discussion and Commentary: The certificate of occupancy is the tool with which the building official can regulate and control the uses and occupancies of the various buildings and structures within the jurisdiction. The code makes it unlawful to use or occupy a building unless a certificate of occupancy has been issued for that specific use.

Certificate of Occupancy
(Address of Structure)

This (applicable portion of structure) has been inspected for compliance with the laws and ordinances of (jurisdiction) and is hereby issued a Certificate of Occupancy

Building permit number _____
Applicable edition of code _____
Use and occupancy _____
Type of construction _____
Design occupant load _____
Sprinkler system required _____
Name and address of owner _____

Special conditions _____

Building Official _____

Sample of Certificate of Occupancy

The building official is permitted to suspend or revoke a certificate of occupancy for any of the following reasons: 1) when the certificate is issued in error, 2) when incorrect information is supplied, or 3) when the building is in violation of the code.

| **Topic:** Alternative Compliance | **Category:** Existing Structures |
| **Reference:** IBC 3401.6 | **Subject:** General Provisions |

Code Text: *Work performed in accordance with the* International Existing Building Code *shall be deemed to comply with the provisions of Chapter 34.*

Discussion and Commentary: The *International Existing Building Code* (IEBC), first published by the International Code Council as the 2003 edition, continues to provide a comprehensive approach to the regulation of existing buildings. Where repairs, alterations, change of occupancy or additions to existing buildings are encountered, such work is initially regulated by IBC Chapter 34. However, the IEBC may be used as an alternative set of provisions where desired by the building owner and designer. Either the provisions of IBC Chapter 34 or those of the IEBC must be applied to the entire scope of work, as it is not permissible to pick and choose requirements.

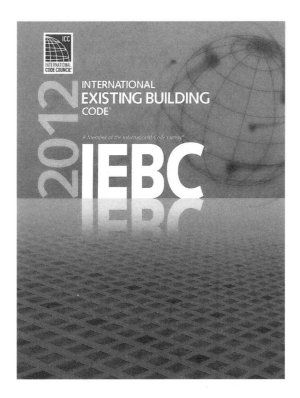

The IEBC is intended to encourage the use and reuse of existing buildings while adequately protecting public health, safety and welfare. Further founding principles require that the provisions do not unnecessarily increase construction costs; do not restrict the use of new materials, products or methods of construction; and do not give preferential treatment to particular types or classes of materials, products or methods of construction.

Topic: Scope
Reference: IBC 3403.1, 3404.1
Category: Existing Structures
Subject: Additions and Alterations

Code Text: *Additions to any building or structure shall comply with the requirements of the IBC for new construction. Except as provided by Section 3401.4* (building materials and systems) *or Section 3404* (alternative compliance), *alterations to any building or structure shall comply with the requirements of the IBC for new construction. Alterations shall be such that the existing building or structure is no less complying with the provisions of the IBC than the existing building or structure was prior to the alteration.* See exceptions for existing stairways and handrails.

Discussion and Commentary: Any construction that takes place on a building will necessitate compliance with the building code in effect at the time of the work. However, only those portions being remodeled or added need comply. Existing portions of the building where no work is being performed are exempt from the current code requirements.

Conformance with the code in effect is also mandated for a change in use or occupancy. When approved by the building official, full conformance is not required where it is determined that the new use is less of a life-safety and fire-safety hazard than the previous use.

Topic: Compliance
Reference: IBC 3412.1, 3412.3
Category: Existing Structures
Subject: Compliance Alternatives

Code Text: *The provisions of Section 3412 are intended to maintain or increase the current degree of public safety, health and general welfare in existing buildings while permitting repair, alteration, addition and a change of occupancy without requiring full compliance with Chapters 2 through 33, or Sections 3401.3, and 3403 through 3409, except where compliance with other provisions of the IBC is specifically required in Section 3412. For repairs, alterations, additions and changes of occupancy to existing buildings that are evaluated in accordance with Section 3412, compliance with Section 3412 shall be accepted by the building official.*

Discussion and Commentary: A thorough investigation and evaluation of an existing building can be the basis for determining the level of compliance when an existing building is altered or repaired, or when the building or portion thereof undergoes a change in use. Based on the results of the evaluation process, it can be determined if the existing building, along with any modifications, is deemed to be in compliance with the IBC.

TABLE 3412.7
SUMMARY SHEET—BUILDING CODE

Existing occupancy: _____   Proposed occupancy: _____

Year building was constructed: _____   Number of stories: _____ Height in feet: _____

Type of construction: _____   Area per floor: _____

Percentage of open perimeter increase: _____%

Completely suppressed: Yes ___ No ___   Corridor wall rating: _____

Compartmentation: Yes ___ No ___   Required door closers: Yes ___ No ___

Fire-resistance rating of vertical opening enclosures: _____

Type of HVAC system: _____, serving number of floors: _____

Automatic fire detection: Yes ___ No ___   Type and location: _____

Fire alarm system: Yes ___ No ___   Type: _____

Smoke control: Yes ___ No ___   Type: _____

Adequate exit routes: Yes ___ No ___   Dead ends: Yes ___ No ___

Maximum exit access travel distance: _____   Elevator controls: Yes ___ No ___

Means of egress emergency lighting: Yes ___ No ___   Mixed occupancies: Yes ___ No ___

SAFETY PARAMETERS	FIRE SAFETY (FS)	MEANS OF EGRESS (ME)	GENERAL SAFETY (GS)
3412.6.1 Building Height 3412.6.2 Building Area 3412.6.3 Compartmentation			
3412.6.4 Tenant and Dwelling Unit Separations 3412.6.5 Corridor Walls 3412.6.6 Vertical Openings			
3412.6.7 HVAC Systems 3412.6.8 Automatic Fire Detection 3412.6.9 Fire Alarm Systems			
3412.6.10 Smoke Control 3412.6.11 Means of Egress Capacity 3412.6.12 Dead Ends	**** **** ****		
3412.6.13 Maximum Exit Access Travel Distance 3412.6.14 Elevator Control 3412.6.15 Means of Egress Emergency Lighting	**** ****		
3412.6.16 Mixed Occupancies 3412.6.17 Automatic Sprinklers 3412.6.18 Standpipes 3412.6.19 Incidental Use		**** ÷ 2 =	
Building score — total value			

* * * *No applicable value to be inserted.

In evaluating an existing building undergoing a change in use, alteration or repair, the process addresses three issues: fire safety, means of egress and general safety. Each category must be deemed satisfactory in order for the existing building to be considered compliant.

Topic: Application
Reference: IBC Chapter 35
Category: Referenced Standards
Subject: Standards

Code Text: Chapter 35 *lists the standards that are referenced in various sections of* the IBC. *The standards are listed herein by the promulgation agency of the standard, the standard identification, the effective date and title, and the section or sections of* the IBC *that reference the standard. The application of the referenced standards shall be as specified in Section 102.4.*

Discussion and Commentary: Limited in their scope, standards define more precisely the general provisions set forth in the code. The *International Building Code* references several hundred different standards, each addressing a specific aspect of building design or construction. In general terms, the standards referenced by the IBC are primarily materials, testing, installation or engineering standards.

Those codes and standards referenced in the IBC are considered an extension of the code, but only to the degree prescribed by the IBC. Where there is a conflict between the provisions in the IBC and any referenced code or standard, the provisions of the IBC apply.

Study Session 1
IBC Chapters 1, 34 and 35

1. The *International Residential Code* is applicable to townhouses a maximum of _____ above grade plane in height and provided with separate means of egress.

    a. 35 feet b. 40 feet

    c. three stories d. four stories

    Reference_____

2. Provisions of the appendix do not apply unless _____.

    a. specified in the code b. applicable to unique conditions

    c. specifically adopted d. relevant to fire or life safety

    Reference_____

3. If there is a conflict in the code between a general requirement and a specific requirement, the _____ requirement shall apply.

    a. general b. specific

    c. least restrictive d. most restrictive

    Reference_____

4. The _____ is considered by the code as the term to describe the individual in charge of the department of building safety.

   a. building official
   b. code official
   c. code administrator
   d. chief building inspector

   Reference_____

5. The building official has the authority to _____ the provisions of the code.

   a. ignore
   b. waive
   c. violate
   d. interpret

   Reference_____

6. Used materials may be utilized under which of the following conditions?

   a. They meet the requirements for new materials.
   b. They are limited to 10 percent of the total materials.
   c. Used materials may never be used in new construction.
   d. A representative sampling is tested for compliance.

   Reference_____

7. The building official has the authority to grant modifications to the code _____.

   a. for only those issues not affecting life safety or fire safety
   b. for individual cases where the strict letter of the code is impractical
   c. where the intent and purpose of the code cannot be met
   d. related only to administrative functions

   Reference_____

8. In order for an alternative material, design or method of construction to be considered acceptable, it must be equivalent to the code based on all but which of the following criteria?

   a. durability
   b. economics
   c. strength
   d. fire resistance

   Reference_____

9. Tests performed by _____ may be required by the building official where there is insufficient evidence of code compliance.

   a. the owner
   b. the contractor
   c. an approved agency
   d. a design professional

   Reference 104.11.2

10. A permit is not required for the construction of a one-story detached accessory structure when it has a maximum floor area of _____ square feet.

    a. 100
    b. 120
    c. 150
    d. 200

    Reference _____

11. Movable fixtures, cases, counters and partitions are exempt from a building permit where they have a maximum height of _____.

    a. 5 feet, 0 inches
    b. 5 feet, 6 inches
    c. 5 feet, 9 inches
    d. 6 feet, 0 inches

    Reference 105.2 B.13

12. Where a building is constructed in a flood hazard area, documentation of the lowest floor elevation shall be submitted to the building official prior to _____.

    a. any framing work above the lowest floor level
    b. approval of the framing inspection
    c. the final inspection
    d. issuance of the certification of occupancy

    Reference 110.3.10.1

13. The building permit, or a copy of the permit, shall be kept _____ until completion of the project.

    a. at the job site
    b. by the permit applicant
    c. by the contractor
    d. by the design professional in responsible charge

    Reference 105.7

Study Session 1

14. When a building permit is issued, the construction documents shall be approved as _____.

   a. "Approved for Construction"
   b. "Conditional Approval"
   c. "Accepted as Reviewed"
   d. "Reviewed for Code Compliance"

   Reference 107.3.1

15. Unless otherwise mandated by state or local laws, the approved construction documents shall be retained by the building official for a minimum of _____ from the date of completion of the permitted work.

   a. 90 days   b. 180 days
   c. 1 year   d. 2 years

   Reference 107.5

16. Which one of the following inspections is not specifically identified by the *International Building Code* as a required inspection?

   a. footing inspection
   b. frame inspection
   c. soil classification
   d. energy efficiency inspection

   Reference 110.3

17. Whose duty is it to notify the building official that the work is ready for inspection?

   a. the permit holder   b. the owner
   c. the contractor   d. the owner's agent

   Reference 110.5

18. The certificate of occupancy shall contain all of the following information except:

   a. the name of the building owner
   b. the name of the building official
   c. the building's type of construction
   d. the building's maximum height and area

   Reference 111.2

19. A temporary certificate of occupancy is valid for what maximum period of time?

    a. 30 days
    b. 60 days
    c. 180 days
    d. a period set by the building official *(circled)*

    Reference 111.3

20. The board of appeals is not authorized to rule on an appeal based on a claim that _____.

    a. the provisions of the code do not fully apply
    b. a code requirement should be waived
    c. the rules have been incorrectly interpreted *(circled)*
    d. a better form of construction is provided

    Reference 113.2

21. Unless in compliance with the code for new structures, alterations to an existing structure are permitted to increase the force in any existing structural element that carries a gravity load by a maximum of _____ percent.

    a. 0 (no increase is permitted)
    b. 5 *(circled)*
    c. 10
    d. 25

    Reference 3403.3

22. Where permitted for an existing building, a fire escape must be designed to support a minimum live load of _____ pounds per square foot.

    a. 40
    b. 50
    c. 100 *(circled)*
    d. 125

    Reference 3406.3

23. Where alterations occur to an existing building, an access ramp with a slope of 1:10 is permitted where necessitated by space limitations, provided the maximum rise of the ramp is _____ inches.

    a. 3
    b. 6 *(circled)*
    c. 12
    d. 30

    Reference 3411.8.5

Study Session 1

24. Which of the following categories is not specifically addressed in the evaluation of existing buildings for compliance alternatives?

    a. fire safety
    b. means of egress
    c. general safety
    d. structural safety

    Reference 3412.5

25. ACI 318-11 is a referenced standard addressing _____.

    a. structural concrete
    b. wood construction
    c. structural steel buildings
    d. gypsum board

    Reference_____

26. A permit is required for a prefabricated above-ground swimming pool, accessory to a Group R-3 occupancy, that is a minimum of _____ in depth or has a capacity of more than _____.

    a. 18 inches; 4,000 gallons
    b. 24 inches; 5,000 gallons
    c. 30 inches; 5,000 gallons
    d. 36 inches; 6,000 gallons

    Reference_____

27. In Group R-3 and U occupancies, the installation of window awnings is exempt from a permit, provided the awnings each project a maximum of _____ inches from the exterior wall and do not require additional support.

    a. 30
    b. 36
    c. 48
    d. 54

    Reference_____

28. A permit may be suspended or revoked for all of the following reasons, except _____.

    a. where it is issued in error
    b. on the basis of incomplete information
    c. where issued in violation of a jurisdictional ordinance
    d. where other permits by the contractor have been voided

    Reference 105.6

29. Unless extended by the building official, what is the maximum time period granted for a permit issued on a temporary structure?

    a. 90 days
    b. 180 days
    c. 1 year
    d. 2 years

    Reference __105.5__

30. The provisions for compliance alternatives in the repair, alteration, addition and change of occupancy in existing buildings are applicable to all of the following occupancy classifications, except _____.

    a. Group A
    b. Group E
    c. Group I
    d. Group R

    Reference __3412.2__

31. A permit is not required for the installation of a self-contained refrigeration system, provided it contains a maximum of _____ pounds of refrigerant and is actuated by a maximum _____-horsepower motor.

    a. 5, 1
    b. 10, 1
    c. 5, 1½
    d. 10, 1½

    Reference __105.2__

32. The means of egress layout required as a part of the construction documents for which of the following occupancies must include the number of occupants to be accommodated?

    a. F-1
    b. I-1
    c. R-2
    d. R-3

    Reference __107.2.3__

33. The final permit valuation shall be determined by the _____.

    a. owner
    b. building official
    c. design professional
    d. general contractor

    Reference __109.3__

Study Session 1

34. Who is responsible for assuring that the work is accessible and exposed for inspection purposes?

    a. owner
    b. contractor
    c. permit applicant
    d. design professional

    Reference _110.1_

35. A stop work order shall be in writing and given to any of the following individuals except the _____.

    a. owner of the property involved
    b. owner's agent
    c. permit holder
    d. person doing the work

    Reference _115.2_

Study Session 2

2012 IBC Chapter 3 and Sections 508 – 509
Use and Occupancy Classification

OBJECTIVE: To gain an understanding of how an occupancy is classified based on its intended use and how a building with incidental uses and/or mixed uses is addressed.

REFERENCE: Chapter 3 and Sections 508 and 509, 2012 *International Building Code*

KEY POINTS:
- What are the 10 general occupancy groups?
- How is a space that is intended to be occupied at different times for different purposes to be addressed?
- How is an occupancy that is not specifically described to be classified?
- Which types of activities are considered assembly uses? What is their general classification?
- How are small assembly uses classified where accessory to a different occupancy?
- What is the classification for restaurants and cafes? Theaters? Places of religious worship, conference rooms and libraries? Arenas? Grandstands?
- What is the primary use classified as Group B?
- Group E occupancies describe educational uses for individuals of what age group?
- Which types of day care are considered Group E occupancies?
- Manufacturing operations fall into what occupancy group? How do the two divisions of factory-use differ from each other?
- What type of operations or materials cause a use to be considered Group H?
- How does the amount of hazardous materials affect the occupancy classification?
- Which occupancies address physical hazards? Health hazards? Semiconductor fabrication facilities?
- Which characteristics are typical of a Group I occupancy?
- In which institutional occupancies are the occupants considered incapable of self-preservation?
- For which types of institutional uses may the *International Residential Code* be utilized?
- What general type of building is considered a Group M occupancy?
- How are residential occupancies classified?
- What is the key difference between a Group R-1 and Group R-2 occupancy?

29

KEY POINTS:
(Cont'd)

- What is a congregate living facility? How should such a facility be classified?
- When is a residential use permitted to be constructed under the provisions of the *International Residential Code*?
- What do storage occupancy classifications have in common with those of manufacturing uses?
- How is a vehicle repair garage classified? An aircraft hangar?
- What is the classification of an enclosed parking garage? An open parking garage?
- What is a utility occupancy? How does its classification differ from that of other occupancies?
- Which three options are available for addressing multiple occupancies within a building?
- What is an accessory occupancy? What benefit is derived from such a designation?
- What is the concept of the nonseparated occupancy provisions? What conditions apply to buildings with nonseparated occupancies?
- What is the basis for separated occupancies? How are the minimum required fire-resistive separations determined?
- How is the fire-resistance rating for an occupancy separation determined? How does the presence of an automatic sprinkler system affect the required rating?
- What is an incidental use? How must such an area be separated from the remainder of the building? When is sprinkler protection required?

Topic: Occupancy Groups
Reference: IBC 302.1
Category: Occupancy Classification
Subject: Classification

Code Text: *Structures or portions of structures shall be classified with respect to occupancy in one or more of the groups listed. Where a structure is proposed for a purpose which is not specifically provided for in* the IBC, *such structure shall be classified in the group which the occupancy most nearly resembles, according to the fire safety and relative hazard involved.*

Discussion and Commentary: The perils contemplated by the occupancy groupings are divided into two general categories: those related to people and those related to content. People-related hazards include the number and density of the occupants, their age and mobility, and their awareness of surrounding conditions. Content-related hazards include the storage and use of hazardous materials, as well as the presence of large quantities of combustible materials.

Assembly	**B**usiness
Educational	**F**actory
Hazardous	**I**nstitutional
Mercantile	**R**esidential
Storage	**U**tility

Proper occupancy classification is critical in making appropriate code determinations throughout a project. In the classification process, the building official must use judgment in the determination of the potential hazards of an affected occupancy.

Study Session 2

Topic: Assembly Uses
Reference: IBC 303.1
Category: Occupancy Classification
Subject: Group A Occupancies

Code Text: *Assembly Group A occupancy includes, among others, the use of a building or structure, or a portion thereof, for the gathering of persons for purposes such as civic, social or religious functions, recreation, food or drink consumption or awaiting transportation.* See exceptions for assembly buildings and assembly spaces with an occupant load of less than 50, accessory assembly spaces less than 750 square feet in floor area, and those assembly spaces associated with Group E occupancies.

Discussion and Commentary: The conditions related to a typical Group A occupancy suggest a moderate hazard use. This use often includes sizable numbers of people who are generally mobile and aware of the surrounding conditions. The extremely high occupant density level often present in an assembly occupancy is what distinguishes Group A from other occupancies. Where the occupant load of an assembly building or tenant space does not exceed 50, a Group B classification is typically more appropriate owing to the lesser hazard.

Group A-1	Group A-3	Group A-4
Motion picture theaters Theaters Symphony and concert halls	Amusement arcades Art galleries Bowling alleys Places of worship Community halls Conference rooms Exhibition halls Lecture halls Libraries Museums Passenger stations	Arenas Skating rinks Swimming pools Tennis courts

Group A-2		Group A-5
Banquet halls Casinos Night clubs Restaurants Taverns		Amusement park structures Bleachers Grandstands Stadiums

Unique conditions are represented by the classifications of Groups A-1, A-2, A-4 and A-5. However, the category Group A-3 includes a variety of broad and diverse assembly uses. It is not uncommon to find high combustible loading in Group A-3 occupancies.

Topic: Business Uses
Reference: IBC 304.1
Category: Occupancy Classification
Subject: Group B Occupancies

Code Text: *Business Group B occupancy includes, among others, the use of a building or structure, or a portion thereof, for office, professional or service-type transactions, including storage of records and accounts.*

Discussion and Commentary: Business occupancies typically have a low to moderate fire load, a moderate density level, and occupants who are usually mobile and have a general awareness of the surrounding conditions. As such, business occupancies are grouped into a classification based upon a relatively moderate fire hazard level. Group B occupancies are not restricted by occupant load, as the number of people in a business use, such as an office, can range from one person to thousands of people.

Group B

- Ambulatory care facilities
- Animal hospitals, kennels and ponds
- Banks
- Barber and beauty shops
- Car wash
- Civil administration
- Clinic-outpatient
- Educational occupancies above the 12th grade
- Laboratories; testing and research
- Motor vehicle showrooms
- Post offices
- Print shops
- Professional services
- Radio and television stations
- Training and skill development

As is the case for many of the occupancy groups, a review of the building's intended uses is necessary to determine the amount of hazardous materials that may be stored, handled or used. If the amounts exceed a specified quantity, then a Group H classification will be in order.

Topic: Educational Uses
Reference: IBC 305.1, 305.2
Category: Occupancy Classification
Subject: Group E Occupancies

Code Text: *Educational Group E occupancy includes, among others, the use of a building or structure, or a portion thereof, by six or more persons at any one time for educational purposes through the 12th grade. This group includes buildings and structures or portions thereof occupied by more than five children older than $2^1/_2$ years of age who receive educational, supervision or personal care services for fewer than 24 hours per day.*

Discussion and Commentary: Educational occupancies include classroom uses for students of high school age and younger. Education facilities limited to use by older students, such as college classrooms, are classified as Group B occupancies; however, a Group A classification should be considered for lecture halls and similar large occupant load spaces.

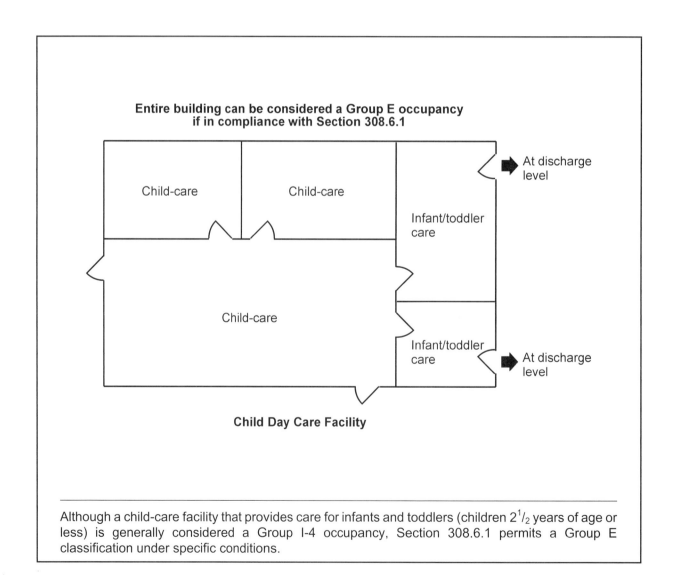

Child Day Care Facility

Although a child-care facility that provides care for infants and toddlers (children $2^1/_2$ years of age or less) is generally considered a Group I-4 occupancy, Section 308.6.1 permits a Group E classification under specific conditions.

Topic: Factory Uses
Reference: IBC 306.1
Category: Occupancy Classification
Subject: Group F Occupancies

Code Text: *Factory Industrial Group F occupancy includes, among others, the use of a building or structure, or a portion thereof, for assembling, disassembling, fabricating, finishing, manufacturing, packaging, repair or processing operations that are not classified as a Group H hazardous or Group S storage occupancy.*

Discussion and Commentary: Although the potential hazard and fire severity varies among the many uses categorized as Group F occupancies, the uses still share elements in common. The occupants are adults who are awake and who generally have enough familiarity with the premises to be able to exit the building with reasonable efficiency. The presence of combustible materials in the industrial process causes a classification of Group F-1, which is by far the most common factory use.

Group F-1	Group F-2
Aircraft	Brick and masonry
Appliances	Ceramic products
Automobiles	Foundries
Bakeries	Glass products
Business machines	Gypsum
Carpets and rugs	Ice
Clothing	Metal products
Electric generation Plants	
Electronics	
Food processing	
Furniture	
Laundries	
Millwork	
Paper mills or products	
Plastic products	
Printing or publishing	
Refuse incineration	
Textiles	
Woodworking	

Classification as a Group F-2 occupancy is strictly limited because of the restrictions placed on such uses. The fabrication or manufacture of noncombustible materials, as well as their finishing, packaging or processing operations, cannot involve a significant fire hazard.

Topic: High-Hazard Uses
Reference: IBC 307.1

Category: Occupancy Classification
Subject: Group H Occupancies

Code Text: *High-hazard Group H occupancy includes, among others, the use of a building or structure, or a portion thereof, that involves the manufacturing, processing, generation or storage of materials that constitute a physical or health hazard in quantities in excess of those allowed in control areas complying with Section 414, based on the maximum allowable quantity limits for control areas set forth in Tables 307.1(1) and 307.1(2).*

Discussion and Commentary: There is only one fundamental type of Group H occupancy—that which is designated based solely on excessive quantities of hazardous materials contained therein. The quantities of hazardous materials that necessitate a Group H classification vary, based on the type, quantity, condition (use or storage) and environment of the materials. Where the use does not exceed the maximum allowable quantities set forth in the code, a classification other than Group H is appropriate.

Where hazardous materials and processes are involved

References for detailed provisions

Although the *International Building Code* is limited to general construction regulations and occupancy-specific requirements, the *International Fire Code*® (IFC®) sets forth special detailed provisions relating to hazardous materials and the specific conditions of their storage, use and handling.

Topic: Exceptions to Group H **Category:** Occupancy Classification
Reference: IBC 307.1, Exceptions **Subject:** Group H Occupancies

Code Text: *The following shall not be classified as Group H, but shall be classified as the occupancy that they most nearly resemble.* See a listing of 13 conditions under which a Group H occupancy is not warranted. *Hazardous materials in any quantity shall conform to the requirements of the IBC, including Section 414, and the* International Fire Code.

Discussion and Commentary: Although some degree of hazardous materials is found in most buildings, the occupancy is designated as Group H only where the quantities are excessive and the hazards are not adequately addressed. The most common condition for a non-H classification is where the amount of hazardous materials contained in the building does not exceed the maximum allowable quantities shown in Table 307.1(1) for physical hazards and Table 307.1(2) for health hazards. Footnotes to both tables can be used to increase the permitted quantities.

[F] TABLE 307.1(1)
MAXIMUM ALLOWABLE QUANTITY PER CONTROL AREA OF HAZARDOUS MATERIALS POSING A PHYSICAL HAZARD[a, j, m, n, p]

MATERIAL	CLASS	GROUP WHEN THE MAXIMUM ALLOWABLE QUANTITY IS EXCEEDED	STORAGE[b]			USE-CLOSED SYSTEMS[b]			USE-OPEN SYSTEMS[b]	
			Solid pounds (cubic feet)	Liquid gallons (pounds)	Gas (cubic feet at NTP)	Solid pounds (cubic feet)	Liquid gallons (pounds)	Gas (cubic feet at NTP)	Solid pounds (cubic feet)	Liquid gallons (pounds)
Combustible dust	N/A	H-2	Note q	N/A	N/A	Note q	N/A	N/A	Note q	N/A
Combustible liquid[c, i]	II	H-2 or H-3	N/A	120[d, e]	N/A	N/A	120[d]	N/A	N/A	30[d]
	IIIA	H-2 or H-3		330[d, e]			330[d]			80[d]
	IIIB	N/A		13,200[e, f]			13,200[f]			3,300[f]
Combustible fiber	Loose Baled[o]	H-3	(100) (1,000)	N/A	N/A	(100) (1,000)	N/A	N/A	(20) (200)	N/A
Consumer fireworks	1.4G	H-3	125[d, e, l]	N/A	N/A	N/A	N/A	N/A	N/A	N/A
Cryogenics, flammable	N/A	H-2	N/A	45[d]	N/A	N/A	45[d]	N/A	N/A	10[d]
Cryogenics, inert	N/A	N/A	N/A	N/A	NL	N/A	N/A	NL	N/A	N/A
Cryogenics, oxidizing	N/A	H-3	N/A	45[d]	N/A	N/A	45[d]	N/A	N/A	10[d]
Explosives	Division 1.1	H-1	1[e, g]	(1)[e, g]	N/A	0.25[g]	(0.25)[g]	N/A	0.25[g]	(0.25)[g]
	Division 1.2	H-1	1[e, g]	(1)[e, g]	N/A	0.25[g]	(0.25)[g]	N/A	0.25[g]	(0.25)[g]
	Division 1.3	H-1 or H-2	5[e, g]	(5)[e, g]	N/A	1[g]	(1)[g]	N/A	1[g]	(1)[g]
	Division 1.4	H-3	50[e, g]	(50)[e, g]	N/A	50[g]	(50)[g]	N/A	N/A	N/A
	Division 1.4G	H-3	125[d, e, l]	N/A	N/A	N/A	N/A	N/A	N/A	N/A
	Division 1.5	H-1	1[e, g]	(1)[e, g]	N/A	0.25[g]	(0.25)[g]	N/A	0.25[g]	(0.25)[g]
	Division 1.6	H-1	1[d, e, g]	N/A	N/A	N/A	N/A	N/A	N/A	N/A
Flammable gas	Gaseous	H-2	N/A	N/A	1,000[d,e]	N/A	N/A	1,000[d,e]	N/A	N/A
	Liquefied			(150)[d,e]	N/A		(150)[d,e]	N/A		
Flammable liquid[c]	1A	H-2 or H-3	N/A	30[d, e]	N/A	N/A	30[d]	N/A	N/A	10[d]
	1B and 1C			120[d, e]			120[d]			30[d]
Flammable liquid, combination (1A, 1B, 1C)	N/A	H-2 or H-3	N/A	120[d, e, h]	N/A	N/A	120[d, h]	N/A	N/A	30[d, h]

(continued)

Where one of the 13 exceptions is applied, the classification is based on the general use. For example, a warehouse containing quantities below the maximum allowable would simply be classified as Group S-1. A manufacturing facility would be classified as a Group F-1 occupancy.

Study Session 2

Topic: Institutional Uses
Reference: IBC 308.1
Category: Occupancy Classification
Subject: Group I Occupancies

Code Text: *Institutional Group I occupancy includes, among others, the use of a building or structure, or a portion thereof, in which care or supervision is provided to persons who are or are not capable of self preservation without physical assistance or in which persons are detained for penal or correctional purposes or in which the liberty of the occupants is restricted. Institutional occupancies shall be classified as Group I-1, I-2, I-3 or I-4.*

Discussion and Commentary: The institutional uses classified as Group I occupancies are of three broad types. The first is a facility in which care is provided for the very young, sick or injured. The second category includes those facilities in which the personal liberties of the inmates or residents are restricted. Thirdly, supervised care facilities are regulated. Though the hazard due to combustible contents is quite low in institutional uses, the occupants' lack of mobility limits their egress ability.

Group I-1
- Alcohol and drug centers
- Assisted living facilities
- Congregate care facilities
- Convalescent facilities
- Group homes
- Halfway houses
- Residential board and custodial care facilities
- Social rehabilitation facilities

Group I-2
- Foster care facilities
- Detoxification facilities
- Hospitals
- Nursing homes
- Psychiatric hospitals

Group I-3
- Correctional centers
- Detention centers
- Jails
- Prerelease centers
- Prisons
- Reformatories

Group I-4
- Adult day care
- Child day care

Where the number of children, patients or residents in institutional uses is five or less, the hazards are similar in nature to a residential use. In most cases, an institutional facility with such a low occupant load would be considered a Group R-3 occupancy or regulated under the IRC.

Topic: Mercantile Uses
Reference: IBC 309.1

Category: Occupancy Classification
Subject: Group M Occupancies

Code Text: *Mercantile Group M occupancy includes, among others, the use of a building or a structure or a portion thereof, for the display and sale of merchandise, and involves stocks of goods, wares or merchandise incidental to such purposes and accessible to the public.*

Discussion and Commentary: A Group M occupancy is a retail or wholesale facility, or a store. An entire building can be classified as a Group M occupancy, such as a department store, or a portion of a building can be considered a mercantile use, such as the sales room in a manufacturing facility. A service station, including a canopy over the pump islands, is also classified as a Group M occupancy. In limited instances, a sales operation is designated as a Group B occupancy, as in the case of automobile showrooms.

Group M

- Department stores
- Drug stores
- Markets
- Motor fuel-dispensing facilities
- Retail or wholesale stores
- Sales rooms

When classifying the occupancy of a storage area accessory to the sales area in a retail store, it is appropriate to apply the provisions that address the specific hazards of the use. In most situations, it is appropriate to classify the incidental storage area as a Group S-1 occupancy.

Study Session 2

Topic: Residential Uses
Reference: IBC 310.1
Category: Occupancy Classification
Subject: Group R Occupancies

Code Text: *Residential Group R occupancy includes, among others, the use of a building or structure, or a portion thereof, for sleeping purposes when not classified as an Institutional Group I or when not regulated by the* International Residential Code.

Discussion and Commentary: Residential occupancies are characterized by: 1) their use by people for living and sleeping purposes, 2) a relatively low potential fire severity, and 3) the worst fire record of all structure types. Because occupants of these types of buildings spend up to one-third of each day sleeping, there is a high potential of a fire to rage out of control before the occupants awaken. After awakening, the residents will typically be disoriented for a short period of time, further decreasing the opportunity for immediate egress. A major difference between the Group R-1 and R-2 occupancy classifications is the transient nature of the use. "Transient" is defined as *occupancy of a dwelling unit or sleeping unit for not more than 30 days.*

Group R-1
Boarding houses (transient)
 > 10 occupants
Congregate living
 facilities (transient) > 10 occupants
Hotels (transient)
Motels (transient)

Group R-2
Apartment houses
Boarding houses (nontransient)
 > 16 occupants
Congregate living
 facilities (nontransient)
 > 16 occupants
Convents
Dormitories
Fraternities and sororities
Hotels (nontransient)
Live/work units
Monasteries
Motels (nontransient)
Vacation timeshare properties

Group R-3
Buildings with ≤ two dwelling units
Boarding houses
 (nontransient) ≤ 16 occupants
Boarding houses
 (transient) ≤ 10 occupants
Care facilities ≤ 5 persons
 receiving care
Congregate living facilities
 (nontransient) ≤ 16 occupants
Congregate living facilities
 (transient) ≤ 10 occupants

Group R-4
Alcohol and drug centers
Assisted living facilities
Congregate care facilities
Convalescent facilities
Group homes
Halfway houses
Residential board and
 custodial care facilities
Social rehabilitation facilities

Detached one- and two-family dwellings, as well as townhouses, are not regulated by the *International Building Code* when limited to the conditions of the exception to Section 101.2. They are to be designed and constructed in accordance with the *International Residential Code*.

Topic: Storage Uses
Reference: IBC 311.1
Category: Occupancy Classification
Subject: Group S Occupancies

Code Text: *Storage Group S occupancy includes among others, the use of a building or structure, or a portion thereof, for storage that is not classified as a hazardous occupancy.*

Discussion and Commentary: Where a warehouse or other storage facility does not contain significant amounts of hazardous commodities (as determined by Section 307), it should be considered a Group S occupancy. A facility used for the storage of combustible goods is classified as Group S-1, whereas a Group S-2 occupancy shall be used only for the storage of noncombustible materials. If it is reasonable to believe that a storage building will house combustible goods for any significant period of time, it would be appropriate to consider the structure a Group S-1 occupancy, designed and constructed accordingly. Motor-vehicle-related uses are also included in the Group S category, with repair garages classified as Group S-1 and parking garages (both open and enclosed) as Group S-2 occupancies.

Group S-1	Group S-2
Aerosols, Level 2 and Level 3	Asbestos
Aircraft hangar	Cement in bags
Bags; cloth, burlap, paper	Chalk and crayons
Belting; canvas, leather	Dairy products
Books	Dry cell batteries
Paper in rolls	Electric motors
Cardboard and cardboard boxes	Food products
Clothing	Fresh fruits and vegetables
Furniture	Frozen foods
Grains	Glass
Lumber	Gypsum board
Motor vehicle repair garages	Meats
Tires, bulk storage of	Metals
Tobacco, cigars, cigarettes	Open parking garages
Upholstery and mattresses	Enclosed parking garages

Although the goods being stored in a Group S-2 occupancy must be noncombustible, the code permits a limited amount of combustibles in the packaging or support materials. Wood pallets, paper cartons, paper wrappings, plastic trim and film wrapping are permitted for such purposes.

Topic: Utility and Miscellaneous Uses
Reference: IBC 312.1
Category: Occupancy Classification
Subject: Group U Occupancies

Code Text: *Buildings and structures of an accessory character and miscellaneous structures not classified in any specific occupancy shall be constructed, equipped and maintained to conform to the requirements of* the IBC *commensurate with the fire and life hazard incidental to their occupancy.*

Discussion and Commentary: Those structures not ordinarily occupied by the general public are typically classified as Group U occupancies. The fire load in these structures varies considerably but is usually not excessive. Because these types of uses are not normally occupied, the concern for fire severity is not very great, and as a group they constitute a low hazard. Several of the structures regulated as Group U occupancies are never occupied, such as fences, towers and tanks.

Group U

- Agricultural buildings
- Barns
- Carports
- Fences more than 6 feet in height
- Greenhouses
- Livestock shelters
- Private garages
- Retaining walls
- Sheds
- Stables
- Tanks
- Towers

Private garages and carports classified as Group U occupancies are generally limited to 1,000 square feet and one story in height. However, such structures are permitted to be 3,000 square feet where no repair work is done and no fuel is dispensed.

Topic: Mixed Occupancies
Reference: IBC 508.1
Category: General Building Heights and Areas
Subject: Mixed Use and Occupancy

Code Text: *Each portion of a building shall be individually classified in accordance with Section 302.1. Where a building contains more than one occupancy group, the building or portion thereof shall comply with the applicable provisions of Section 508.2* (Accessory Occupancies), *508.3* (Nonseparated Occupancies), *or 508.4* (Separated Occupancies), *or a combination of these sections.* See exceptions for 1) occupancies separated in accordance with Section 510 (Special Provisions), 2) Group H-1, H-2 and H-3 occupancies required by Table 415.5.2 to be located in a separate and detached building, and 3) uses within live/work units complying with Section 419.

Discussion and Commentary: It is not uncommon for two or more distinct occupancy classifications to occur in the same building. Where such conditions exist, the code requires that such multiple occupancies be either 1) isolated from each other using fire-resistive separation elements (fire barriers and/or horizontal assemblies), or 2) imposed with special provisions that eliminate the need for such fire separations.

Although compliance with only one of the three mixed-occupancy methods is required, it is acceptable to utilize two or even all three methods within the same building.

Study Session 2

Topic: Accessory Occupancies
Reference: IBC 508.2, 508.2.1

Category: General Building Heights and Areas
Subject: Mixed Use and Occupancy

Code Text: *Accessory occupancies are those occupancies that are ancillary to the main occupancy of the building or portion thereof. Aggregate accessory occupancies shall not occupy more than 10 percent of the area of the story in which they are located and shall not exceed the tabular values in Table 503, without building area increases in accordance with Section 506 for such accessory occupancies.*

Discussion and Commentary: The mixed-occupancy method of "Accessory Occupancies" is one of the three design options that the code provides when dealing with mixed-occupancy buildings. This approach is only applicable where one or more of the occupancies is quite small in relationship to the major occupancy in the building. The aggregate floor area of all accessory occupancies is limited to 10 percent of the floor area of the story in which the accessory occupancies are located. In addition, the aggregate floor area of the accessory occupancies cannot exceed the allowable floor area taken directly from Table 503.

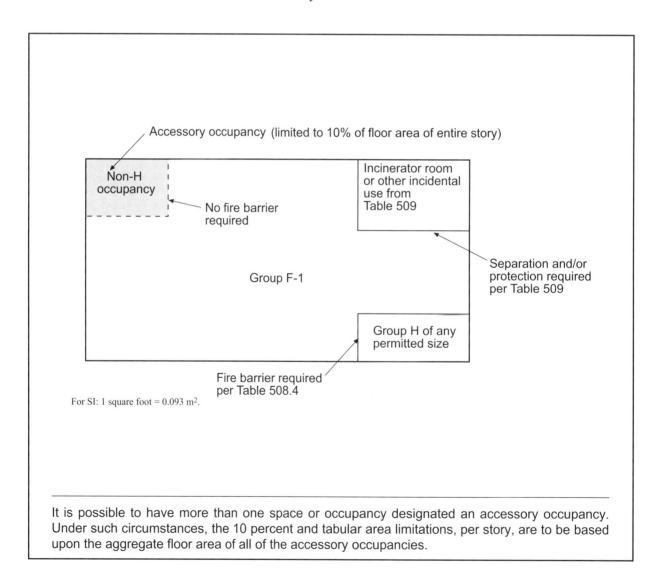

For SI: 1 square foot = 0.093 m².

It is possible to have more than one space or occupancy designated an accessory occupancy. Under such circumstances, the 10 percent and tabular area limitations, per story, are to be based upon the aggregate floor area of all of the accessory occupancies.

Topic: Occupancy Classification
Reference: IBC 508.2.2

Category: Mixed Use and Occupancy
Subject: Accessory Occupancies

Code Text: *Accessory occupancies shall be individually classified in accordance with Section 302.1. The requirements of the IBC shall apply to each portion of the building based on the occupancy classification of that space.*

Discussion and Commentary: The occupancy classification of a use that is regulated under the provisions for accessory occupancies is based solely upon the specific use of that area. Although the size of the occupancy may be quite small in comparison with the remainder of the building, the accessory occupancy has its own unique hazards that must be adequately addressed.

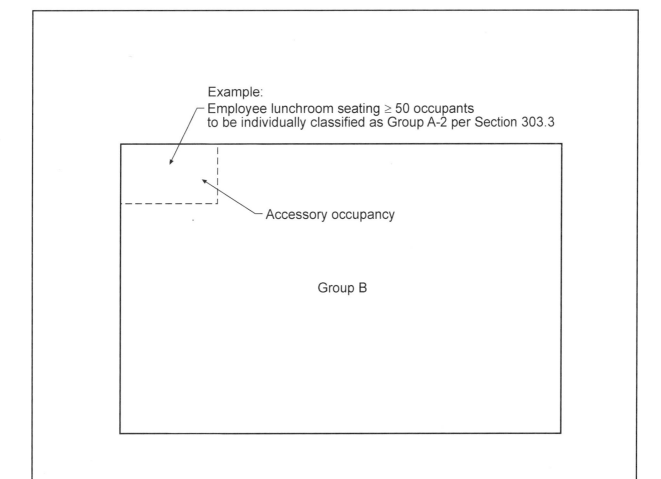

Unless modified by Section 508.2, requirements of the code apply to each occupancy independently based on their specific uses. For example, the ventilation requirements applicable to the accessory occupancy may not be appropriate for the major occupancy. Each of the occupancies would be regulated based upon their unique classifications.

Topic: Height, Area and Separation
Reference: IBC 508.2.3, 508.2.4
Category: Mixed Use and Occupancy
Subject: Accessory Occupancies

Code Text: *The allowable building area and height of the building shall be based on the allowable building area and height for the main occupancy in accordance with Section 503.1. No separation is required between accessory occupancies and the main occupancy.* See exceptions for 1) Group H-2, H-3, H-4 and H-5 occupancies, and 2) dwelling and sleeping units in Groups I-1, R-1, R-2 and R-3.

Discussion and Commentary: Where the methodology of "Accessory Occupancies" is utilized, the allowable height and area of the accessory occupancies, as well as that of the major occupancy, is based solely on the building's major occupancy. There is no mandate to apply the more restrictive height and area provisions of each of the occupancies involved, as required under the "Nonseparated Occupancies" method, nor to go through calculations based on the unity formula as required for "Separated Occupancies."

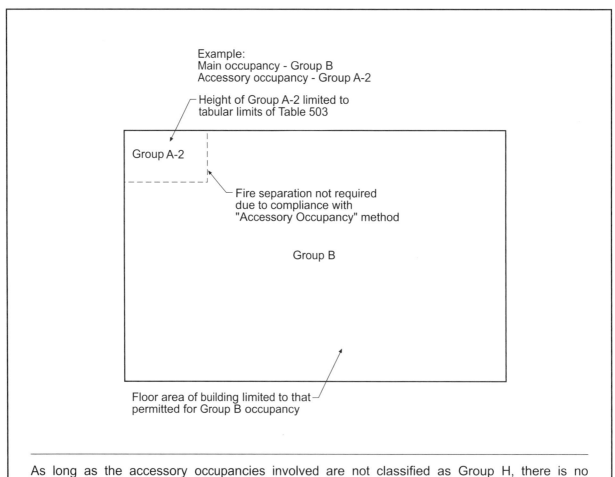

Example:
Main occupancy - Group B
Accessory occupancy - Group A-2

Height of Group A-2 limited to tabular limits of Table 503

Group A-2

Fire separation not required due to compliance with "Accessory Occupancy" method

Group B

Floor area of building limited to that permitted for Group B occupancy

As long as the accessory occupancies involved are not classified as Group H, there is no requirement to separate the assembly occupancies from the major occupancy of the building. In addition, where two or more accessory occupancies are present, they do not need to be separated from each other.

Topic: Occupancy Classification
Reference: IBC 508.3.1
Category: Mixed Use and Occupancy
Subject: Nonseparated Occupancies

Code Text: *Nonseparated occupancies shall be individually classified in accordance with Section 302.1. The requirements of the IBC shall apply to each portion of the building based on the occupancy classification of that space. In addition, the most restrictive provisions of Chapter 9 that apply to the nonseparated occupancies shall apply to the total nonseparated occupancy area.*

Discussion and Commentary: The allowance for "Nonseparated Occupancies," one of the alternatives to the physical separation of different occupancies, is based on the most limiting requirements for building size (height and area) and fire-protection features such as sprinklers, standpipes and alarm systems. Where occupancies are regulated by the nonseparated occupancy provisions, a physical separation is permitted, but it is not required to be fire-resistance rated.

- Type of construction limited by:
    - Lesser height limit of Occupancy 1 or 2
    - Lesser floor area limit of Occupancy 1 or 2
- Most restrictive fire-protection system requirements of Occupancy 1 and 2

Where the provisions for nonseparated occupancies are utilized in a high-rise building as defined by Section 403, the special high-rise provisions that apply to the nonseparated occupancies are applicable to the entire building.

Study Session 2

Topic: Height, Area and Separation
Reference: IBC 508.3.2, 508.3.3
Category: Mixed Use and Occupancy
Subject: Nonseparated Occupancies

Code Text: *The allowable building area and height of the building or portion thereof shall be based on the most restrictive allowances for the occupancy groups under consideration for the type of construction of the building in accordance with Section 503.1. No separation is required between nonseparated occupancies. See exceptions for 1) Group H-2, H-3, H-4 and H-5 occupancies, and 2) dwelling and sleeping units in Groups I-1, R-1, R-2 and R-3.*

Discussion and Commentary: Where the option for nonseparated occupancies is utilized to address a mixed-occupancy building, it is necessary to determine the maximum building size for each of the occupancies that are not appropriately separated. The maximum allowable height and area for each occupancy would be based upon the building's type of construction. The most restrictive height and area of those nonseparated occupancies would then be the limiting size for the combination of such occupancies.

Example:

Given: A Type VB building contains both Group B and Group E occupancies.

Determine: The height and area limitations if the occupancies are not separated under the nonseparated occupancies provisions of Section 508.3.

OCCUPANCY	ALLOWABLE HEIGHT[1]	ALLOWABLE AREA[1]
Group B[2]	2 stories	9,000 square feet
Group E[2]	1 story	9,500 square feet

[1] Based on Table 503 assuming no permitted increases.

[2] Most restrictive fire protection requirements of Chapter 9 also applicable to entire building.

∴ Thus, for nonseparated occupancies, the maximum building size would be 1 story and 9,000 square feet.

Nonseparated Occupancies

The use of the nonseparated occupancies method is not applicable to high-hazard occupancies. Those areas or spaces classified as Group H occupancies must be isolated from other occupancies within the building by fire barriers and/or horizontal assemblies in accordance with Table 508.4 for occupancy separations.

Topic: Occupancy Classification
Reference: IBC 508.4
Category: Mixed Use and Occupancy
Subject: Separated Occupancies

Code Text: *Separated occupancies shall be individually classified in accordance with Section 302.1. Each separated space shall comply with* the IBC *based on the occupancy classification of that portion of the building.*

Discussion and Commentary: Under the provisions for "Separated Occupancies," each of the distinct uses is to be individually classified as to occupancy. This approach is consistent with that for accessory occupancies and nonseparated occupancies. The concept of separated occupancies provides for a fire-resistance-rated separation in order to isolate the hazards associated with a specific occupancy from other portions of the building.

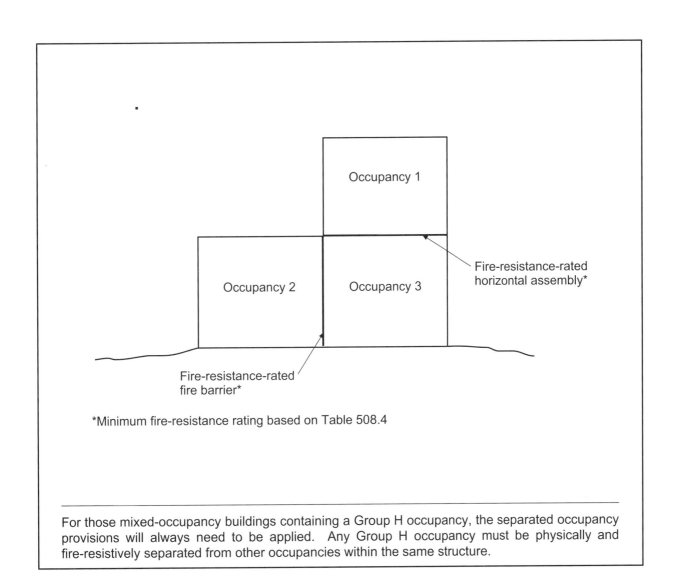

For those mixed-occupancy buildings containing a Group H occupancy, the separated occupancy provisions will always need to be applied. Any Group H occupancy must be physically and fire-resistively separated from other occupancies within the same structure.

Topic: Allowable Height and Area
Reference: IBC 508.4.2, 508.4.3

Category: Mixed Use and Occupancy
Subject: Separated Occupancies

Code Text: *In each story, the building area shall be such that the sum of the ratios of the actual building area of each separated occupancy divided by the allowable area of each separated occupancy shall not exceed 1. Each separated occupancy shall comply with the building height limitations based on the type of construction of the building in accordance with Section 503.1.*

Discussion and Commentary: The approach to separated occupancies mandates that the ratios of the actual and allowable floor areas be calculated in order to determine compliance. Often known as the "unity formula," this calculation recognizes the relationship between the permitted sizes of the various occupancies involved. The unity formula is only applicable where the separated occupancy method is utilized and does not apply to accessory occupancies or nonseparated occupancies.

Example:

Fire-resistance-rated occupancy separations per Table 508.4

B / R-1 / A-2

Allowable height per Section 503.1 for each individual occupancy

$$\frac{\text{Actual area A-2}}{\text{Allowable area A-2}} + \frac{\text{Actual area B}}{\text{Allowable area B}} + \frac{\text{Actual area R-1}}{\text{Allowable area R-1}} \leq 1.0$$

The height limitations for separated occupancies are based upon the general provisions of Section 503.1. The height limit, in both feet and stories, is to be measured from the grade plane, and the measurement must include all intervening fire areas.

Topic: Separation and Construction
Reference: IBC 508.4.4, 508.4.4.1
Category: Mixed Use and Occupancy
Subject: Separated Occupancies

Code Text: *Individual occupancies shall be separated from adjacent occupancies in accordance with Table 508.4. Required separations shall be fire barriers constructed in accordance with Section 707 or horizontal assemblies constructed in accordance with Section 711, or both, so as to completely separate adjacent occupancies.*

Discussion and Commentary: A matrix, Table 508.4, has been established to identify any required fire-resistance-rated separation between various occupancies. The table is based on the perceived degree of dissimilarity between the occupancies involved. Where Table 508.4 requires a level of fire-resistance between the adjoining occupancies, fire barriers and/or horizontal separations are to be used. The intended result is that the hazards associated with one occupancy be completely isolated from those present in the remainder of the building.

TABLE 508.4
REQUIRED SEPARATION OF OCCUPANCIES (HOURS)

OCCUPANCY	A, E		I-1, I-3, I-4		I-2		R[a]		F-2, S-2[b], U		B, F-1, M, S-1		H-1		H-2		H-3, H-4		H-5	
	S	NS	S	NS	S	NS	S	NS	S	NS	S	NS	S	NS	S	NS	S	NS	S	NS
A, E	N	N	1	2	2	NP	1	2	N	1	1	2	NP	NP	3	4	2	3	2	NP
I-1, I-3, I-4	—	—	N	N	2	NP	1	NP	1	2	1	2	NP	NP	3	NP	2	NP	2	NP
I-2	—	—	—	—	N	N	2	NP	2	NP	2	NP	NP	NP	3	NP	2	NP	2	NP
R[a]	—	—	—	—	—	—	N	N	1[c]	2[c]	1	2	NP	NP	3	NP	2	NP	2	NP
F-2, S-2[b], U	—	—	—	—	—	—	—	—	N	N	1	2	NP	NP	3	4	2	3	2	NP
B, F-1, M, S-1	—	—	—	—	—	—	—	—	—	—	N	N	NP	NP	2	3	1	2	1	NP
H-1	—	—	—	—	—	—	—	—	—	—	—	—	N	NP	NP	NP	NP	NP	NP	NP
H-2	—	—	—	—	—	—	—	—	—	—	—	—	—	—	N	NP	1	NP	1	NP
H-3, H-4	—	—	—	—	—	—	—	—	—	—	—	—	—	—	—	—	1[d]	NP	1	NP
H-5	—	—	—	—	—	—	—	—	—	—	—	—	—	—	—	—	—	—	N	NP

S = Buildings equipped throughout with an automatic sprinkler system installed in accordance with Section 903.3.1.1.
NS = Buildings not equipped throughout with an automatic sprinkler system installed in accordance with Section 903.3.1.1.
N = No separation requirement.
NP = Not permitted.
a See Section 420.
b. The required separation from areas used only for private or pleasure vehicles shall be reduced by 1 hour but to not less than 1 hour.
c. See Section 406.3.4.
d. Separation is not required between occupancies of the same classification.

Although the title of Section 508.4 implies that the method described requires a physical and/or fire-resistive separation, that is not always the case. A number of the occupancies do not require such a separation due to the lack of hazard dissimilarity between the uses that occur.

Topic: Classification and Area Limits
Reference: IBC 509.1 – 509.3
Category: General Building Heights and Areas
Subject: Incidental Uses

Code Text: *Incidental uses are ancillary functions associated with a given occupancy that generally pose a greater level of risk to that occupancy and are limited to those uses listed in Table 509. Incidental uses shall not be individually classified in accordance with Section 302.1. Incidental uses shall be included in the building occupancies within which they are located. Incidental uses shall not occupy more than 10 percent of the building area of the story in which they are located.*

Discussion and Commentary: It is common to find uses that are typical of the general occupancy classification of the building, yet which create a hazard different from the other hazards found in the occupancy. An example would be a chemistry laboratory classroom in a high school building. The code addresses such conditions by requiring incidental uses to be separated from the remainder of the building with fire-resistance-rated construction, or to be protected with an automatic sprinkler system. An incidental use should be assigned an occupancy classification consistent with the portion of the building in which it is located.

TABLE 509
INCIDENTAL USES

ROOM OR AREA	SEPARATION AND/OR PROTECTION
Furnace room where any piece of equipment is over 400,000 Btu per hour input	1 hour or provide automatic sprinkler system
Rooms with boilers where the largest piece of equipment is over 15 psi and 10 horsepower	1 hour or provide automatic sprinkler system
Refrigerant machinery room	1 hour or provide automatic sprinkler system
Hydrogen cutoff rooms, not classified as Group H	1 hour in Group B, F, M, S and U occupancies; 2 hours in Group A, E, I and R occupancies.
Incinerator rooms	2 hours and automatic sprinkler system
Paint shops, not classified as Group H, located in occupancies other than Group F	2 hours; or 1 hour and provide automatic sprinkler system
Laboratories and vocational shops, not classified as Group H, located in a Group E or I-2 occupancy	1 hour or provide automatic sprinkler system
Laundry rooms over 100 square feet	1 hour or provide automatic sprinkler system
Group I-3 cells equipped with padded surfaces	1 hour
Waste and linen collection rooms located in either Group I-2 occupancies or ambulatory care facilities	1 hour
Waste and linen collection rooms over 100 square feet	1 hour or provide automatic sprinkler system
Stationary storage battery systems having a liquid electrolyte capacity of more than 50 gallons for flooded lead-acid, nickel cadmium or VRLA, or more than 1,000 pounds for lithium-ion and lithium metal polymer used for facility standby power, emergency power or uninterruptable power supplies	1 hour in Group B, F, M, S and U occupancies; 2 hours in Group A, E, I and R occupancies.

For SI: 1 square foot = 0.0929 m², 1 pound per square inch (psi) = 6.9 kPa, 1 British thermal unit (Btu) per hour = 0.293 watts, 1 horsepower = 746 watts, 1 gallon = 3.785 L.

The separation and protection of incinerator rooms is unique in that both methods are required. Although an automatic sprinkler system is required within the incinerator room, it is still necessary to provide a fire-resistance-rated separation utilizing minimum 2-hour fire barriers.

Topic: Construction and Protection
Reference: IBC 509.4
Category: General Building Heights and Areas
Subject: Incidental Uses

Code Text: *Where Table 509 specifies a fire-resistance-rated separation, the incidental uses shall be separated from the remainder of the building by a fire barrier constructed in accordance with Section 707 or a horizontal assembly constructed in accordance with Section 711, or both. Where Table 509 permits an automatic sprinkler system without a fire barrier, the incidental uses shall be separated from the remainder of the building by construction capable of resisting the passage of smoke.*

Discussion and Commentary: In utilizing Table 509, it is common that two options are available for addressing rooms or areas considered incidental uses. A fire barrier may often be used to isolate the specific hazard from the remainder of the building. As an alternative, a sprinkler system may be used to limit any fire in the incidental use to that space only. By incorporating smoke containment construction, little if any smoke created would be transferred to other portions of the building.

Note: Doors shall:
• be self-closing or automatic-closing upon detection of smoke
• have no air transfer openings
• have no excessive undercuts

Construction to resist the passage of smoke

Where an automatic sprinkler system is used to provide protection for an incidental use, the area must still be separated from the remainder of the building. The separation must be constructed such that smoke will be contained. Enclosure doors also are regulated.

Study Session 2
IBC Chapter 3 and Sections 508 – 509

1. An institutional occupancy is typically considered Group _____.

    a. A
    b. B
    c. I ⟵ (circled)
    d. R

    Reference __308__

2. A Group _____ occupancy is the general classification for miscellaneous and utility structures.

    a. A
    b. M
    c. S
    d. U ⟵ (circled)

    Reference __312.1__

3. Accessory religious educational rooms need not be considered separate occupancies where the occupant load is less than _____ occupants.

    a. 100 ⟵ (circled)
    b. 150
    c. 200
    d. 300

    Reference __305.1.1__

4. An accessory assembly area may be classified as a Group B occupancy where the floor area is a maximum of _____ square feet.

    a. 120   b. 399

    (c.) 749   d. 1,000

Reference __303.1.2__

5. In a Group E middle school, an assembly area associated with the Group E shall be separated at what minimum level from the remainder of the school building?

    (a.) no separation is required (considered same occupancy)

    b. 1-hour fire barrier

    c. 2-hour fire barrier

    d. 3-hour fire barrier

Reference __303.1.3__

6. Which of the following uses is typically considered a Group A-4 occupancy?

    a. restaurant with a dance floor   b. school library

    c. outdoor football stadium   (d.) indoor hockey arena

Reference __303.5__

7. Which of the following uses is not considered a Group B occupancy?

    (a.) convenience store   b. motor vehicle showroom

    c. car wash   d. college classroom building

Reference __304.1__

8. The gaming floor of a casino shall be classified as a Group _____ occupancy.

    (a.) A-2   b. A-3

    c. B   d. M

Reference __303.3__

9. A Group _____ occupancy classification is to be assigned to a facility where combustible dusts are generated in a manner that creates a fire or explosion hazard.

   a. H-1   (b.) H-2
   c. H-4   d. H-5

   Reference 307.4

10. Buildings containing materials that present a detonation hazard are typically considered _____ occupancies.

    (a.) Group H-1   b. Group H-2
    c. Group H-3   d. Group H-5

    Reference 307.3

11. A foster-care facility providing care on a 24-hour basis to six or more infants/toddlers ($2^{1}/_{2}$ years of age or less) is classified as a Group _____ occupancy.

    a. E   b. I-1
    (c.) I-2   d. R-4

    Reference 308.4

12. Prior to any permitted increases, the maximum allowable quantity per control area of a Class IB flammable liquid permitted in a storage condition in a one-story Group F-1 occupancy is _____ gallons.

    a. 15   b. 30
    c. 60   (d.) 120

    Reference 307.1

13. A facility used for supervised custodial care and housing more than 16 persons is classified as a Group _____ occupancy.

    a. (I-1)   b. I-4
    c. R-3   d. R-4

    Reference 308.3

14. Which of the following uses is not considered a Group M occupancy?

    a. wholesale store b. sales room

    c. motor vehicle showroom d. motor vehicle service station

    Reference __309.1__

15. A sorority house with 40 occupants is considered a Group _____ occupancy.

    a. R-1 b. R-2

    c. R-3 d. R-4

    Reference __310.4__

16. Which one of the following facilities with 12 residents receiving custodial care would not be considered a Group R-4 occupancy?

    a. halfway house b. drug abuse center

    c. convalescent facility d. detoxification facility

    Reference __310.6__

17. Which of the following uses is not considered a Group S-2 occupancy?

    a. open parking garage b. enclosed parking garage

    c. dry cell battery storage d. stable

    Reference __311.3__

18. A public library is typically classified as a _____ occupancy.

    a. Group A-3 b. Group A-4

    c. Group B d. Group M

    Reference __303.4__

19. A manufacturing facility utilizing highly toxic materials exceeding the maximum allowable quantities set forth in Table 307.1(2) is considered a _____ occupancy.

    a. Group F-1 b. Group F-2

    c. Group H-3 d. Group H-4

    Reference __307.6__

Study Session 2

20. Aerosol storage buildings shall be classified as Group _____ occupancies when constructed in accordance with the *International Fire Code*.

    a. H-2   b. H-3
    c. S-1   d. S-2

    Reference 311.2

21. What is the largest occupant load permitted in an assembly building for it to be classified as a Group B occupancy?

    a. 15   b. 30
    c. 49   d. 99

    Reference 303.1.1

22. Where a training and skill development use occurs in other than a school or academic program, it is classified as a(n) Group _____ occupancy.

    a. A-3   b. B
    c. E   d. M

    Reference 304.1

23. A congregate living facility (transient) with an occupant load of 8 persons is to be classified as a Group _____ occupancy.

    a. R-1   b. R-2
    c. R-3   d. R-4

    Reference 310.5

24. By definition, a transient residential dwelling unit or sleeping unit has a maximum occupancy period of _____ days.

    a. 14   b. 30
    c. 90   d. 180

    Reference DEF

25. Unless protected by an automatic sprinkler system, what is the minimum required fire separation between a 350-square-foot laundry room and the manufacturing building it is incidental to?

    a. no fire separation is required

    b. 1-hour fire partition

    c. 1-hour fire barrier and/or horizontal assembly

    d. smoke barrier

    Reference 509.4.1

26. What minimum level of protection is required for an incinerator room located in a manufacturing facility?

    a. 1-hour fire partition only

    b. automatic sprinkler system only

    c. 1-hour fire barrier and an automatic sprinkler system

    d. separation of two hours and an automatic sprinkler system

    Reference 509

27. What minimum level of protection is required for a chemistry laboratory/classroom in a high school?

    a. 1-hour fire barrier/horizontal assembly only

    b. automatic sprinkler system only

    c. both a 1-hour fire barrier/horizontal assembly and an automatic sprinkler system

    d. either a 1-hour fire barrier/horizontal assembly or an automatic sprinkler system

    Reference 509

28. Where an automatic sprinkler system without a fire barrier is utilized for the protection of incidental use areas, what minimum level of separation is required?

    a. 1-hour fire partition

    b. 1-hour fire barrier/horizontal assembly

    c. 1-hour smoke barrier walls and horizontal assemblies

    d. construction capable of resisting the passage of smoke

    Reference _____

Study Session 2

29. Where an accessory occupancy is not required to be separated from the major use by a fire barrier, it is limited to a maximum floor area of _____ of the area of the story in which it is located.

    a. 10 percent   b. 15 percent
    c. 25 percent   d. 33 percent

    Reference 509.4

30. Any accessory occupancy area classified as a Group _____ occupancy must always be separated by a fire barrier horizontal assembly from other occupancies in the building.

    a. A   b. E
    c. H   d. I

    Reference 508.2.4

31. Where the provisions for nonseparated occupancies are used for a mixed-occupancy building, the most restrictive _____ requirements shall apply to the nonseparated uses.

    a. fire protection system   b. means of egress
    c. accessibility   d. interior finish

    Reference 508.3.1

32. Where the provisions for separated occupancies are used for a nonsprinklered mixed-occupancy building, the minimum separation between a Group A-2 and Group B occupancy shall be a _____.

    a. 1-hour fire partition
    b. 1-hour fire barrier/horizontal assembly
    c. 2-hour fire barrier/horizontal assembly
    d. 2-hour fire wall

    Reference 508.4

33. Where the provisions for separated occupancies are used for a sprinklered mixed-occupancy building, the minimum separation between a Group H-2 and Group F-1 occupancy shall be a _____ .

    a. 1-hour fire partition
    b. 1-hour fire barrier/horizontal assembly
    c. 2-hour fire barrier/horizontal assembly
    d. 2-hour fire wall

    Reference 508.4

34. Where the provisions for separated occupancies are utilized for a fully sprinklered building housing both a Group A-2 occupancy and a Group R-1 occupancy, the minimum required separation between the two occupancies shall be a _____ .

    a. 1-hour fire partition
    b. 1-hour fire barrier/horizontal assembly
    c. 2-hour fire barrier/horizontal assembly
    d. 2-hour fire wall

    Reference 508.4

35. Under the provisions for separated occupancies, what is the minimum hourly fire-resistance-rated separation required between a Group F-1 occupancy and a Group S-1 occupancy where located in a nonsprinklered building?

    a. 0 hours (no separation requirement)
    b. 1-hour
    c. 2-hour
    d. 3-hour

    Reference 508.4

Study Session 2

Study Session

3

2012 IBC Chapter 6
Types of Construction

OBJECTIVE: To gain an understanding of how a building is classified as a specific type of construction, based on the construction materials and the various building elements' resistance to fire.

REFERENCE: Chapter 6, 2012 *International Building Code*

KEY POINTS:
- What do the various types of construction indicate?
- How are the required fire-resistance ratings of building elements determined?
- Why are exterior walls regulated by additional criteria?
- Why are exterior walls protected differently based on fire separation distance?
- At what minimum distance is the protection of exterior walls unnecessary?
- Which types of materials are required to be used as building elements of a Type I or II building?
- How do the two different categories of Type I construction differ in fire protection? Type II construction?
- Which types of materials are required for use in the exterior walls of a Type III structure? In the interior building elements?
- What is another name for Type IV construction?
- How shall exterior walls be constructed? Interior building elements?
- What are the minimum construction details for columns used in a building of Type IV construction?
- In Type IV buildings, what is the minimum size of heavy-timber members used in the floor and roof framing? Floors? Roofs? Partitions?
- Where the minimum dimensions for Type IV solid sawn members are prescribed, how are the equivalent sizes established for glued laminated members?
- Type V buildings may be constructed of which building materials?
- How does a Type VA building differ from a Type VB building?
- In noncombustible Type I and II buildings, where may fire-retardant-treated wood be used?
- Which specific allowances are provided for combustible materials in Type I and Type II buildings?

KEY POINTS:
(Cont'd)

- What are the limitations for the use of fire-retardant-treated wood in the roof construction of noncombustible buildings? In nonbearing partitions? In nonbearing exterior walls?
- Which building elements are considered primary structural frame elements for the determination of fire resistance? Secondary members?
- When are bracing members considered part of the structural frame?
- Under which conditions may the required fire resistance of roof supports be reduced?
- At what height may the required fire resistance of roof construction be eliminated? In which occupancies is the elimination not applicable?
- For which building elements are heavy-timber members and 1-hour fire-resistance-rated construction interchangeable?
- How can a sprinkler system affect a building's type of construction classification?
- How are interior nonbearing walls regulated for fire resistance based on construction type? Exterior nonbearing walls?

Topic: Construction Types
Reference: IBC 602.1
Category: Types of Construction
Subject: Construction Classification

Code Text: *Buildings and structures erected or to be erected, altered or extended in height or area shall be classified in one of the five construction types defined in Sections 602.2 through 602.5.*

Discussion and Commentary: There are two major groupings based on the construction materials: noncombustible construction (Types I and II) and noncombustible or combustible construction (Types III, IV and V). These groupings are divided into two more categories: protected, where the major structural elements are provided with some degree of fire resistance, and unprotected, where no fire protection of the building elements is typically mandated. Protected construction is further distinguished in Type I buildings where the required protection for many structural elements exceeds a 1-hour fire-resistance rating.

Noncombustible	Exterior and interior (bearing or nonbearing) walls, floors, roofs and structural elements are to be of noncombustible materials	I	A	B
		II	A	B
Noncombustible or combustible	Exterior walls are to be of noncombustible materials	III	A	B
		IV	HT	
		V	A	B

It is the intent of the *International Building Code* that each building be classified as a single type of construction. The construction materials and the degree to which such materials are protected determine the classification based on the criteria of Table 601 and Chapter 6.

Topic: Fire Resistance Ratings
Reference: IBC 602.1, Table 601
Category: Types of Construction
Subject: Construction Classification

Code Text: *The building elements shall have a fire-resistance rating not less than that specified in Table 601. The protection of openings, ducts and air transfer openings in building elements shall not be required unless required by other provisions of the IBC.*

Discussion and Commentary: The building elements regulated by Table 601 for types of construction include primary structural frame members, such as columns, girders and trusses; bearing walls, both interior and exterior; floor construction, including supporting beams and joists; and roof construction, consisting of supporting beams, joists, rafters and other members. The required fire-resistance rating for each of these elements is based on the specific type of construction assigned to the building. The required fire-resistance rating can be as high as a 3-hour or as little as a 0-hour (no fire-resistance rating required).

TABLE 601
FIRE-RESISTANCE RATING REQUIREMENTS FOR BUILDING ELEMENTS (HOURS)

BUILDING ELEMENT	TYPE I A	TYPE I B	TYPE II A^d	TYPE II B	TYPE III A^d	TYPE III B	TYPE IV HT	TYPE V A^d	TYPE V B
Primary structural frame^g (see Section 202)	3^a	2^a	1	0	1	0	HT	1	0

Bearing walls
  Exterior$^{f,\,g}$
  Interior
	3
	3^a
		2
		2^a
			1
			1
				0
				0
					2
					1
						2
						0
							2
							1/HT
								1
								1
									0
									0

Nonbearing walls and partitions Exterior	See Table 602								
Nonbearing walls and partitions Interior^e	0	0	0	0	0	0	See Section 602.4.6	0	0
Floor construction and associated secondary members (see Section 202)	2	2	1	0	1	0	HT	1	0
Roof construction and associated secondary members (see Section 202)	$1\frac{1}{2}^b$	$1^{b,c}$	$1^{b,c}$	0^c	$1^{b,c}$	0	HT	$1^{b,c}$	0

For SI: 1 foot = 304.8 mm.

a. Roof supports: Fire-resistance ratings of primary structural frame and bearing walls are permitted to be reduced by 1 hour where supporting a roof only.
b. Except in Group F-1, H, M and S-1 occupancies, fire protection of structural members shall not be required, including protection of roof framing and decking where every part of the roof construction is 20 feet or more above any floor immediately below. Fire-retardant-treated wood members shall be allowed to be used for such unprotected members.
c. In all occupancies, heavy timber shall be allowed where a 1-hour or less fire-resistance rating is required.
d. An approved automatic sprinkler system in accordance with Section 903.3.1.1 shall be allowed to be substituted for 1-hour fire-resistance-rated construction, provided such system is not otherwise required by other provisions of the code or used for an allowable area increase in accordance with Section 506.3 or an allowable height increase in accordance with Section 504.2. The 1-hour substitution for the fire resistance of exterior walls shall not be permitted.
e. Not less than the fire-resistance rating required by other sections of this code.
f. Not less than the fire-resistance rating based on fire separation distance (see Table 602).
g. Not less than the fire-resistance rating as referenced in Section 704.10

Where a structure is separated by one or more fire walls, the code treats those individual compartments created by the fire walls as separate buildings. Thus, each separate compartment would be considered a distinct building for the purpose of classification by type of construction.

Topic: Fire Separation Distance
Reference: IBC 602.1, Table 602
Category: Types of Construction
Subject: Exterior Walls

Code Text: *Exterior walls shall have a fire-resistance rating not less than that specified in Table 602.*

Discussion and Commentary: The rationale behind exterior wall protection is that an owner has no control over what occurs on an adjacent lot. The lot line concept provides a convenient means of protecting one building from another insofar as radiant heat could potentially be transmitted from one building to another during a fire. The requirements are based on "fire separation distance," which must be considered for all exterior walls. Where such walls are also bearing walls, the provisions of Table 601 also apply, governed by the more restrictive of the hourly ratings. Additional provisions for exterior walls are found in Section 705.

Table 602 Regulates Exterior Walls Only

- Table 602 used in conjunction with Table 601 for fire resistance of exterior bearing walls
- Only Table 602 used for nonbearing exterior walls
- Based primarily on occupancy type
- Highest required rating for exterior wall is 3 hours
- Final threshold at ≥ 30'
- Additional provisions for exterior walls and openings in Section 705

TABLE 602
FIRE-RESISTANCE RATING REQUIREMENTS FOR EXTERIOR WALLS BASED ON FIRE SEPARATION DISTANCE[a, e, h]

FIRE SEPARATION DISTANCE = X (feet)	TYPE OF CONSTRUCTION	OCCUPANCY GROUP H[f]	OCCUPANCY GROUP F-1, M, S-1[g]	OCCUPANCY GROUP A, B, E, F-2, I, R, S-2[g], U[b]
X < 5[c]	All	3	2	1
5 ≤ X < 10	IA	3	2	1
	Others	2	1	1
10 ≤ X < 30	IA, IB	2	1	1[d]
	IIB, VB	1	0	0
	Others	1	1	1[d]
X ≥ 30	All	0	0	0

For SI: 1 foot = 304.8 mm.
a. Load-bearing exterior walls shall also comply with the fire-resistance rating requirements of Table 601.
b. For special requirements for Group U occupancies, see Section 406.3.
c. See Section 706.1.1 for party walls.
d. Open parking garages complying with Section 406 shall not be required to have a fire-resistance rating.
e. The fire-resistance rating of an exterior wall is determined based upon the fire separation distance of the exterior wall and the story in which the wall is located.
f. For special requirements for Group H occupancies, see Section 415.5.
g. For special requirements for Group S aircraft hangars, see Section 412.4.1.
h. Where Table 705.8 permits nonbearing exterior walls with unlimited area of unprotected openings, the required fire-resistance rating for the exterior walls is 0 hours.

The "fire separation distance" is defined in Section 202 as the distance measured from the building face to the closest interior lot line, to the centerline of a street, alley or public way, or to an assumed imaginary line between two buildings on the same lot.

Topic: Fire Separation Distance
Reference: IBC Table 602, 202
Category: Type of Construction
Subject: Exterior Walls

Code Text: Fire separation distance *is the distance measured from the building face to one of the following: 1) the closest interior lot line; 2) to the centerline of a street, alley or public way; or 3) to an imaginary line between two buildings on the property. The distance shall be measured at right angles from the face of the wall.*

Discussion and Commentary: The atmospheric separation provided between a building and an adjoining structure provides resistance to fire spread due to radiant heat transfer. Many provisions throughout the IBC, such as those regulating projections and parapets, are based upon the degree of separation provided. A measurement at a right angle from the building face addresses heat transfer from the building of fire incident toward other structures and properties.

Protection of walls adjacent to lot lines

The measurement of fire separation distance is theoretically taken at an infinite number of points along the exterior wall. In reality, zones are created adjacent to the building under consideration, with higher degrees of regulation mandated for those zones closest to the building.

Topic: Fire-Resistance Ratings **Category:** Type of Construction
Reference: IBC T601, Note f; T602, Note a **Subject:** Exterior Walls

Code Text: The minimum fire-resistance rating for exterior bearing walls shall be *not less than the fire-resistance rating based on fire separation distance*. In addition to the fire-resistance rating requirements for exterior walls based on fire separation distance, *load-bearing exterior walls shall also comply with the fire-resistance rating requirements of Table 601.*

Discussion and Commentary: When analyzing an exterior wall for its required level of fire resistance, it is necessary to use both Tables 601 and 602. Table 601 addresses the potential need for structural stability of exterior bearing walls under fire conditions. Exterior nonbearing walls are not regulated by this table. The concern of radiant heat transfer from an adjoining burning building results in Table 602 regulating the exterior wall rating based upon its fire separation distance (typically the distance from the building face to the lot line). This concern exists for both bearing and nonbearing exterior walls.

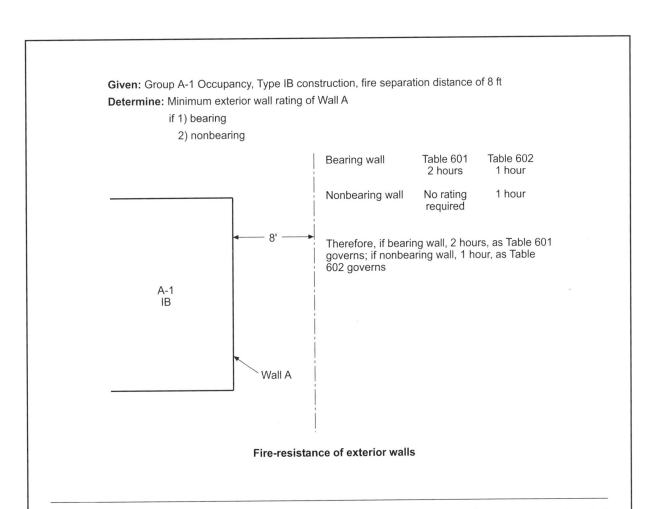

Fire-resistance of exterior walls

The provisions of Section 704.10 regulate the fire resistance of structural frame members located within nonbearing exterior walls. The required fire rating is based on the highest of ratings found in Table 601 (structural frame, exterior bearing wall) and Table 602 (fire separation distance).

Study Session 3

Topic: Types I and II Construction
Reference: IBC 602.2
Category: Types of Construction
Subject: Construction Classification

Code Text: *Types I and II construction are those types of construction in which the building elements listed in Table 601 are of noncombustible materials, except as permitted in Section 603 and elsewhere in* the IBC.

Discussion and Commentary: Type I buildings are noncombustible, and the building elements are also provided with a mandated degree of fire resistance. This type of construction requires the highest level of fire protection specified in the code. Type II buildings are also of noncombustible construction; however, the level of fire resistance is usually less than that required for Type I structures. Buildings of Type II construction may have a limited degree of fire resistance (Type IIA) or no fire resistance whatsoever (Type IIB). There are limited allowances for the use of fire-retardant-treated wood in nonbearing partitions, nonbearing exterior walls and roof construction.

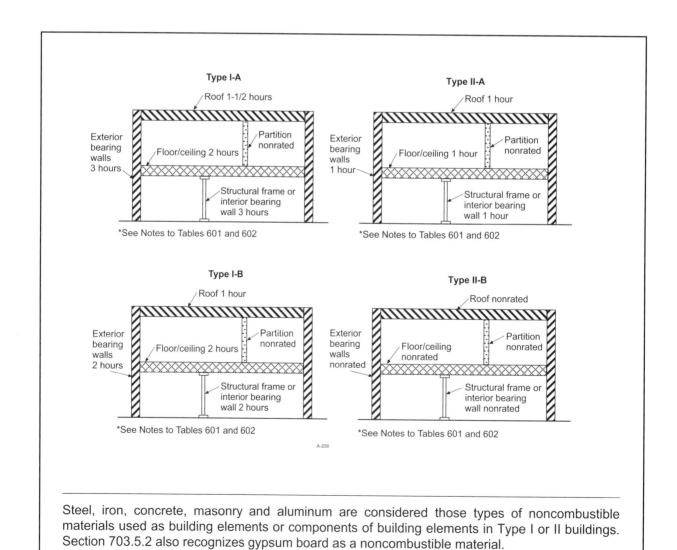

Steel, iron, concrete, masonry and aluminum are considered those types of noncombustible materials used as building elements or components of building elements in Type I or II buildings. Section 703.5.2 also recognizes gypsum board as a noncombustible material.

Topic: Type III Construction
Reference: IBC 602.3
Category: Types of Construction
Subject: Construction Classification

Code Text: *Type III construction is that type of construction in which the exterior walls are of noncombustible materials and the interior building elements are of any material permitted by the IBC. Fire-retardant-treated wood framing complying with Section 2303.2 shall be permitted within exterior wall assemblies of a 2-hour rating or less.*

Discussion and Commentary: Type III buildings are considered combustible buildings and are either protected or unprotected. This building type was developed out of the necessity to prevent conflagrations in heavily built-up areas where buildings were erected side-by-side in congested downtown business districts. To limit the spread of fire from building to building, exterior walls were required to be of both noncombustible and fire-resistant construction.

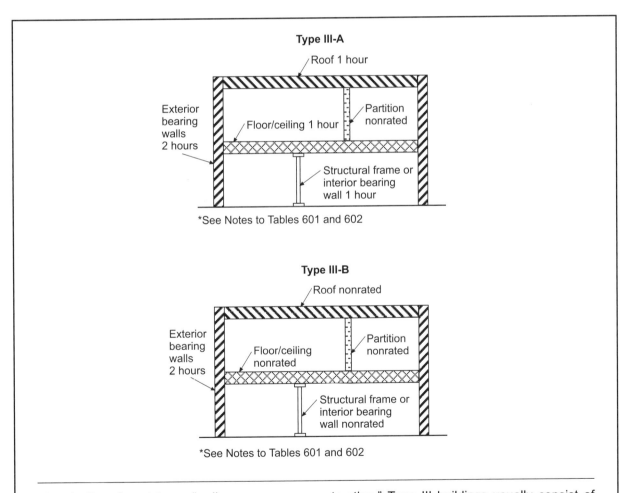

Historically referred to as "ordinary masonry construction," Type III buildings usually consist of masonry exterior walls with wood floor and roof systems. However, the IBC permits such walls to contain fire-retardant-treated wood as an element of the exterior wall construction.

Topic: Type IV Construction
Reference: IBC 602.4
Category: Types of Construction
Subject: Construction Classification

Code Text: *Type IV construction (Heavy Timber, HT) is that type of construction in which the exterior walls are of noncombustible materials and the interior building elements are of solid or laminated wood without concealed spaces. The details of Type IV construction shall comply with the provisions of Section 602.4. Fire-retardant-treated wood framing complying with Section 2303.2 shall be permitted within exterior wall assemblies with a 2-hour rating or less.*

Discussion and Commentary: Referred to as "heavy-timber," buildings of Type IV construction are essentially Type III buildings with an interior of timber members. To conform to Type IV construction, building members must be of substantial thickness. Given the characteristics of massive wood members, there is little chance for sudden structural collapse during or after a fire.

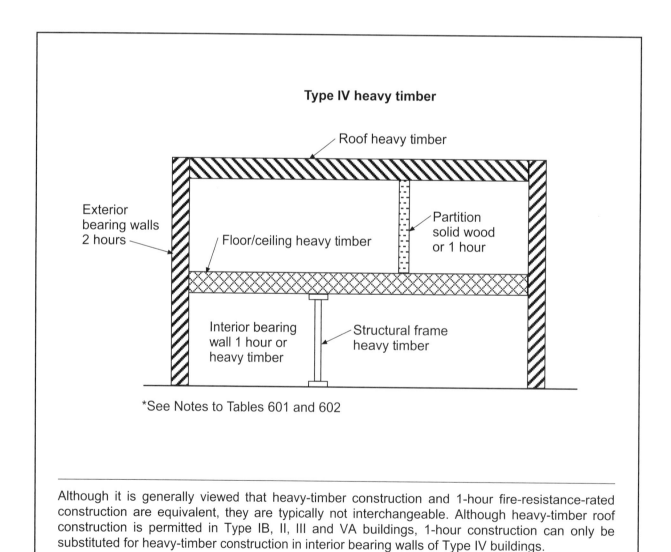

Although it is generally viewed that heavy-timber construction and 1-hour fire-resistance-rated construction are equivalent, they are typically not interchangeable. Although heavy-timber roof construction is permitted in Type IB, II, III and VA buildings, 1-hour construction can only be substituted for heavy-timber construction in interior bearing walls of Type IV buildings.

Topic: Glued-Laminated Members
Reference: IBC 602.4
Category: Types of Construction
Subject: Type IV Construction

Code Text: *Minimum solid-sawn nominal dimensions are required for structures built using Type IV construction (HT). For glued-laminated members the equivalent net finished width and depths corresponding to the minimum nominal width and depths of solid sawn lumber are required as specified in Table 602.4.*

Discussion and Commentary: Solid-sawn wood members and glued-laminated timbers are manufactured with different methods and procedures: therefore; they do not have the same dimensions. However, they both have the same inherent fire-resistive capability that has been long recognized in the code. A comparison of the varied widths and depths indicates how the reduced width of glued-laminated members is offset by an increase in the depth, providing for similar cross-sectional areas.

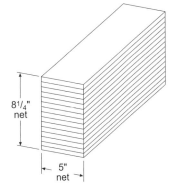

For SI: 1 inch = 25.4 mm

Solid-sawn member   Glued-laminated member

**TABLE 602.4
WOOD MEMBER SIZE**

MINIMUM NOMINAL SOLID SAWN SIZE		MINIMUM GLUED-LAMINATED NET SIZE	
Width, inch	Depth, inch	Width, inch	Depth, inch
8	8	6 3/4	8 1/4
6	10	5	10 1/2
6	8	5	8 1/4
6	6	5	6
4	6	3	6 7/8

For SI: 1 inch = 25.4 mm.

Table 602.4 in the IBC, which sets forth the minimum net sizes for glued-laminated members necessary to comply as Type IV construction elements, is a simplified version of a similar table in American Institute of Timber Construction (AITC) Standard 113.

Topic: Type V Construction
Reference: IBC 602.5

Category: Types of Construction
Subject: Construction Classification

Code Text: *Type V construction is that type of construction in which the structural elements, exterior walls and interior walls are of any materials permitted by* the IBC.

Discussion and Commentary: Type V buildings are essentially construction systems that will not fit into any of the other higher types of construction specified by the IBC. Although the construction normally considered Type V is the conventional light-frame wood building, any combination of approved materials can be considered Type V construction. Section 602.1.1 indicates that a building is not required to conform to the details of a type of construction higher than the type that meets the minimum requirements based on occupancy, even though certain features of such a building actually conform to a higher construction type.

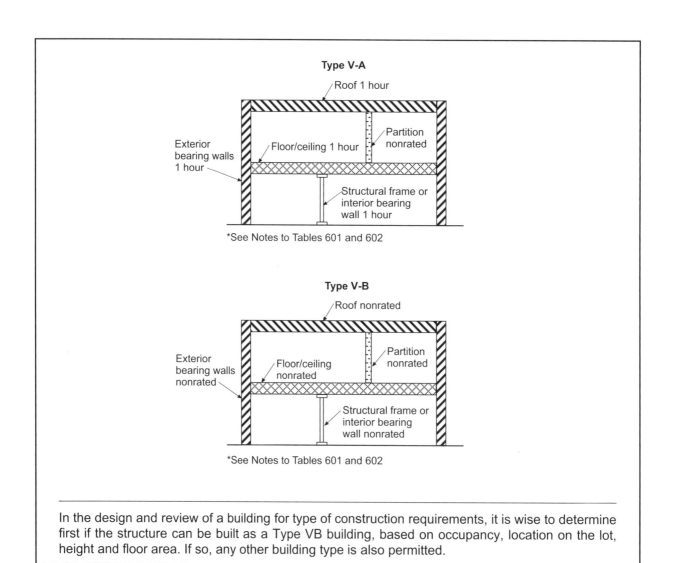

In the design and review of a building for type of construction requirements, it is wise to determine first if the structure can be built as a Type VB building, based on occupancy, location on the lot, height and floor area. If so, any other building type is also permitted.

Topic: Allowable Uses
Reference: IBC 603.1

Category: Types of Construction
Subject: Combustible Material in Types I and II

Code Text: *Combustible materials shall be permitted in buildings of Type I and II construction in the following applications:* (25 applications listed).

Discussion and Commentary: Materials used in the construction of buildings classified as either Type I or Type II are intended to be noncombustible, thereby not increasing the potential fire loading (fuel contribution). There are, however, a number of applications where the presence of combustible building materials is desirable in otherwise noncombustible structures. Such materials are typically permitted where they are adequately protected, limited in use or amount, or installed in accordance with the *International Fire Code*, *International Mechanical Code®*, or other provisions of the *International Building Code*.

Combustible materials permitted in buildings of Type I and Type II construction in the following applications:

- Fire-retardant-treated wood in:
  - Roof construction of most buildings.
  - Nonbearing partitions with fire-resistance rating ≤ 2 hours.
  - Nonbearing exterior walls requiring no fire rating.
- Thermal and acoustical insulation with limited flame spread.
- Foam plastics per Chapter 26.
- A, B or C roof coverings.
- Interior floor finish, trim, millwork such as, doors, frames, etc.
- Stages and platforms per Section 410.
- Blocking for handrails, fixtures, windows and door frames, etc.
- Light-transmitting plastics per Chapter 26.
- Nailing or furring strips per Section 803.11.
- Heavy timber for specific components.
- Additional applications as specified.

In Type I and II construction, the use of fire-retardant-treated wood is permitted in roof construction of all Type IB and II buildings and those buildings of Type IA construction that do not exceed two stories or have a top-story height of at least 20 feet.

Study Session 3

Topic: Ducts, Piping and Electrical
Reference: IBC 603.1.1, 603.1.2, 603.1.3
Category: Types of Construction
Subject: Combustible Material in Types I and II

Code Text: *Combustible materials shall be permitted in buildings of Type I or II construction in accordance with Sections 603.1.1 through 603.1.3. The use of nonmetallic ducts shall be permitted when installed in accordance with the limitations of the* International Mechanical Code. *The use of combustible piping materials shall be permitted when installed in accordance with the limitations of the* International Mechanical Code *and the* International Plumbing Code®. *The use of electrical wiring methods with combustible insulation, tubing, raceways and related components shall be permitted when installed in accordance with the limitations of* the IBC.

Discussion and Commentary: The IMC contains requirements for nonmetallic ducts that address the issues of flammability, flame spread and smoke development. The IPC regulates the use of combustible piping materials, such as plastic, and also addresses those same characteristics applicable to nonmetallic ducts. Similar regulations apply to combustible wiring materials. These provisions in Chapter 6 clarify that such combustible materials are acceptable for installation in buildings of Type I and II construction, provided they meet the limitations set forth in the appropriate code.

Additional allowances for combustible mechanical and plumbing materials in Type I and II buildings are established by Exception 24 of Section 603. Specified combustible materials are permitted in concealed spaces under the provisions of Section 718.5.

Topic: Primary Structural Frame
Reference: IBC Table 601, 202
Category: Types of Construction
Subject: Building Elements

Code Text: *The primary structural frame shall include all of the following structural members: 1) the columns; 2) structural members having direct connections to the columns, including girders, beams, trusses and spandrels; 3) members of the floor construction and roof construction having direct connections to the columns; and 4) bracing members that are essential to the vertical stability of the primary structural frame under gravity loading shall be considered part of the primary structural frame whether or not the bracing member carries gravity loads.*

Discussion and Commentary: To maintain stability of the building as a whole, the major structural elements are regulated for endurance when subjected to a fire. In addition to the columns, beams and girders, both interior bearing walls and exterior bearing walls are regulated to a level of fire resistance equal to or greater than that of other structural elements. Secondary members, such as floor joists, roof joists or rafters, are protected within the rated floor-ceiling or roof-ceiling assemblies.

Components of primary structural frame

Lateral force bracing is not considered part of the structural frame where it serves no other purpose than to resist the lateral loads. For example, lateral load bracing within exterior nonbearing walls or interior partitions would be protected by the wall or partition construction. Such bracing elements would be considered secondary members.

Topic: Roof Construction
Reference: IBC Table 601, Note b
Category: Types of Construction
Subject: Building Elements

Code Text: *Except in Group F-1, H, M and S-1 occupancies, fire protection of structural members shall not be required, including protection of roof framing and decking, where every part of the roof construction is 20 feet or more above any floor immediately below. Fire-retardant-treated wood members shall be allowed to be used for such unprotected members.*

Discussion and Commentary: Where there is limited potential for a fire to be of a severe nature at the roof structure due to its height above the floor below, an elimination of the required fire-resistance rating of the roof construction is permitted in construction of Types I, IIA, IIIA and VA. Elimination of the required fire resistance is not allowed where combustible or hazardous materials are located adjacent to the roof.

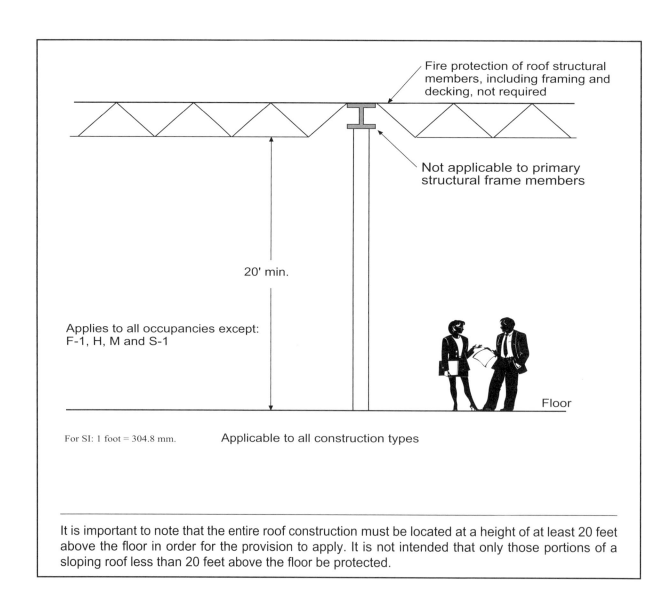

It is important to note that the entire roof construction must be located at a height of at least 20 feet above the floor in order for the provision to apply. It is not intended that only those portions of a sloping roof less than 20 feet above the floor be protected.

Topic: Fire-Resistant Substitution
Reference: IBC Table 601, Note d
Category: Types of Construction
Subject: Building Elements

Code Text: *An approved automatic sprinkler system in accordance with Section 903.3.1.1 shall be allowed to be substituted for 1-hour fire-resistance-rated construction, provided such system is not otherwise required by other provisions of the IBC or used for an allowable area increase in accordance with Section 506.3 or an allowable height increase in accordance with Section 504.2. The 1-hour substitution for the fire resistance of exterior walls shall not be permitted.*

Discussion and Commentary: In buildings of Type IIA, IIIA or VA construction, the code allows for a reduction in the "built-in" fire protection, due to the presence of a sprinkler system that would not otherwise be required. This reduction does not apply to exterior walls, nor does it apply when a sprinkler system is used for an increase in allowable height or allowable area.

Type II-B Construction + Approved Automatic Sprinkler System* = Type II-A Construction

Type III-B Construction + Approved Automatic Sprinkler System* = Type III-A Construction

Type V-B Construction + Approved Automatic Sprinkler System* = Type V-A Construction

***Unless** otherwise required, or used for allowable height and area increase*

The application of this note to Table 601 may be extremely limited based on the language of Section 901.2, which states that *any fire protection system for which an exception or reduction to the provisions of the code has been granted shall be considered to be a required system.*

Study Session 3

Quiz

Study Session 3
IBC Chapter 6

1. What types of construction are considered "noncombustible"?

    (a.) I, II   b. I, II, III, IV

    c. III, IV   d. III, IV, V

    Reference _602.2_

2. Type III buildings are constructed with _____ exterior walls and interior elements _____.

    a. fire-resistant, of noncombustible materials

    b. noncombustible, of noncombustible materials

    (c.) noncombustible, of any material permitted by the code

    d. combustible, of any material permitted by the code

    Reference _602.3_

3. A Type IV building is also described as _____ construction.

    a. ordinary   (b.) heavy timber

    c. combustible   d. provincial

    Reference _602.4_

4. In buildings of Type III and IV construction, under what condition is fire-retardant-treated wood framing permitted within exterior wall assemblies?

    a. the wall has a 2-hour rating or less
    b. the fire separation distance exceeds 10 feet
    c. wood columns of heavy-timber sizes are used
    d. the wall is a nonbearing element

    Reference 602.3

5. Where supporting floor loads, wood columns of Type IV construction shall be of what minimum nominal size?

    a. 5 inches by 5 inches
    b. 6 inches by 6 inches
    c. 6 inches by 8 inches
    d. 8 inches by 8 inches

    Reference 602.4.1

6. Where used in floor framing, wood beams of Type IV construction shall be of what minimum nominal size?

    a. 4 inches by 8 inches
    b. 4 inches by 10 inches
    c. 6 inches by 10 inches
    d. 8 inches by 10 inches

    Reference 602.4.2

7. Roofs shall be without concealed spaces in buildings of Type _____ construction.

    a. I
    b. III
    c. IV
    d. V

    Reference 602.4.5

8. Which of the following materials is permitted in a building of Type VB construction?

    a. wood
    b. steel
    c. masonry
    d. all of the above

    Reference 602.5

Study Session 3

9. In a building of Type IB construction, what is the minimum required fire-resistance rating of the floor construction?

   a. 3 hours
   b. 2 hours
   c. 1 hour
   d. 0 hours (no rating required)

   Reference 601

10. Which one of the following members is not considered to be a part of the primary structural frame?

    a. columns
    b. girders
    c. bracing members carrying gravity loads
    d. floor joists

    Reference 202

11. In a building of Type IIB construction, what is the minimum required fire-resistance rating of the roof construction?

    a. 0 hours (no rating required)
    b. 1 hour
    c. 1½ hours
    d. 2 hours

    Reference 601

12. In a building of Type VA construction, which of the following building elements does not require a minimum 1-hour fire-resistance rating?

    a. primary structural frame
    b. exterior bearing wall
    c. interior nonbearing wall
    d. roof construction

    Reference 601

13. In a one-story Type IA building, what is the minimum required fire-resistance rating for the interior bearing walls supporting the roof only?

    a. 3 hours
    b. 2 hours
    c. 1½ hours
    d. 1 hour

    Reference 601

14. In Type IA buildings limited to two stories in height, what building element is permitted to be constructed of fire-retardant-treated wood?

    a. structural frame
    b. bearing walls
    c. floor construction
    d. roof construction ✓

    Reference 603.1

15. In a Type IB building housing a Group A-4 occupancy, fire protection of the roof structural members, framing and decking is not required where every portion of the roof construction is a minimum of _____ feet above the floor below.

    a. 18
    b. 20 ✓
    c. 25
    d. 35

    Reference 601

16. In a Type IIB building housing a Group I-2 occupancy, what is the minimum required rating for an exterior nonbearing wall located with a fire separation distance of 8 feet?

    a. 3 hours
    b. 2 hours
    c. 1 hour ✓
    d. 0 hours (no rating required)

    Reference 602

17. In a Type IIA building housing a Group R-2 occupancy, what is the minimum required rating for an exterior nonbearing wall located with a fire separation distance of 3 feet?

    a. 3 hours
    b. 2 hours
    c. 1 hour ✓
    d. 0 hours (no rating required)

    Reference 602

18. In a Type IIIA building housing a Group A-1 occupancy, what is the minimum required rating for an exterior bearing wall located with a fire separation distance of 10 feet?

    a. 3 hours
    b. 2 hours ✓
    c. 1 hour
    d. 0 hours (no rating required)

    Reference 601

19. In a Type VB building housing a Group B occupancy, what is the minimum required rating for an exterior bearing wall located with a fire separation distance of 10 feet?

    a. 3 hours
    b. 2 hours
    c. 1 hour
    d. 0 hours (no rating required) ✓

    Reference __602__

20. Considering all occupancies and construction types, for which minimum fire separation distance is no fire rating required for a nonbearing exterior wall?

    a. 20 feet
    b. 30 feet ✓
    c. 40 feet
    d. 60 feet

    Reference __602__

21. In a building of Type I or II construction, nonbearing partitions having a maximum fire-resistance rating of _____ are permitted to be constructed of fire-retardant-treated wood.

    a. 0 hours (no rating required)
    b. ½ hour
    c. 1 hour
    d. 2 hours ✓

    Reference __603__

22. A detached garage classified as a Group U occupancy shall not be required to have a fire-resistance rating where the fire separation distance is a minimum of _____ feet.

    a. 0, no rating is required
    b. 3
    c. 5 ✓
    d. 10

    Reference __406.3.2__

23. In a building of Type I or II construction, nonbearing exterior walls having a maximum fire-resistance rating of _____ are permitted to be constructed of fire-retardant-treated wood.

    a. 0 hours (no rating required) ✓
    b. ½ hour
    c. 1 hour
    d. 2 hours

    Reference __603.1.2__

24. Show windows may be of combustible construction in Type I and II construction where located a maximum of _____ above grade.

    a. 15 feet
    b. 35 feet
    c. one story
    d. three stories

    Reference 603.1 #9

25. Which type of roof covering is not permitted on an office building of Type IIB construction?

    a. Class A
    b. Class B
    c. Class C
    d. nonclassified

    Reference 603. #4

26. Which of the following methods of construction is not permitted for partitions in Type IV structures?

    a. two layers of ½-inch fire-retardant-treated structural wood panels
    b. 4 inches of laminated solid-wood construction
    c. two layers of 1-inch matched boards
    d. one-hour fire-resistance-rated construction

    Reference 602.4.6

27. What is the minimum fire-resistive rating required for interior metal stud partitions in Type IIB construction?

    a. 0, no rating is required
    b. 1 hour
    c. 2 hours
    d. 3 hours

    Reference 601

28. What is the minimum required vertical distance between the upper floor and the roof to allow the use of fire-retardant-treated wood in the roof construction of an eight-story Type IA building?

    a. 20 feet
    b. 25 feet
    c. there is no minimum distance required
    d. FRT wood is never permitted in such a case

    Reference 601

Study Session 3

29. For an office building of Type IB construction, what is the minimum required fire-resistance rating for an exterior bearing wall located on an interior lot line?

    a. 0, no rating is required   b. 1 hour
    c. 2 hours   d. 3 hours

    Reference _____

30. In a building of Type IIA construction, heavy timber members may be used in lieu of one-hour fire-resistance-rated construction for which building element?

    a. structural frame members   b. interior bearing walls
    c. floor construction   d. roof construction

    Reference 601

31. In a building of Type IV construction, 1-hour combustible construction is permitted in lieu of heavy-timber construction for which of the following building elements?

    a. floor construction   b. exterior bearing walls
    c. interior bearing walls   d. roof construction

    Reference _____

32. Where an exterior bearing wall of a Group M occupancy of Type IIB construction is located 3 feet from an interior lot line, the wall must have a minimum fire-resistance rating of _____ hour(s).

    a. 0 (no rating required)   b. 1
    c. 2   d. 3

    Reference _____

33. A nonbearing exterior wall of a Type IB open parking garage complying with Section 406 shall have a minimum fire-resistance rating of _____ hour(s) where the fire separation distance is 10 feet.

    a. 0 (no rating required)   b. 1
    c. 2   d. 3

    Reference _____

34. A glued-laminated beam, where utilized in a Type IV building requiring a 6-inch by 10-inch solid-sawn member of nominal size, shall have a minimum net finished size of _____.

    a. $5^1/_4$ inches by $9^1/_4$ inches
    b. 6 inches by 10 inches
    c. $6^3/_4$ inches by $10^1/_2$ inches
    d. 5 inches by $10^1/_2$ inches

    Reference _____

35. Under general conditions, thermal and acoustical insulation other than foam plastic is permitted to be installed in buildings of Type I or II construction, provided the insulation has a maximum flame spread index of _____.

    a. 25
    b. 75
    c. 200
    d. 450

    Reference _____

2012 IBC Chapter 5
General Building Heights and Areas

OBJECTIVE: To gain an understanding of how a building is classified and regulated based on its floor area, height and number of stories.

REFERENCE: Chapter 5, 2012 *International Building Code*

KEY POINTS:
- How and why must buildings be identified by their address?
- How is the maximum basic floor area of a building determined? The basic height in feet and number of stories?
- How is the tabular allowable building area determined?
- How must multiple buildings located on the same lot be handled?
- What types of special industrial occupancies are exempt from the height and area limitations of Table 503?
- Under which conditions may the basic allowable height of a building be increased? How much of an increase is permitted?
- Which special provisions address the construction of towers, spires, steeples and other roof structures?
- What is a mezzanine?
- Under which conditions may a floor level be considered a mezzanine?
- Under what conditions is a mezzanine permitted to be enclosed or in some manner separated from the room in which it is located?
- Which conditions provide for an increase in the allowable floor areas specified in Table 503?
- What is the minimum width of a yard or public way that can provide for a floor area increase?
- How much of a building's perimeter must be considered "open" for the area increase to apply?
- What amount of area increase is permitted for sprinklered single-story buildings? For sprinklered multistory structures?
- How is a basement viewed in the calculation of maximum allowable floor area?
- How is the maximum total combined floor area for a multistory building determined?

KEY POINTS:
(Cont'd)

- Which occupancy groups are eligible for unlimited floor area in a one-story nonsprinklered building?
- Which criteria must be met for buildings of Groups A-4, B, F, M or S to be unlimited in floor area? Groups A-1 and A-2?
- What is the minimum width required for open space surrounding an unlimited area building?
- Under which conditions may that width be reduced?
- Under what limitations is a Group A-3 occupancy permitted to be of unlimited area?
- How may high-hazard occupancies be accommodated in an unlimited area building?
- Which limitations are placed on motion picture theaters of unlimited area?
- Where the special provisions for a horizontal building separation allowance are applied, what criteria must be met?
- Which special provisions address the situation where an enclosed or open parking garage is located below another occupancy group? Below an open parking garage?
- What are the height limitations for apartment houses and other Group R-2 occupancies where the special provisions are met?
- How is an open parking garage regulated where located below Groups A, B, I, M or R?

Topic: Address Identification
Reference: IBC 501.2
Category: Building Heights and Areas
Subject: General Provisions

Code Text: *New and existing buildings shall be provided with approved address numbers or letters. Each character shall be not less than 4 inches (102 mm) in height and not less than 0.5 inch (12.7 mm) in width. They shall be installed on a contrasting background and be plainly visible from the street or road fronting the property. When required by the fire code official, address numbers shall be provided in additional approved locations to facilitate emergency response. Where access is by means of a private road and the building address cannot be viewed from the public way, a monument, pole or other approved sign or means shall be used to identify the structure. Address numbers shall be maintained.*

Discussion and Commentary: Buildings must be provided with plainly visible and legible address numbers posted on the building or in such a place on the property that the building may be identified by emergency services such as fire, medical and police.

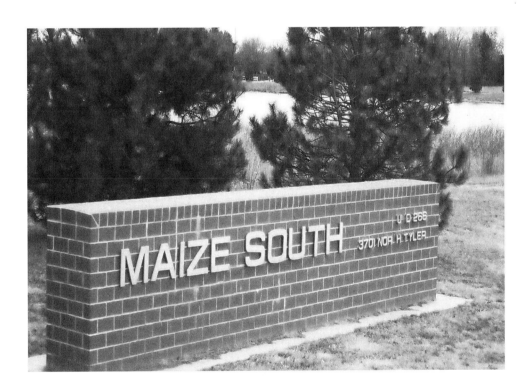

As a fundamental requirement, the approved street numbers are to be placed in a location readily visible from the street fronting the property. The fire code official has the authority to require that the address numbers be posted in more than one location to help eliminate any confusion or delay in identifying the location of the emergency.

Study Session 4

Topic: Building Area
Reference: IBC 503.1, 202

Category: Building Heights and Areas
Subject: Definitions

Code Text: *Building area is the area included within surrounding exterior walls (or exterior walls and fire walls) exclusive of vent shafts and courts. Areas of the building not provided with surrounding walls shall be included in the building area if such areas are included within the horizontal projection of the roof or floor above.*

Discussion and Commentary: The building area must be determined in order to verify that it does not exceed the maximum allowable area as determined by Section 503.1. The building area is considered, in very general terms, the "footprint" of the building, excluding those unroofed areas and any projections that may extend beyond the exterior walls. Where complying mezzanines are located within a building, they are not assumed to contribute to the building area.

Definition of Building Area

Basements are usually considered to be unoccupied spaces and, as such, do not pose a high degree of hazard. Therefore, under all conditions, a single basement does not contribute to a building's allowable height or area. However, it cannot exceed the size permitted for a single story.

Topic: Story Above Grade/Basement
Reference: IBC 503, 202
Category: Building Heights and Areas
Subject: Definitions

Code Text: *A "story above grade plane" is any story having its finished floor surface entirely above grade plane, or in which the finished surface of the floor next above the basement is 1) more than 6 feet above grade plane, or 2) more than 12 feet above the finished ground level at any point. A "basement" is a story that is not a story above grade plane.*

Discussion and Commentary: A number of provisions in the IBC are applicable based on the location of the floor under consideration, relative to the exterior ground level. Therefore, it is necessary to define specifically the circumstances under which a floor level is considered a story above grade plane.

The "grade plane" is defined as a reference plane representing the average of finished ground level adjoining the building at exterior walls. It is measured at the lowest point between the building and the lot line, though never more than 6 feet from the building.

Topic: Use of Table 503
Reference: IBC 503.1, Table 503
Category: Building Heights and Areas
Subject: Height and Area Limitations

Code Text: *The building height and area shall not exceed the limits specified in Table 503 based on the type of construction as determined by Section 602 and the occupancies as determined by Section 302 except as modified hereafter.*

Discussion and Commentary: Table 503 is the foremost code provision used in establishing equivalent risk (offsetting a building's inherent fire hazard—represented by occupancy group—with materials and construction features). Sections 504 and 506 give height and area increases to the limits of Table 503 for buildings with certain features. Table 503 has three components. The left column lists all of the occupancy classifications, the top row lists the various construction types, and each cell of the matrix contains the specific height (in stories above grade plane) and area (per story) limitations for the occupancy group/type of construction combination in question.

TABLE 503
ALLOWABLE BUILDING HEIGHTS AND AREAS[a]
Building height limitations shown in feet above grade plane. Story limitations shown as stories above grade plane.
Building area limitations shown in square feet, as determined by the definition of "Area, building," per story

		TYPE OF CONSTRUCTION								
		TYPE I		TYPE II		TYPE III		TYPE IV	TYPE V	
		A	B	A	B	A	B	HT	A	B
	HEIGHT(feet)	UL	160	65	55	65	55	65	50	40
GROUP		STORIES(S) AREA (A)								
A-1	S	UL	5	3	2	3	2	3	2	1
	A	UL	UL	15,500	8,500	14,000	8,500	15,000	11,500	5,500
A-2	S	UL	11	3	2	3	2	3	2	1
	A	UL	UL	15,500	9,500	14,000	9,500	15,000	11,500	6,000
A-3	S	UL	11	3	2	3	2	3	2	1
	A	UL	UL	15,500	9,500	14,000	9,500	15,000	11,500	6,000
A-4	S	UL	11	3	2	3	2	3	2	1
	A	UL	UL	15,500	9,500	14,000	9,500	15,000	11,500	6,000
A-5	S	UL	UL	UL	UL	UL	UL	UL	UL	UL
	A	UL	UL	UL	UL	UL	UL	UL	UL	UL
B	S	UL	11	5	3	5	3	5	3	2
	A	UL	UL	37,500	23,000	28,500	19,000	36,000	18,000	9,000
E	S	UL	5	3	2	3	2	3	1	1
	A	UL	UL	26,500	14,500	23,500	14,500	25,500	18,500	9,500
F-1	S	UL	11	4	2	3	2	4	2	1
	A	UL	UL	25,000	15,500	19,000	12,000	33,500	14,000	8,500
F-2	S	UL	11	5	3	4	3	5	3	2
	A	UL	UL	37,500	23,000	28,500	18,000	50,500	21,000	13,000

(Continued)

The maximum height in feet above grade plane is shown along a top row and is based solely on construction type. Buildings must meet both the height in stories and the height in feet criteria to be considered in compliance with that type of construction.

Topic: Special Industrial Occupancies
Reference: IBC 503.1.1
Category: Building Heights and Areas
Subject: Height and Area Limitations

Code Text: *Buildings and structures designed to house special industrial processes that require large areas and unusual building heights to accommodate craneways or special machinery and equipment, including among others, rolling mills; structural metal fabrication shops and foundries; or the production and distribution of electric, gas or steam power, shall be exempt from the building height and area limitations of Table 503.*

Discussion and Commentary: A limited number of buildings that house special industrial processes need extensive heights and/or areas for their operations. The activities that occur are generally of moderate to low hazard, and the buildings are not typically accessible to the public. Therefore, it has been deemed appropriate that no type of construction limitations should be placed on these unique structures.

The special processes addressed by this provision are limited to those having a low or moderate degree of hazard, commonly designated as Group F-1 or S-1 occupancies. Where the processes under consideration would necessitate a Group H classification because of the high hazards involved, the application of this provision is inappropriate.

Study Session 4

Topic: Buildings on the Same Lot
Reference: IBC 503.1.2
Category: Building Heights and Areas
Subject: Height and Area Limitations

Code Text: *Two or more buildings on the same lot shall be regulated as separate buildings or shall be considered as portions of one building if the building height of each building and the aggregate building area of buildings are within the limitations of Table 503 as modified by Sections 504 and 506. The provisions of* the IBC *applicable to the aggregate building shall be applicable to each building.*

Discussion and Commentary: In general, the provisions of Section 705.3 require an assumed imaginary line to be located between two buildings on the same site to regulate exterior wall and opening protection, as well as projection and roof-covering requirements. This method would provide protection equivalent to that of buildings on adjoining lots.

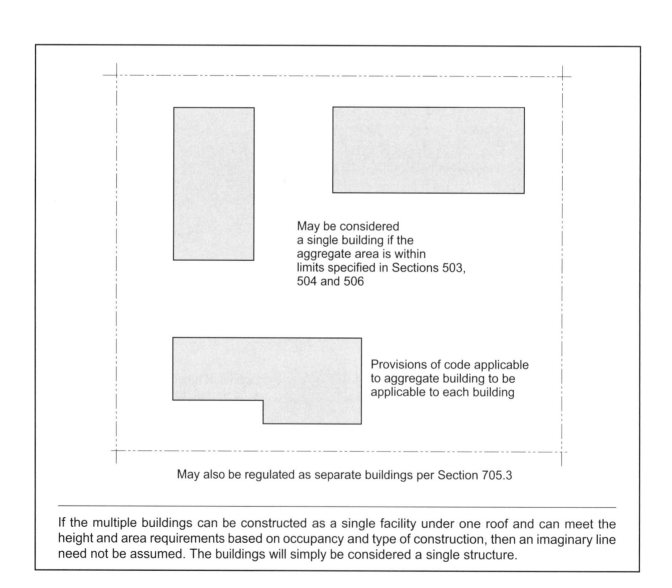

May be considered a single building if the aggregate area is within limits specified in Sections 503, 504 and 506

Provisions of code applicable to aggregate building to be applicable to each building

May also be regulated as separate buildings per Section 705.3

If the multiple buildings can be constructed as a single facility under one roof and can meet the height and area requirements based on occupancy and type of construction, then an imaginary line need not be assumed. The buildings will simply be considered a single structure.

Topic: Automatic Sprinkler Increase
Reference: IBC 504.2
Category: Building Heights and Areas
Subject: Height Modifications

Code Text: *Where a building is equipped throughout with an approved automatic sprinkler system installed in accordance with Section 903.3.1.1, the value specified in Table 503 for maximum building height is increased by 20 feet (6096 mm) and the maximum number of stories is increased by one. These increases are permitted in addition to the building area increase in accordance with Sections 506.2 and 506.3. See exceptions for specific Group I and H occupancies and for buildings utilizing the fire-resistance-rating substitution method.*

Discussion and Commentary: The installation of an automatic sprinkler system improves the fire-safety aspects of a building to the degree that an increase in height is justified. The presence of a sprinkler system also can be used to provide the sizable area increase permitted in Section 506.3.

EXAMPLE

Given: A Type VA office building is permitted to be 3 stories and 50 feet in height per Table 503.

If sprinkler system installed per Section 504.2, the story limit may be increased by one story, and the height can exceed the limit in Table 503 by 20 feet.

Increase does not apply when sprinkler installed under following conditions:
1. Group I-2 occupancies of Type IIB, III, IV or V construction
2. Group H-1, H-2, H-3 or H-5 occupancies
3. Fire-resistance-rating substitution per Table 601, note d

For SI: 1 foot = 304.8 mm

In a residential building protected with an NFPA 13R sprinkler system, the increase is permitted, provided the building does not exceed four stories or 60 feet. The provision does not intend for a residential sprinkler system to be installed in buildings over four stories in height.

Study Session 4

Topic: Roof Structures
Reference: IBC 504.3

Category: Building Heights and Areas
Subject: Height Modifications

Code Text: *Towers, spires, steeples and other roof structures shall be constructed of materials consistent with the required type of construction of the building except where other construction is permitted by Section 1509.2.5. Such structures shall not be used for habitation or storage. The structures shall be unlimited in height if of noncombustible materials and shall not extend more than 20 feet (6096 mm) above the allowable building height if of combustible materials (see Chapter 15 for additional requirements).*

Discussion and Commentary: The types of structures addressed by this provision are intended to be unoccupied with no significant fire loading. It would seem logical that the height of such structures could be increased over that required for typical buildings. The only limitation occurs where the structure is of combustible materials, which would create a higher hazard.

For SI: 1 foot = 304.8 mm.

The provisions of Section 1509 regulate the type of construction for towers, spires and similar rooftop structures based on the height above the roof surface, the height above grade and the largest cross-sectional dimension. Penthouses and equipment screening are also addressed.

Topic: Definition
Reference: IBC 505.2.1, 202
Category: Building Heights and Areas
Subject: Mezzanines

Code Text: *A mezzanine is an intermediate level or levels between the floor and ceiling of any story and in accordance with Section 505. The aggregate area of a mezzanine or mezzanines within a room shall not exceed one-third of the floor area of that room or space in which they are located.* See exceptions which allow for increased mezzanine sizes in 1) special industrial occupancies of Type I or II construction, and 2) fully sprinklered Type I or II buildings provided with an approved emergency voice/alarm communication system.

Discussion and Commentary: Because of size limitation and openness (a mezzanine is open to the room in which it is located, with exceptions), an intermediate floor level within a room adds minimal hazard to the building and its occupants. The occupants of the mezzanine by means of sight, smell or hearing will be able to determine if there is some emergency or fire taking place either on the mezzanine or in the room in which the mezzanine is located.

The maximum permitted floor area of a mezzanine is based on the floor area of the room in which it is located. Only those portions of the lower room that are unenclosed may be considered in the calculation of maximum mezzanine size.

Study Session 4

Topic: Scope
Reference: IBC 505.2

Category: Building Heights and Areas
Subject: Mezzanines

Code Text: *A mezzanine or mezzanines in compliance with Section 505.2 shall be considered a portion of the story below. Such mezzanines shall not contribute to either the building area or number of stories as regulated by Section 503.1. The area of a mezzanine shall be included in determining the fire area.*

Discussion and Commentary: There are two distinct benefits derived from the qualification of a floor level as a mezzanine. One, the mezzanine is not considered in the allowable number of stories, and two, for allowable area purposes, the mezzanine floor area does not increase the building area of the story in which it is located. However, in the determination of fire area size for sprinkler requirements, the floor area must be considered. The requirements for sprinkler systems are generally based on the fire load expected in an occupancy; thus, an increased floor area would increase the potential fire loading.

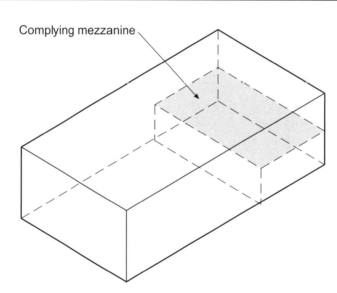

Mezzanine:
- Does <u>not</u> contribute to floor area for maximum allowable area
- Does <u>not</u> contribute as an additional story
- Does contribute to floor area for fire area size determination

Example:
For 8,000 sq ft first floor as shown with 2,000 sq ft mezzanine, building area is 8,000 sq ft, building is one story in height, and fire area is 10,000 sq ft

For SI: 1 square foot = 0.93 m^2.

Although it is quite possible that an individual floor level within a building can meet all of the provisions of the IBC and qualify as a mezzanine, its actual designation is the choice of the designer. It may be more advantageous to treat the floor level simply as an additional story.

Topic: Openness
Reference: IBC 505.2.3

Category: Building Heights and Areas
Subject: Mezzanines

Code Text: *A mezzanine shall be open and unobstructed to the room in which such mezzanine is located except for walls not more than 42 inches high, columns and posts.* See exceptions addressing mezzanines 1) where the enclosed area has a maximum occupant load of 10, 2) having two or more means of egress with at least one providing direct access to an exit, 3) where the aggregate floor area of the enclosed space does not exceed 10 percent of the mezzanine area, 4) in industrial facilities, and 5) in fully sprinklered buildings where two or more means of egress are provided.

Discussion and Commentary: By definition, a mezzanine is intended to be open to the room or space below. This common environment allows individuals on either floor level to be aware of the conditions and hazards that may affect their safety. The IBC, through the application of one of five exceptions, permits the mezzanine to be enclosed when it has been determined that the enclosure creates little, if any concern.

A common exception used to permit the enclosing of a mezzanine is based on egress conditions. If a minimum of two means of egress are provided from the mezzanine level, with at least one such egress path leading directly to an exit, then the mezzanine is not required to be open to the room or space below.

Topic: General Provisions
Reference: IBC 506.1

Category: Building Heights and Areas
Subject: Area Modifications

Code Text: *The building areas limited by Table 503 shall be permitted to be increased due to frontage (I_f) and automatic sprinkler system protection (I_s) in accordance with the following:*

$A_a = \{A_t + [A_t \times I_f] + [A_t \times I_s]\}$ *where:*
A_a = *Allowable building area per story (square feet)*
A_t = *Tabular building area per story in accordance with Table 503 (square feet)*
I_f = *Area increase factor due to frontage as calculated in accordance with Section 506.2*
I_s = *Area increase factor due to sprinkler protection as calculated in accordance with Section 506.3*

Discussion and Commentary: The tabular allowable building areas set forth in Table 503 are limited on a story-by-story basis. The presence of sufficient open space adjacent to a building provides for an increase above the tabular value. The protection afforded by an automatic sprinkler system justifies a significant allowable area increase. The IBC permits both increases to be used to provide the maximum building area permitted per story.

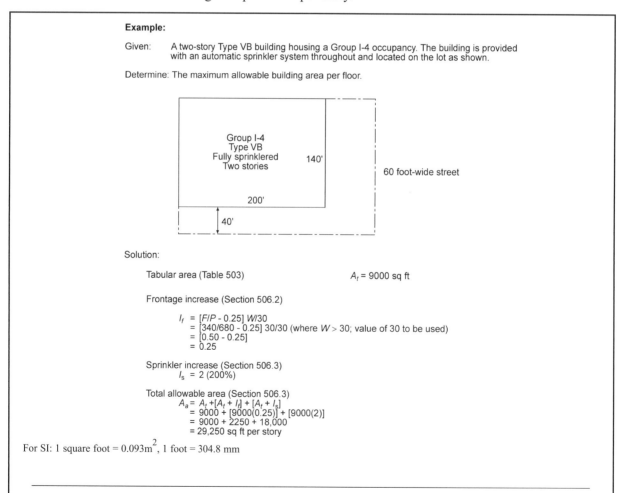

Example:

Given: A two-story Type VB building housing a Group I-4 occupancy. The building is provided with an automatic sprinkler system throughout and located on the lot as shown.

Determine: The maximum allowable building area per floor.

Solution:

Tabular area (Table 503) A_t = 9000 sq ft

Frontage increase (Section 506.2)
I_f = [F/P − 0.25] W/30
   = [340/680 − 0.25] 30/30 (where W > 30; value of 30 to be used)
   = [0.50 − 0.25]
   = 0.25

Sprinkler increase (Section 506.3)
I_s = 2 (200%)

Total allowable area (Section 506.3)
A_a = A_t +[A_t + I_f] + [A_t + I_s]
   = 9000 + [9000(0.25)] + [9000(2)]
   = 9000 + 2250 + 18,000
   = 29,250 sq ft per story

For SI: 1 square foot = 0.093m², 1 foot = 304.8 mm

A single basement need not be considered in the determination of the total allowable building area, provided the basement does not exceed the area permitted for a one-story building.

Topic: Frontage Increase
Reference: IBC 506.2
Category: Building Heights and Areas
Subject: Area Modifications

Code Text: *Every building shall adjoin or have access to a public way to receive a building area increase for frontage. Where a building has more than 25 percent of its perimeter on a public way or open space having a minimum width of 20 feet, the frontage increase shall be determined in accordance with the following: $I_f = [F/P - 0.25] \, W/30$.*

Discussion and Commentary: It is assumed that every building will adjoin a street, alley or yard on at least one side. Therefore, no frontage increase is given for the first 25 percent of a building's perimeter that is open. Credit is provided, however, where additional frontage is considered open (20 feet or more in width). The benefit of increased allowable building area is accrued based on better access for the fire department, as well as decreased exposure to adjoining properties. *F/P* represents the percentage of the building's perimeter that has adequate open space, whereas *W* is width of the open space. The value of *W* must be at least 20 feet and typically cannot exceed 30 feet.

Entire perimeter considered for frontage increase

Min. 20'-wide fire lane*

Street

Street

Open space to be on same lot or dedicated for public use, and accessed from a street or approved fire lane

*Fire lane need only be provided to within 150 feet of exterior wall per the IFC.

For SI: 1 foot = 304.8 mm

Access must be provided from a street or an approved fire lane for any open space that is used for a frontage increase in allowable floor area. The *International Fire Code* mandates that a fire lane for fire apparatus be maintained with an unobstructed width of at least 20 feet.

Topic: Sprinkler System Increase
Reference: IBC 506.3
Category: Building Heights and Areas
Subject: Area Modifications

Code Text: *Where a building is equipped throughout with an approved automatic sprinkler system in accordance with Section 903.3.1.1, the building area limitation in Table 503 is permitted to be increased by an additional 200 percent ($I_s = 2$) for buildings with more than one story above grade plane and an additional 300 percent ($I_s = 3$) for buildings with no more than one story above grade plane. See exceptions for Group H-1, H-2 and H-3 occupancies, and for buildings utilizing the fire-resistance-rated substitution allowance.*

Discussion and Commentary: Because of its excellent record of in-place fire suppression, an automatic sprinkler system provides for a sizable allowable area increase. The presence of the sprinkler system also allows for a height increase as addressed in Section 504.2. However, sprinkler systems are considered an absolute necessity in occupancies associated with high hazard levels, specifically Groups H-1, H-2 and H-3; thus, no size increases are permitted for such occupancies.

Examples

Given: Group B occupancy single-story
Type VB construction
No open yards available

Find: Total allowable area

Basic allowable area = 9,000 sq ft (Table 503)
Sprinkler increase (I_s) = <u>27,000 sq ft</u> (300%)
Total allowable area = 36,000 sq ft

Given: Same situation, however two stories in height
Find: Total allowable area

Basic allowable area = 9,000 sq ft (Table 503)
Sprinkler increase (I_s) = <u>18,000 sq ft</u> (200%)
Total allowable area per story = 27,000 sq ft

For SI: 1 foot = 304.8 mm

It is assumed that in many cases, fire department suppression activities will supplement an automatic sprinkler system. Because a multistory building presents more problems to the fire department than a single-story structure, a smaller increase in story area is justified.

Topic: Multistory Area Determination
Reference: IBC 506.4.1
Category: Building Heights and Areas
Subject: Area Modifications

Code Text: *The total allowable building area of a single occupancy building with more than one story above grade plane shall be determined by multiplying the allowable building area per story (A_a) by the number of stories above grade plane as listed below: 1) for buildings with two stories above grade plane, multiply by 2; 2) for buildings with three or more stories above grade plane, multiply by 3; and 3) no story shall exceed the allowable building area per story (A_a) for the occupancies on that story.* See exceptions for 1) unlimited area buildings complying with Section 507, and 2) Group R buildings equipped with an NFPA 13R sprinkler system.

Discussion and Commentary: In the analysis for allowable building area in a multistory building, each individual floor, as well as the building as a whole, must be in compliance. This is of particular importance for those buildings four or more stories in height above grade plane.

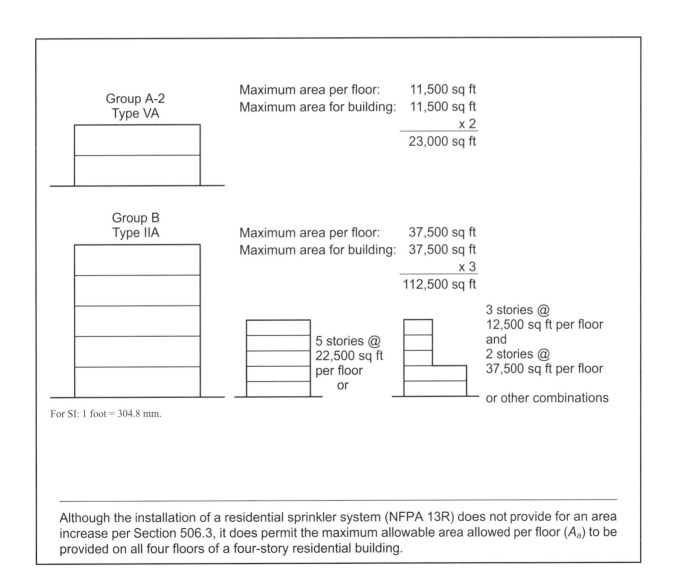

Although the installation of a residential sprinkler system (NFPA 13R) does not provide for an area increase per Section 506.3, it does permit the maximum allowable area allowed per floor (A_a) to be provided on all four floors of a four-story residential building.

Topic: One-Story, Sprinklered Buildings
Reference: IBC 507.3
Category: Building Heights and Areas
Subject: Unlimited Area Buildings

Code Text: *The area of a Group B, F, M or S building no more than one story above grade plane of any construction type, or a Group A-4 building no more than one story above grade plane of other than Type V construction, shall not be limited when the building is provided with an automatic sprinkler system throughout in accordance with Section 903.3.1.1, and is surrounded and adjoined by public ways or yards not less than 60 feet (18 288 mm) in width. Provisions also apply to two-story buildings of such occupancies other than Group A-4.*

Discussion and Commentary: It is often beneficial to have very large, undivided floor areas for facilities such as arenas, office buildings, factories, retail centers and warehouses. The unlimited area provisions allow for an alternative to the higher types of construction that would normally be required. The installation of a sprinkler system and sufficient open space around the building reduce the potential fire severity to a reasonable level in these moderate-hazard occupancies.

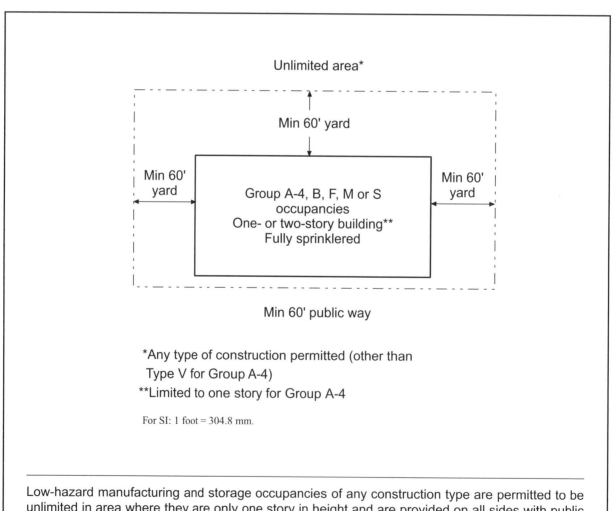

Low-hazard manufacturing and storage occupancies of any construction type are permitted to be unlimited in area where they are only one story in height and are provided on all sides with public ways or yards at least 60 feet in width. Installation of an automatic sprinkler system is not required.

Topic: Reduced Open Space	**Category:** Building Heights and Areas
Reference: IBC 507.5	**Subject:** Unlimited Area Buildings

Code Text: *The public ways or yards of 60 feet (18 288 mm) in width required in Sections 507.2, 507.3, 507.4, 507.6 and 507.11 shall be permitted to be reduced to not less than 40 feet (12 192 mm) in width provided all of the following requirements are met: 1) the reduced open space shall not be allowed for more than 75 percent of the perimeter of the building, 2) the exterior walls facing the reduced open space shall have a minimum fire-resistance rating of 3 hours, and 3) openings in the exterior walls facing the reduced open space shall have opening protectives with a minimum fire-resistance rating of 3 hours.*

Discussion and Commentary: When it is necessary or desirable to reduce the open space around the perimeter of an unlimited area building, the code provides an alternative. An equivalent level of protection can be provided by increasing the level of exterior wall and opening protection.

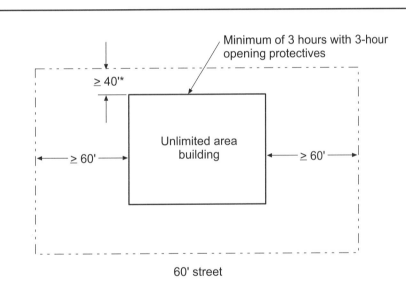

*Reduced open space permitted:

- Up to 75% of building perimeter
- Where exterior wall facing reduced open space has minimum 3-hour fire-resistance rating
- Openings in such walls are protected for 3 hours

For SI: 1 foot = 304.8 mm.

This provision is designed for warehouses, factories, retail stores, office buildings, Group A-3 uses and movie theaters where fire resistance at the exterior wall is easily accomplished. The reduction does not apply to other buildings permitted to be unlimited in area, such as educational uses and aircraft paint hangars.

Topic: Group A-3 Buildings
Reference: IBC 507.6, 507.7
Category: Building Heights and Areas
Subject: Unlimited Area Buildings

Code Text: *The area of a Group A-3 building no more than one story above grade plane used as a place of religious worship, community hall, dance hall, exhibition hall, gymnasium, lecture hall, indoor swimming pool or tennis court of Type II construction, shall not be limited when all of the following criteria are met:* See three conditions for allowance of unlimited area. The provisions are also applicable to buildings of Type III and IV construction, provided four conditions are met.

Discussion and Commentary: The Group A-3 occupancy classification includes the most diverse types of assembly uses assigned by the code. Traditionally, the allowable area of Group A occupancies is greatly limited as compared to most other occupancy groups. However, those assembly uses expected to have a relatively low fire load are permitted in unlimited area buildings subject to the special conditions prescribed by the code.

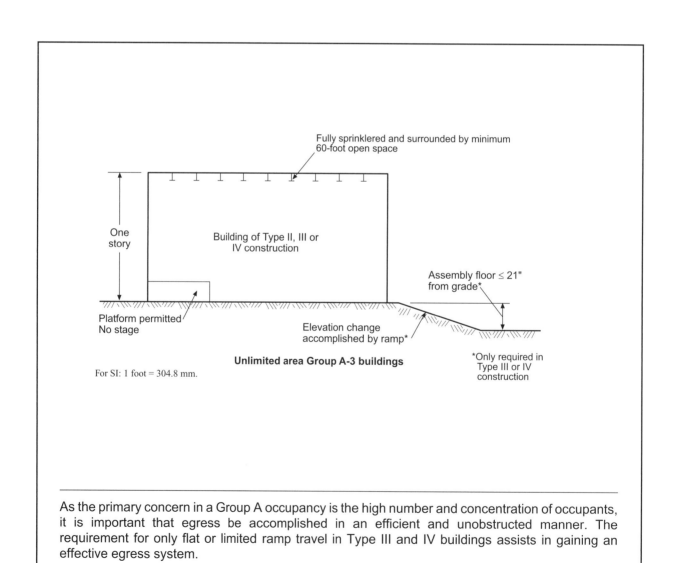

Unlimited area Group A-3 buildings

For SI: 1 foot = 304.8 mm.

As the primary concern in a Group A occupancy is the high number and concentration of occupants, it is important that egress be accomplished in an efficient and unobstructed manner. The requirement for only flat or limited ramp travel in Type III and IV buildings assists in gaining an effective egress system.

Topic: High-Hazard Use Groups
Reference: IBC 507.8

Category: Building Heights and Areas
Subject: Unlimited Area Buildings

Code Text: *Group H-2, H-3 and H-4 occupancies shall be permitted in unlimited area buildings containing Group F and S occupancies, in accordance with Sections 507.3 and 507.4 and the provisions of Sections 507.8.1 through 507.8.4. The aggregate floor area of the Group H occupancies located in an unlimited area building shall not exceed 10 percent of the area of the building nor the area limitations for the Group H occupancies as specified in Table 503 as modified by Section 506.2, based upon the perimeter of each Group H floor area that fronts on a public way or open space. The aggregate floor area of Group H occupancies not located at the perimeter of the building shall not exceed 25 percent of the area limitations for the Group H occupancies as specified in Table 503.*

Discussion and Commentary: The aggregate allowable area of the permitted Group H occupancies in a factory or warehouse regulated as an unlimited area building is dependent on the type of construction of the building and the location of the Group H occupancies in the building.

Group H-2 at the corner of an unlimited area Group F or S building

More ready access to the Group H from the exterior of the building provides the fire department with an opportunity to respond more effectively to an incident. As such, the allowable floor area of the Group H can be far greater than where completely surrounded by the Group F or S use.

Topic: Horizontal Building Separation
Reference: IBC 510.2

Category: Building Heights and Areas
Subject: Special Provisions

Code Text: *A building shall be considered as separate and distinct buildings for the purpose of determining area limitations, continuity of fire walls, limitation of number of stories and type of construction where all of the following conditions are met:* See list of seven criteria.

Discussion and Commentary: The special provisions of Section 510 are intended to modify the specific requirements of Chapter 5 regarding allowable heights and areas of buildings. The allowances granted in Section 510.2 address those structures typically referred to as "podium" or "pedestal" buildings. Compliance with the multiple conditions results in consideration of the structure as two separate and distinct buildings for four distinct issues. Similar provisions are established in Sections 510.3, 510.4, 510.7 and 510.8 where complying horizontal separations, along with other requirements, permit modifications to the general allowable height and area limitations.

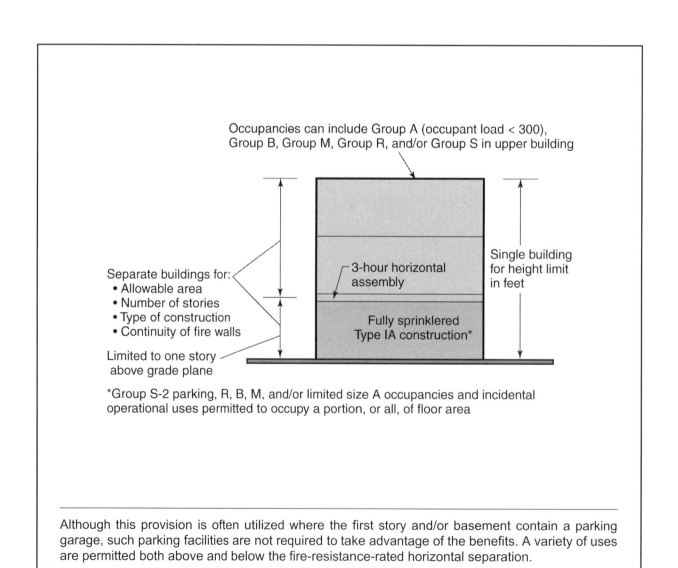

Although this provision is often utilized where the first story and/or basement contain a parking garage, such parking facilities are not required to take advantage of the benefits. A variety of uses are permitted both above and below the fire-resistance-rated horizontal separation.

Topic: Open Parking Garages
Reference: IBC 510.7

Category: Building Heights and Areas
Subject: Special Provisions

Code Text: *Open parking garages constructed under Groups A, I, B, M and R shall not exceed the height and area limitations permitted under Section 406.3. The height and area of the portion of the building above the open parking garage shall not exceed the limitations in Section 503 for the upper occupancy. The height, in both feet and stories, of the portion of the building above the open parking garage shall be measured from grade plane and shall include both the open parking garage and the portion of the building above the parking garage.*

Discussion and Commentary: In the more common types of occupancies, it is desirable at times to provide tiers of parking below the major use of the building. In this special mixed-use condition, two different types of construction are permitted for determining the maximum allowable height and area. This special allowance is just one of several special provisions established in Section 509 that modify the general requirements of the code.

It is important to address the appropriate fire separation issues when using this provision. In addition, the structural members supporting the upper occupancy must be protected by the more restrictive fire-resistant assemblies of all of the occupancies involved.

Quiz

Study Session 4
IBC Chapter 5

1. Premises must be identified by numbers or addresses visible from the street, with a minimum character height of at least _____ inches.

    a. three   b. four

    c. six   d. eight

    Reference_____

2. Building height is measured to the _____.

    a. average height of the highest roof surface

    b. highest point of the highest roof surface

    c. average height of all of the roof surfaces

    d. highest point of the lowest roof surface

    Reference_____

3. The tabular allowable height and building area per floor for a Group I-2 occupancy of Type IB construction is _____ stories and _____ square feet.

    a. 3, 10,000   b. 2, 15,000

    c. 4, unlimited   d. unlimited, unlimited

    Reference_____

4. What is the maximum tabular allowable height, in feet, for a Group B occupancy of Type IIA construction?

   a. 50 b. 55
   c. 65 d. unlimited

   Reference_____

5. A single basement does not need to be included in the total allowable area of a single-occupancy building, provided it does not exceed _____.

   a. one-third the floor area permitted for any single story
   b. the area permitted for a one-story building
   c. twice the area permitted for a single story
   d. the tabular area based on construction type and occupancy group

   Reference_____

6. A story is considered a story above grade plane where the finished surface of the floor above is more than _____ feet above grade plane.

   a. 3 b. 4
   c. 5 d. 6

   Reference_____

7. The maximum building area of a six-story single-occupancy building is limited to _____ times the allowable area permitted per story.

   a. two b. three
   c. four d. six

   Reference_____

8. An allowable height increase of _____ and _____ is permitted in a building housing a Group B occupancy and protected throughout with an automatic sprinkler system.

   a. 15 feet, one story b. 20 feet, one story
   c. 15 feet, two stories d. 20 feet, two stories

   Reference_____

9. In a Type IIB building, an increase in allowable height is not permitted if the building is classified as which one of the following occupancy groups?

   a. Group A-1   b. Group H-4
   c. Group I-2   d. Group R-2

   Reference_____

10. Combustible steeples are limited to a maximum height of _____ feet above the allowable building height.

    a. 15   b. 20
    c. 30   d. 40

    Reference_____

11. The area of a mezzanine is not to be included in the determination of the _____.

    a. fire area   b. building area
    c. occupant load   d. plumbing fixture count

    Reference_____

12. A minimum clear height of _____ is required above and below mezzanine floor construction.

    a. 6 feet, 8 inches   b. 7 feet
    c. 7 feet, 6 inches   d. 8 feet

    Reference_____

13. In general, the aggregate area of mezzanines within a room is limited to _____ of the area of the room in which the mezzanines are located.

    a. 10 percent   b. 25 percent
    c. $33^1/_3$ percent   d. 50 percent

    Reference_____

14. Portions of a mezzanine need not be open to the room in which the mezzanine is located, provided the enclosed space is limited in size to a maximum of _____ of the mezzanine area.

    a. 10 percent   b. 25 percent
    c. 33 1/3 percent   d. 50 percent

    Reference_____

15. A mezzanine is not required to be open to the room in which it is located where the occupant load of the enclosed space does not exceed _____.

    a. 10   b. 20
    c. 30   d. 50

    Reference_____

16. An allowable area increase for frontage is not permitted where a maximum of _____ of the building perimeter is sufficiently open.

    a. 10 percent   b. 25 percent
    c. 33 1/3 percent   d. 40 percent

    Reference_____

17. In order to be considered as sufficiently open for an allowable area increase for frontage, the public way or open space must have a minimum width of _____ feet.

    a. 10   b. 20
    c. 25   d. 30

    Reference_____

18. What is the maximum tabular height, in number of stories, for a Group R-1 occupancy of Type IIB construction?

    a. 3   b. 4
    c. 5   d. 6

    Reference _____

Study Session 4

19. An increase in allowable height is permitted for a Group _____ occupancy in a building fully sprinklered in accordance with NFPA 13.

    a. H-2   b. H-3
    c. H-4   d. H-5

    Reference _____

20. The allowable building area may be increased by _____ in a multistory building protected throughout with an approved automatic sprinkler system installed in accordance with NFPA 13.

    a. 50 percent   b. 100 percent
    c. 200 percent   d. 300 percent

    Reference _____

21. In a fully-sprinklered single-story building, the allowable building area may be increased by _____ .

    a. 50 percent   b. 100 percent
    c. 200 percent   d. 300 percent

    Reference _____

22. In order for a Group E building to comply with the unlimited area allowances of Section 507, the building's construction type cannot be Type _____ .

    a. IIA   b. IIB
    c. IIIA   d. IIIB

    Reference _____

23. In order to reduce the required open space surrounding certain unlimited area buildings from 60 feet to 40 feet, walls facing the reduced open space shall have a minimum fire-resistance rating of _____ .

    a. 45 minutes   b. 1 hour
    c. 2 hours   d. 3 hours

    Reference _____

24. A complying Group H-2 aircraft paint hangar may be unlimited in floor area when limited to one story, provided the hangar is surrounded by yards or public ways having a minimum width of _____ .

    a. 40 feet

    b. 60 feet

    c. twice the height of the hangar

    d. one and one-half times the height of the hangar

    Reference_____

25. Where using the special provisions of Section 510.2 for the horizontal building separation allowance, the horizontal separation between buildings shall be provided with a horizontal assembly having a minimum _____ fire-resistance rating.

    a. 1-hour
    b. 2-hour
    c. 3-hour
    d. 4-hour

    Reference_____

26. What is the maximum allowable height permitted for a noncombustible communications tower located on the roof of a Type IIB office building?

    a. 55 feet
    b. 65 feet
    c. 75 feet
    d. unlimited

    Reference_____

27. In order for a Type IIIB office building to be considered for unlimited area, it must be limited to a maximum height of _____ .

    a. one story
    b. two stories
    c. 50 feet
    d. 40 feet

    Reference_____

28. A one-story Group A-3 gymnasium building of Type IIIA construction may be considered under the unlimited area provisions, provided the gymnasium floor is located a maximum of _____ above or below grade level.

    a. 21 inches
    b. 24 inches
    c. 30 inches
    d. 48 inches

    Reference_____

29. A Group H-3 storage room located within a 100,000-square foot unlimited area manufacturing building of Type IIB construction is limited to _____ square feet where not located on the building's perimeter.

    a. 3,125   b. 3,500
    c. 12,500   d. 14,000

    Reference_____

30. A nine-story apartment building of Type IIA construction shall be located a minimum of _____ feet from any other building on the lot and from all lot lines.

    a. 10 feet   b. 30 feet
    c. 50 feet   d. 60 feet

    Reference_____

31. In buildings of Type I or II construction housing special industrial occupancies in accordance with Section 503.1.1, the aggregate floor area of mezzanines is limited to a maximum of _____ of the area of the room in which they are located.

    a. 1/4   b. 1/3
    c. 1/2   d. 2/3

    Reference_____

32. Where the aggregate area of all equipment platforms within a room is limited to a maximum of _____ of the area of the room in which they are located, the equipment platforms shall not be considered a portion of the floor below.

    a. 1/4   b. 1/3
    c. 1/2   d. 2/3

    Reference_____

33. The area of a complying covered mall building and any anchor stores is not limited where their maximum height is _____ stories.

    a. 2   b. 3
    c. 4   d. 5

    Reference_____

34. A Group E building of unlimited area shall be surrounded and adjoined by public ways and yards a minimum of _____ feet in width.

   a. 30   b. 40

   c. 60   d. 75

   Reference _____

35. In a fully sprinklered building, an allowable area increase is not permitted for a building area classified as a Group _____ occupancy.

   a. H-3   b. I-2

   c. I-3   d. R-2

   Reference _____

Study Session 5

2012 IBC Sections 701 – 705
Fire and Smoke Protection Features I

OBJECTIVE: To gain an understanding of the fundamentals of fire-resistance-rated construction, the methods for the determination of fire resistance, and the regulation of exterior walls for fire-resistance rating and opening protection.

REFERENCE: Sections 701 through 705, 2012 *International Building Code*

KEY POINTS:
- Why are fire-resistance-rated materials and systems used in the construction of buildings?
- What is a fire-resistance rating? How is such a rating determined?
- What referenced standards are the basis for determining fire-resistance ratings?
- What is nonsymmetrical wall construction?
- How must interior walls and partitions of nonsymmetrical construction be tested? Exterior walls?
- Which alternative methods are available for determining the fire-resistance rating of different building elements?
- What is a prescriptive design of fire-resistance-rated building elements? What specific types of elements are addressed?
- Which types of materials can be evaluated for fire-resistance ratings through calculations?
- When is a material considered noncombustible? Is gypsum board considered a noncombustible material?
- Under what conditions is the use of fire-resistance-rated glazing permitted? How must such glazing be identified?
- What fire-resistance-rated walls must be identified by signs or stenciling? Where is the identification required to be located? What minimum lettering size is required?
- Under what conditions must fire-resistance-rated columns be individually protected? Other primary structural frame members?
- How are secondary members required to be protected for fire-resistance purposes?
- How must the fire protective covering of a structural member be protected from impact that is due to moving vehicles?
- How is the minimum fire-resistance rating determined for structural members located within exterior walls?

KEY POINTS:
(Cont'd)
- What are the general requirements applicable to sprayed fire-resistant materials?
- How is a projection defined? What types of building elements are considered projections?
- What limits the extent of a projection?
- Which types of projections are permitted from walls of Type I and II buildings? Type III, IV and V buildings?
- When must combustible projections be protected? What other options are available?
- How are multiple buildings on the same lot addressed in regard to exterior wall and opening protection?
- Under which conditions must fire-resistance-rated exterior walls be rated for fire exposure from both sides?
- Under which conditions are openings in exterior walls prohibited? What are the limitations where protected openings are provided? Unprotected openings?
- How are protected and unprotected openings regulated in the same exterior wall?
- How does the presence of a sprinkler system affect the amount of unprotected openings?
- Which special provisions apply to openings in the first story of exterior walls?
- When must exterior openings in adjacent stories be protected? Where required, what methods of protection are available?
- What is a parapet? Where are parapets required?
- What fire-resistance rating is mandated for required parapets?
- What is the minimum required height of a parapet? How does the slope of the roof affect the minimum required height?

Topic: Scope	Category: Fire and Smoke Protection Features
Reference: IBC 701.1	Subject: Materials and Assemblies

Code Text: *The provisions of Chapter 7 shall govern the materials, systems and assemblies used for structural fire resistance and fire-resistance-rated construction separation of adjacent spaces to safeguard against the spread of fire and smoke within a building and the spread of fire to or from buildings.*

Discussion and Commentary: There are basically two reasons for the protection of various building elements with construction resistant to fire. One, structural elements such as columns, girders, bearing walls and other load-bearing members are often required by the code to maintain their structural integrity under fire conditions for a prescribed time period. Two, horizontal and vertical assemblies are used to create compartments, including control areas, or to isolate portions of the building, such as exitways, through fire-resistant construction.

FIRE RESISTANCE. That property of materials or their assemblies that prevents or retards the passage of excessive heat, hot gases or flames under conditions of use.

In addition to limiting or resisting the spread of fire and heat within a building, certain provisions are intended to provide protection for adjoining structures. The code also addresses construction utilized to restrict the passage of smoke to specific areas.

Topic: Materials and Systems
Reference: IBC 703.2, 202
Category: Fire and Smoke Protection Features
Subject: Fire-resistance Ratings

Code Text: *Fire-resistance rating is the period of time a building element, component or assembly maintains the ability to confine a fire, continues to perform a given structural function, or both as determined by the tests, or the methods based on tests, prescribed in Section 703. The fire-resistance rating of building elements, components or assemblies shall be determined in accordance with the test procedures set forth in ASTM E 119 or UL 263 or in accordance with Section 703.3 (Alternative methods for determining fire resistance).*

Discussion and Commentary: ASTM E 119 is the referenced standard, *Standard Test Methods for Fire Tests of Building Construction and Materials*. These test methods are used for the great majority of building components or assemblies that are mandated by the code to have a fire-resistance rating. Assemblies tested under the criteria of UL 263 are also considered to have the fire-resistance rating as assigned.

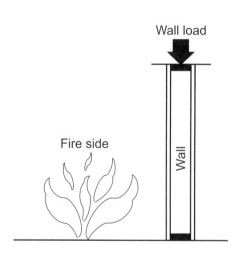

Assembly must:

sustain applied load,

have no passage of flame or gases hot enough to ignite cotton waste,

have average temperature rise on unexposed surface not more than 250°F above initial temperature or more than 325°F at any point, and

have no water pass through during hose-stream test.

Conditions of acceptance - wall fire test

For nonsymmetrical wall construction, where interior walls and partitions are provided with differing membranes on opposing sides, the IBC mandates that tests be performed from both sides. The side with the shortest test duration is the basis for the fire-resistance rating.

Topic: Alternative Methods
Reference: IBC 703.3

Category: Fire and Smoke Protection Features
Subject: Fire-Resistance Ratings

Code Text: *The application of any of the alternative methods listed in Section 703.3 shall be based on the fire exposure and acceptance criteria specified in ASTM E 119 or UL 263. The required fire resistance of a building element, component or assembly shall be permitted to be established by any of the following methods or procedures: (1) fire-resistance designs documented in approved sources, (2) prescriptive designs of fire-resistance-rated building elements as prescribed in Section 721, (3) calculations in accordance with Section 722, (4) engineering analysis based on a comparison of building element, component or assembly designs having fire-resistance ratings as determined by the test procedures set forth in ASTM E 119 or UL 263, or (5) alternative protection methods as allowed by Section 104.11.*

Discussion and Commentary: Prescriptive details of fire-resistance-rated building elements are contained in Section 721. Generic listings for structural parts, walls, partitions, floor systems and roof systems are addressed.

For SI: 1 foot = 304.8 mm.

Wall test furnace

Section 722 provides methods of calculated fire resistance for concrete, masonry, steel, and wood assemblies or members. The procedures and calculations are limited to the specific information set forth in this section and are not to be used in any other manner.

Study Session 5

Topic: Scope
Reference: IBC 721.1

Category: Fire and Smoke Protection Features
Subject: Prescriptive Fire Resistance

Code Text: *The provisions of Section 721 contain prescriptive details of fire-resistance-rated building elements. The materials of construction listed in Tables 721.1(1), 721.1(2) and 721.1(3) shall be assumed to have the fire-resistance ratings prescribed therein. Where materials that change the capacity for heat dissipation are incorporated into a fire-resistance-rated assembly, fire test results or other substantiating data shall be made available to the building official to show that the required fire-resistance rating time period is not reduced.*

Discussion and Commentary: The tables in Section 721 provide the details for obtaining desired fire-resistance ratings for structural parts, walls and partitions, and floor and roof systems. The methods and materials found in the tables are to be used in the same manner as any listed assembly.

Table 721.1(2)
Interior partition: Item 14-1.5

2 in. by 4 in. wood studs 16 in. on center with two layers $5/8$ in. Type X gypsum wallboard each side. Base layers applied vertically and nailed with 6d cooler or wallboard nails at 9 in. on center. Face layer applied vertically or horizontally and nailed with 8d cooler or wallboard nails at 7 in. on center. For nail-adhesive application, base layers are nailed 6 in. on center. Face layers applied with coating of approved wallboard adhesive and nailed 12 in. on center.

2 hour rated

For SI: 1 inch = 25.4 mm.

When insulation or a similar material is added to a fire-resistance-rated assembly, it may change the assembly's capacity to dissipate heat. Particularly in noncombustible horizontal assemblies, the fire-resistance rating may be diminished to some degree.

Topic: Scope
Reference: IBC 722.1
Category: Fire and Smoke Protection Features
Subject: Calculated Fire Resistance

Code Text: *The provisions of Section 722 contain procedures by which the fire resistance of specific materials or combinations of materials is established by calculations. The procedures apply only to the information contained in Section 722 and shall not be otherwise used. The calculated fire resistance of concrete, concrete masonry, and clay masonry assemblies shall be permitted in accordance with ACI 216.1/TMS 0216. The calculated fire resistance of steel assemblies shall be permitted in accordance with Chapter 5 of ASCE 29.*

Discussion and Commentary: Another method used to obtain the necessary fire-resistance ratings mandated by the code is calculation. The provisions for calculating fire resistance are applicable to concrete assemblies, concrete masonry, clay brick and tile masonry, steel assemblies and wood assemblies.

TABLE 722.6.2(1)
TIME ASSIGNED TO WALLBOARD MEMBRANES[a, b, c, d]

DESCRIPTION OF FINISH	TIME*(minutes)
$3/8$-inch wood structural panel bonded with exterior glue	5
$15/32$-inch wood structural panel bonded with exterior glue	10
$19/32$-inch wood structural panel bonded with exterior glue	15
$3/8$-inch gypsum wallboard	10
$1/2$-inch gypsum wallboard	15
$5/8$-inch gypsum wallboard	30
$1/2$-inch Type X gypsum wallboard	25
$5/8$-inch Type X gypsum wallboard	40
Double $3/8$-inch gypsum wallboard	25
$1/2$-inch + $3/8$-inch gypsum wallboard	35
Double $1/2$-inch gypsum wallboard	40

For SI: 1 inch = 25.4 mm.
a. These values apply only when membranes are installed on framing members which are spaced 16 inches o.c. or less.
b. Gypsum wallboard installed over framing or furring shall be installed so that all edges are supported, except $5/8$-inch Type X gypsum wallboard shall be permitted to be installed horizontally with the horizontal joints staggered 24 inches each side and unsupported but finished.
c. On wood frame floor/ceiling or roof/ceiling assemblies, gypsum board shall be installed with the long dimension perpendicular to framing members and shall have all joints finished.
d. The membrane on the unexposed side shall not be included in determining the fire resistance of the assembly. When dissimilar membranes are used on a wall assembly, the calculation shall be made from the least fire-resistant (weaker) side.
e. The time assigned is not a finished rating.

TABLE 722.6.2(2)
TIME ASSIGNED FOR CONTRIBUTION OF WOOD FRAME[a, b, c]

DESCRIPTION	TIME ASSIGNED TO FRAME (minutes)
Wood studs 16 inches o.c.	20
Wood floor and roof joists 16 inches o.c.	10

For SI: 1 inch = 25.4 mm.
a. This table does not apply to studs or joists spaced more than 16 inches o.c.
b. All studs shall be nominal 2 × 4 and all joists shall have a nominal thickness of at least 2 inches.
c. Allowable spans for joists shall be determined in accordance with Sections 2308.8, 2308.10.2 and 2308.10.3.

Applicable to both load-bearing and nonload-bearing assemblies, the calculated fire resistance for wood-framed walls, floor/ceiling assemblies and roof/ceiling assemblies is limited to a 1-hour rating.

Topic: Noncombustibility Tests
Reference: IBC 703.5
Category: Fire and Smoke Protection Features
Subject: Fire-Resistance Ratings

Code Text: *Materials required to be noncombustible shall be tested in accordance with ASTM E 136. The term "noncombustible" does not apply to the flame spread characteristics of interior finish or trim materials. A material shall not be classified as a noncombustible building construction material if it is subject to an increase in combustibility or flame spread beyond the limitations herein established through the effects of age, moisture, or other atmospheric conditions.*

Discussion and Commentary: In buildings of Types I, II, III and IV construction, specific elements are required to be constructed of noncombustible materials. Such materials are desirable because they do not aid combustion, nor do they add appreciable heat to an ambient fire. Under conditions of the test, a material may have a limited amount of combustible content and still qualify as noncombustible.

Gypsum wallboard and similar products are also acceptable as noncombustible materials. They must have a structural base of noncombustible materials, a surface material no more than $1/8$-inch in thickness and a maximum flame-spread index of 50.

Topic: Fire-Resistance-Rated Glazing
Category: Fire and Smoke Protection Features
Reference: IBC 703.6
Subject: Fire Resistance Ratings

Code Text: *Fire-resistance-rated glazing, when tested in accordance with ASTM E 119 or UL 263 and complying with the requirements of Section 707, shall be permitted. Fire-resistance-rated glazing shall bear a label marked in accordance with Table 716.3 issued by an agency and shall be permanently identified on the glazing.*

Discussion and Commentary: Under the provisions of Table 716.6, fire windows are not permitted in fire walls and most fire barriers having a required fire-resistance rating of 1 hour or greater. Therefore, glazing in such fire separations must either be protected by complying fire shutters or be in compliance with the requirements of Section 703.6. The fire-resistance-rated glazing permitted by this section is acceptable because it is tested to the same criteria as any fire wall or fire barrier.

*XXX = fire-resistance rating in minutes

In the identifier, the "W" indicates that the glazing meets the requirements of ASTM E 119 or UL 263, thus qualifying the glazing to be used as a wall element. It also indicates that the glazing meets the fire-resistance, hose-stream and temperature-rise requirements of the test standard.

Study Session 5

Topic: Marking and Identification
Reference: IBC 703.7
Category: Fire and Smoke Protection Features
Subject: Fire-Resistance Ratings

Code Text: *Fire walls, fire barriers, fire partitions, smoke barriers and smoke partitions or any other wall required to have protected openings or penetrations shall be effectively and permanently identified with signs or stenciling. Such identification shall be located in accessible concealed floor, floor-ceiling or attic spaces.* See exception for Group R-2 occupancies.

Discussion and Commentary: The integrity of fire and/or smoke separation walls is subject to compromise during the life of a building. During maintenance and remodel activities, it is not uncommon for new openings and penetrations to be installed in a fire separation wall without recognition that the integrity of the construction must be maintained or that some type of fire or smoke protective is required. Provisions mandating the appropriate identification of such walls under certain conditions have been established to better ensure that tradespeople, maintenance workers and inspectors will recognize the required level of protection that must be maintained.

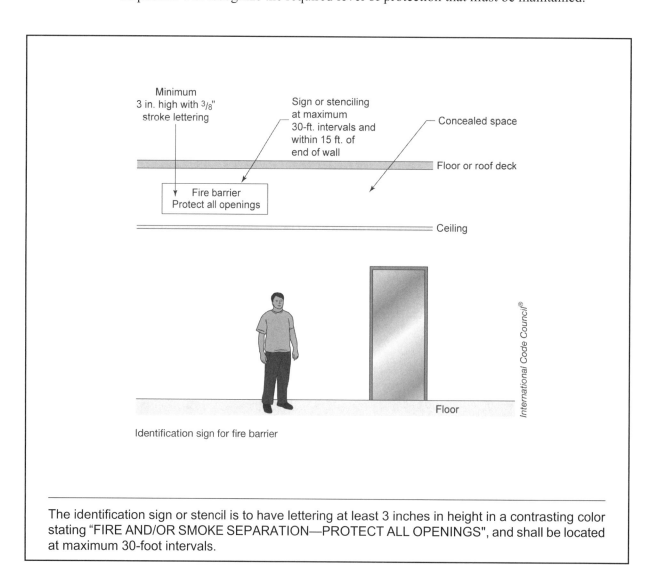

Identification sign for fire barrier

The identification sign or stencil is to have lettering at least 3 inches in height in a contrasting color stating "FIRE AND/OR SMOKE SEPARATION—PROTECT ALL OPENINGS", and shall be located at maximum 30-foot intervals.

Topic: Individual Protection
Reference: IBC 704.3, 704.4
Category: Fire and Smoke Protection Features
Subject: Structural Members

Code Text: *Members of the primary structural frame other than columns that are required to have protection to achieve a fire-resistance rating and support more than two floors or one floor and roof, or support a load-bearing wall or a nonload-bearing wall more than two stories high, shall be provided individual encasement protection by protecting them on all sides for their full length, including connections to other structural members, with materials having the required fire-resistance rating. Secondary members that are required to have a fire-resistance rating shall be protected by individual encasement, by a membrane or ceiling protection as specified in Section 711, or by a combination of both.*

Discussion and Commentary: Because of the differences in both the testing procedure and the conditions of acceptance, primary structural frame members carrying significant portions of the structure cannot simply be protected by enclosure within a fire-resistance-rated wall, floor/ceiling or roof/ceiling assembly. Therefore, under specific conditions, individual encasement is

Code requirements:

1. Individual protection for horizontal primary structural frame members is required if supporting loads are from more than two floors or from one floor and roof.

2. Otherwise, protection may be by individual encasement, membrane or ceiling protection per Sec. 711, or combination of both.

3. Columns must always be individually encased and protected for full height (Sec. 704.2).

When a column requires a fire-resistance rating, it must always be fully protected by individual encasement, including its connections to beams or girders. If the column extends above a ceiling, the fire protection must continue through the above-ceiling space to the top of the column.

Study Session 5

Topic: Exterior Members
Reference: IBC 704.10

Category: Fire and Smoke Protection Features
Subject: Structural Members

Code Text: *Load-bearing structural members located within the exterior walls or on the outside of a building or structure shall be provided with the highest fire-resistance rating as determined in accordance with the following: 1) as required by Table 601 for the type of building element based on the type of construction of the building; 2) as required by Table 601 for exterior bearing walls based on the type of construction; and 3) as required by Table 602 for exterior walls based on the fire separation distance.*

Discussion and Commentary: Structural frame members such as columns that are placed inside an exterior wall assembly, or are located on the outside of a building, must be evaluated for fire-resistance purposes. Three criteria are to be evaluated, with the minimum required fire-resistance rating of the structural members based on the highest rating of the three criteria.

GIVEN: An exterior nonbearing wall in a Type IIIB building housing a Group M occupancy. The wall has a fire separation distance of 15 feet to an interior lot line.

DETERMINE: The minimum required fire-resistance rating for structural columns located within the exterior wall.

SOLUTION:

Per Table 601 for structural frame members, a minimum of 0 hours
Per Table 601 for exterior bearing walls, a minimum of 2 hours
Per Table 602 for a FSD of 15 feet, a minimum of 1 hour
∴ The columns shall have a minimum fire-resistance rating of 2 hours

Any structural frame members located within, or on the outside of, an exterior wall should never have a lower fire-resistance rating than that required to protect the members from an internal fire. However, if the exposure hazard from an external source is so great as to require exterior wall protection, a higher rating may be required.

Topic: Extent of Projections
Reference: IBC 705.2
Category: Fire and Smoke Protection Features
Subject: Exterior Walls

Code Text: *Cornices, eave overhangs, exterior balconies and similar projections extending beyond the exterior wall shall conform to the requirements of Sections 705 and 1406. Exterior egress balconies and exterior exit stairways and ramps shall also comply with Sections 1019 and 1026, respectively. Projections shall not extend any closer to the line used to determine the fire separation distance than shown in Table 705.2. See exception for projections beyond opposing walls of two buildings on the same lot.*

Discussion and Commentary: Where a building projection is located in close proximity to another building, either on the same lot or an adjoining lot, there is significant potential under fire conditions for convected heat to be trapped. The requirement for some degree of physical separation between any adjacent building and the edge of a projection allows for open space to assist in the heat's dissipation.

TABLE 705.2
MINIMUM DISTANCE OF PROJECTION

FIRE SEPARATION DISTANCE (FSD)	MINIMUM DISTANCE FROM LINE USED TO DETERMINE FSD
0 feet to less than 2 feet	Projections not permitted
2 feet to less than 5 feet	24 inches
5 feet or greater	40 inches

For SI: 1 foot = 304.8 mm; 1 inch = 25.4 mm.

The allowable projection length is based upon the fire separation distance related to the exterior wall. Where the fire separation distance is 2 feet or less, no projection is permitted. If the fire separation distance exceeds 2 feet, some amount of projection is allowed up to the limit established by Table 705.2.

Topic: Construction of Projections
Reference: IBC 705.2.1 – 705.2.3
Category: Fire and Smoke Protection Features
Subject: Exterior Walls

Code Text: *Projections from walls of Type I or II construction shall be of noncombustible materials or combustible materials as allowed by Sections 1406.3 and 1406.4. Projections from walls of Type III, IV or V construction shall be of any approved material. Combustible projections extending to within 5 feet (1524 mm) of the line used to determine the fire separation distance, or located where openings are not permitted or where protection of some openings is required shall be of at least 1-hour fire-resistance-rated construction, Type IV construction, fire-retardant-treated wood or as required by Section 1406.3. See exception for Group R-3 and U occupancies.*

Discussion and Commentary: In noncombustible buildings, those of Type I and II construction, projections must also be combustible in order to maintain the limited fire loading created by the building's construction features. In buildings where combustible construction is permitted—Types III, IV and V—combustible projections typically pose no greater hazard than the other combustible elements in the building.

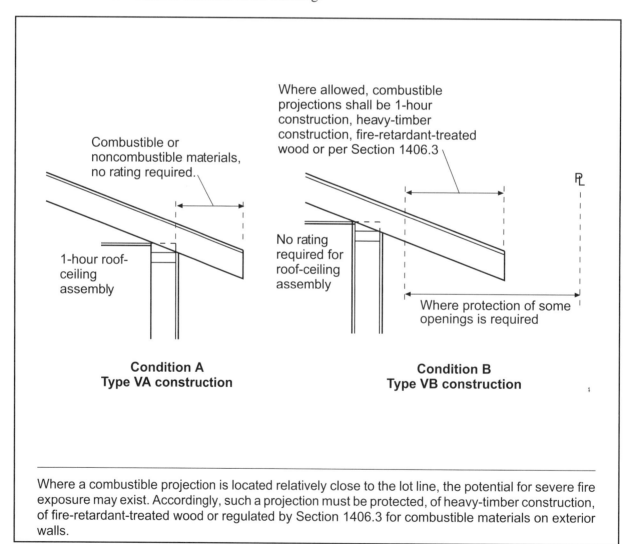

Where a combustible projection is located relatively close to the lot line, the potential for severe fire exposure may exist. Accordingly, such a projection must be protected, of heavy-timber construction, of fire-retardant-treated wood or regulated by Section 1406.3 for combustible materials on exterior walls.

Topic: Buildings on the Same Lot
Reference: IBC 705.3

Category: Fire and Smoke Protection Features
Subject: Exterior Walls

Code Text: *For the purposes of determining the required wall and opening protection, projections and roof-covering requirements, buildings on the same lot shall be assumed to have an imaginary line between them.* See exception where aggregate area of multiple buildings is within limits specified in Chapter 5 for a single building.

Discussion and Commentary: Where two or more buildings are placed on the same piece of property, their exterior walls and openings must be regulated in the same manner as if they were on separate lots. However, if the buildings could be constructed as a single structure under one roof and meet the size requirements based on occupancy and type of construction, then an assumed imaginary line is not required.

Case I: an assumed imaginary line between buildings

    A. Fire resistance and opening protection for walls adjacent to the imaginary line must comply with the code

    B. Imaginary line may be placed to take best advantage of wall and opening protection

Case II: as a single building

    A. Allowable area and type of construction are based on the most restrictive requirements for the occupancies housed

    B. Total floor area may not exceed that allowed for a single building

Buildings on the same lot

Where a new building is to be constructed on the same lot as an existing building, the assumed imaginary line must be placed in a location where it will not cause the exterior wall and opening protection of the existing building to become noncompliant.

Topic: Fire-Resistance Ratings
Reference: IBC 705.5
Category: Fire and Smoke Protection Features
Subject: Exterior Walls

Code Text: *Exterior walls shall be fire-resistance rated in accordance with Tables 601 and 602 and Section 705. The required fire-resistance rating of exterior walls with a fire separation distance of greater than 10 feet (3048 mm) shall be rated for exposure to fire from the inside. The required fire-resistance rating of exterior walls with a fire separation distance of less than or equal to 10 feet (3048 mm) shall be rated for exposure to fire from both sides.*

Discussion and Commentary: Exposure of the exterior wall to an interior fire does not vary based on the distance of the wall from the lot line. However, exterior fire exposure decreases with an increase in the distance between the lot line and the exterior wall (fire separation distance). A fire separation distance of 10 feet is considered by the IBC to be a reasonable limit of flame impingement (direct exterior fire exposure) from an adjacent building.

The determination of the minimum fire-resistance-rating for an exterior wall is typically based on two conditions: 1) the type of construction of the building, and 2) the fire separation distance. The higher of the two ratings regulates the minimum level of fire resistance.

Topic: Allowable Area of Openings
Reference: IBC 705.8.1, 705.8.4
Category: Fire and Smoke Protection Features
Subject: Exterior Walls

Code Text: *The maximum area of unprotected and protected openings permitted in an exterior wall in any story shall not exceed the percentages specified in Table 705.8. See exceptions for 1) openings on the first story, and 2) stories where the exterior wall is not required to have a fire-resistance rating. Where both unprotected and protected openings are located in the exterior wall in any story of a building, the total area of the openings shall comply with the following formula:* $A_p/a_p + A_u/a_u \leq 1$.

Discussion and Commentary: Based on the fire separation distance, the amount of openings in an exterior wall is regulated on a floor-by-floor basis. Where all of the exterior openings are protected, a higher percentage of the exterior wall surface may be provided with openings, whereas a lesser amount is permitted if all openings are unprotected. The IBC also permits both protected and unprotected openings in an exterior wall, provided they comply with the unity formula. In a fully-sprinklered building, the maximum allowable area of unprotected openings is the same as that allowed for protected openings.

TABLE 705.8
MAXIMUM AREA OF EXTERIOR WALL OPENINGS BASED ON FIRE SEPARATION DISTANCE AND DEGREE OF OPENING PROTECTION

FIRE SEPARATION DISTANCE (feet)	DEGREE OF OPENING PROTECTION	ALLOWABLE AREA[a]
0 to less than 3[b, c]	Unprotected, Nonsprinklered (UP, NS)	Not Permitted
	Unprotected, Sprinklered (UP, S)[i]	Not Permitted
	Protected (P)	Not Permitted
3 to less than 5[d, e]	Unprotected, Nonsprinklered (UP, NS)	Not Permitted
	Unprotected, Sprinklered (UP, S)[i]	15%
	Protected (P)	15%
5 to less than 10[e, f]	Unprotected, Nonsprinklered (UP, NS)	10%[h]
	Unprotected, Sprinklered (UP, S)[i]	25%
	Protected (P)	25%
10 to less than 15[e, f, g]	Unprotected, Nonsprinklered (UP, NS)	15%[h]
	Unprotected, Sprinklered (UP, S)[i]	45%
	Protected (P)	45%
15 to less than 20[f, g]	Unprotected, Nonsprinklered (UP, NS)	25%
	Unprotected, Sprinklered (UP, S)[i]	75%
	Protected (P)	75%

(Continued)

If a building's exterior wall is not required to be fire-resistance rated by Table 602, then an unlimited percentage of unprotected openings is permitted regardless of fire separation distance.

Topic: First Story Walls
Reference: IBC 705.8.1, Exc. 1
Category: Fire and Smoke Protection Features
Subject: Exterior Walls

Code Text: *In other than Group H occupancies, unlimited unprotected openings are permitted in the first story above grade either 1.1) where the wall faces a street and has a fire separation distance of more than 15 feet (4572 mm), or 1.2) where the wall faces an unoccupied space. The unoccupied space shall be on the same lot or dedicated for public use, shall not be less than 30 feet (9144 mm) in width and shall have access from a street by a posted fire lane in accordance with the* International Fire Code.

Discussion and Commentary: Because the first story of a building is generally readily available for fire department access and manual suppression efforts, unprotected openings are not restricted, provided a moderate amount of open space is provided adjacent to the exterior wall. It is expected that the fire department can quickly mitigate the potential radiant heat exposure to surrounding buildings or structures.

Unlimited openings in the first story exterior wall

For SI: 1 foot = 304.8 mm.

As the primary allowance for unlimited unprotected openings is the ability of the fire service to easily access and suppress at the ground level, it is critical that adequate open space be available for such purposes. Where not accessed directly from a street, a fire lane must be provided.

Topic: Vertical Separation of Openings
Reference: IBC 705.8.5
Category: Fire and Smoke Protection Features
Subject: Exterior Walls

Code Text: *Openings in exterior walls in adjacent stories shall be separated vertically to protect against fire spread on the exterior of the buildings where the openings are within 5 feet (1524 mm) of each other horizontally and the opening in the lower story is not a protected opening with a fire protection rating of not less than $^3/_4$ hour. Such openings shall be separated vertically at least 3 feet (914 mm) by spandrel girders, exterior walls or other similar assemblies that have a fire-resistance rating of at least 1 hour or by flame barriers than extend horizontally at least 30 inches (762 mm) beyond the exterior wall. Flame barriers shall also have a fire-resistance rating of at least 1 hour.* See exceptions for buildings no more than three stories in height, buildings that are fully sprinklered and open parking garages.

Discussion and Commentary: Where unprotected openings occur in adjacent stories, a fire that breaks out of an opening in a lower story can spread vertically to upper stories of the building. Two methods of flame resistance are available to restrict such fire spread.

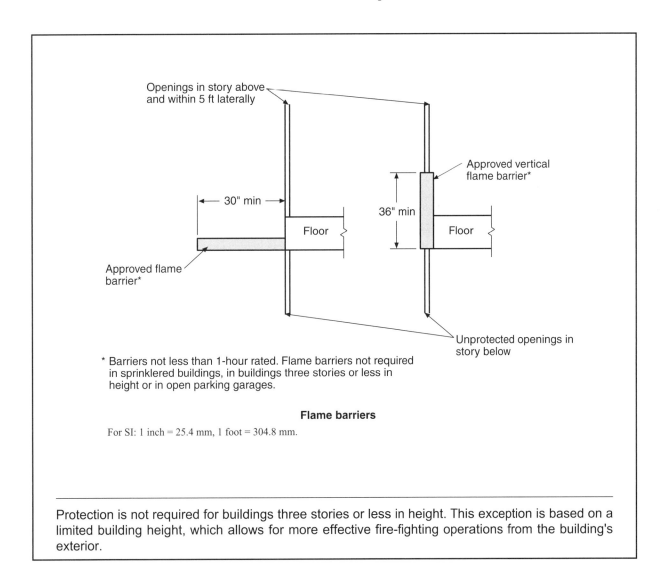

Flame barriers

For SI: 1 inch = 25.4 mm, 1 foot = 304.8 mm.

Protection is not required for buildings three stories or less in height. This exception is based on a limited building height, which allows for more effective fire-fighting operations from the building's exterior.

Topic: Parapets
Reference: IBC 705.11, 705.11.1

Category: Fire and Smoke Protection Features
Subject: Exterior Walls

Code Text: *Parapets shall be provided on exterior walls of buildings.* See six exceptions for construction methods or locations that would eliminate the requirement for parapets. *Parapets shall have the same fire-resistance rating as that required for the supporting wall, and on any side adjacent to a roof surface, shall have noncombustible faces for the uppermost 18 inches (457 mm), including counterflashing and coping materials. The height of the parapet shall not be less than 30 inches (762 mm) above the point where the roof surface and the wall intersect.*

Discussion and Commentary: A parapet wall is defined as *the part of any wall entirely above the roof line*. Its purpose is to prevent the spread of fire from the roof of the subject building to an adjacent building and to protect the roof of a building from exposure to a fire in an adjacent building.

For SI: 1 inch = 25.4 mm

As very few buildings actually provide exterior parapet walls for fire protection, it is evident that the exceptions are widely used. In many situations, a parapet is provided only to hide the roof slope or to screen rooftop equipment, in which case the requirements do not apply.

Topic: Parapet Exceptions
Reference: IBC 705.11, Exception 4
Category: Fire and Smoke Protection Features
Subject: Exterior Walls

Code Text: *A parapet need not be provided on an exterior wall where 1-hour fire-resistance-rated exterior walls terminate at the underside of the roof sheathing, deck or slab, provided:* four conditions are met.

Discussion and Commentary: Limited to walls having a 1-hour fire-resistance rating, Exception 4 permits exterior walls to terminate at the underside of the roof sheathing as an alternative to the use of parapets. Protection of the roof construction is provided from the interior of the building rather than at the exterior side. In addition to the restrictions on roof covering materials and openings in the roof, roof framing elements are addressed where installed parallel or perpendicular to their supporting walls.

Roof/ceiling framing parallel to wall

For SI: 1 foot = 304.8 mm.

Other parapet exceptions are applicable where 1) the wall is not required to be fire-resistance-rated by Table 602; 2) no story exceeds 1,000 square feet in floor area; 3) the wall terminates at a minimum 2-hour fire-resistance-rated roof; 4) the roof is constructed of noncombustible materials, including the deck and supporting construction; 5) the wall is permitted to have a minimum of 25 percent unprotected openings per Section 705.8; or 6) in Groups R-2 and R-3, a number of special conditions are met.

Quiz

Study Session 5
IBC Sections 701 – 705

1. An opening around a penetrating item is a(n) _____.

    a. annular space
    b. penetration
    c. through penetration
    d. joint

    Reference_____

2. A _____ is a listed device installed in a ceiling membrane of a fire-resistance-rated floor/ceiling or roof/ceiling assembly to limit automatically the radiative heat transfer through an air inlet/outlet opening.

    a. horizontal fire damper
    b. ceiling radiation damper
    c. combination fire/smoke damper
    d. horizontal access door

    Reference_____

3. The time period that a through-penetration firestop system limits the spread of fire through a penetration when tested in accordance with ASTM E 814 or UL 1479 is considered _____.

    a. a fire protection rating
    b. an F rating
    c. a T rating
    d. a fire-resistance rating

    Reference_____

4. _____ consists of building materials installed to resist the free passage of flame to other areas of the building through concealed spaces.

    a. Draftstopping   b. Fireblocking

    c. Firestopping   d. Opening protectives

Reference_____

5. The measurement between the face of an exterior wall and the closest interior lot line is described as the _____.

    a. fire separation distance   b. fire exposure setback

    c. clearance to construction   d. exterior fire exposure

Reference_____

6. A _____ must have sufficient structural stability under fire conditions to allow collapse of construction on either side without collapse of the wall.

    a. fire wall   b. fire separation wall

    c. fire barrier   d. smoke barrier

Reference_____

7. A smoke compartment is created by enclosure of a space on all sides, including top and bottom, by _____.

    a. smoke partitions   b. smoke barriers

    c. shaft enclosure construction   d. fire partitions

Reference_____

8. The T rating for a penetration firestop system is based on a maximum temperature rise of _____ °F above its initial temperature through the penetration on the nonfire side when tested in accordance with ASTM E 184 or UL 1479.

    a. 250   b. 325

    c. 375   d. 400

Reference_____

9. Unless alternate methods are used for determining the fire-resistance ratings of building elements, the test procedures set forth in ASTM E 119 or _____ shall be applicable.

    a. ANSI Z 97.1   b. ASCE 5

    c. UL 263   d. UL 555

Reference_____

10. Where a fire barrier is required to be identified by signage in an accessible concealed attic space, the identification shall occur at maximum intervals of _____ feet measured horizontally along the wall.

    a. 15   b. 20

    c. 30   d. 40

Reference_____

11. Composite materials with a noncombustible structural base are considered noncombustible where the surface material is limited to _____ inch in thickness and has a maximum flame spread index of _____.

    a. $^1/_2$, 25   b. $^1/_{16}$, 50

    c. $^1/_8$, 25   d. $^1/_8$, 50

Reference_____

12. A projection is considered a building element that extends beyond the _____ of a building.

    a. exterior wall   b. fire area

    c. building area   d. floor area

Reference_____

13. Projections beyond an exterior wall are not permitted where the exterior wall has a maximum fire separation distance of _____ inches.

    a. 12   b. 24

    c. 30   d. 36

Reference_____

14. In a building of Type III construction, projections may be _____.

    a. of noncombustible construction

    b. of combustible construction

    c. of minimum one-hour fire-resistance-rated construction

    d. of any approved materials

    Reference_____

15. For wall and opening protection, projections and roof covering requirements, an assumed imaginary line shall be placed between _____ unless they are considered a single building.

    a. court walls   b. fire areas

    c. buildings on the same lot   d. different occupancies

    Reference_____

16. The fire-resistance rating for an exterior wall shall be based on both interior and exterior fire exposure where the wall is located a maximum of _____ feet from the lot line.

    a. 3   b. 5

    c. 10   d. 20

    Reference_____

17. In a nonsprinklered Type VB building, what is the maximum allowable area of unprotected exterior wall openings for a fire separation distance of 8 feet?

    a. 10 percent   b. 25 percent

    c. unlimited   d. unprotected openings are prohibited

    Reference_____

18. In a nonsprinklered Type IIIA building, what is the maximum allowable area of protected exterior wall openings for a fire separation distance of 20 feet?

    a. 25 percent   b. 45 percent

    c. 75 percent   d. unlimited

    Reference_____

Study Session 5

19. In a fully-sprinklered Type IIB office building, what is the maximum allowable area of unprotected exterior wall openings for a fire separation distance of 5 feet?

    a. 10 percent
    b. 15 percent
    c. 25 percent
    d. unprotected openings are prohibited

    Reference_____

20. In other than Group H occupancies, unlimited unprotected exterior openings are permitted in the first story above grade of exterior walls facing an unoccupied space of a least _____ feet in width.

    a. 15
    b. 20
    c. 30
    d. 40

    Reference_____

21. Where flame barriers are required for the vertical separation of exterior openings in adjacent stories, horizontal barriers must extend at least _____ inches beyond the exterior wall.

    a. 12
    b. 24
    c. 30
    d. 36

    Reference_____

22. Flame barriers protecting openings in exterior walls in adjacent stories shall have a minimum fire-resistance rating of _____.

    a. 20 minutes
    b. 45 minutes
    c. 1 hour
    d. 2 hours

    Reference_____

23. A parapet is not required at an exterior wall of a building having a maximum floor area of _____ square feet on any floor.

    a. 400
    b. 1,000
    c. 1,500
    d. 3,000

    Reference_____

24. Where required at exterior walls, parapets must extend a minimum of _____ inches above the roof.

   a. 30
   b. 32
   c. 36
   d. 42

   Reference_____

25. The uppermost portion of parapet walls must have noncombustible faces for a minimum of _____ inches.

   a. 12
   b. 18
   c. 24
   d. 30

   Reference_____

26. To obtain a fire-resistance rating, all nonsymmetrical walls except for _____ shall be tested with both faces exposed to the furnace.

   a. fire walls
   b. fire barriers
   c. smoke barriers
   d. exterior walls

   Reference_____

27. A roof eave, where projecting into an area where protection of some openings is required, may be constructed of all of the following materials, except _____ .

   a. heavy-timber construction
   b. noncombustible construction
   c. fire-retardant-treated wood
   d. one-half-hour fire-resistance-rated construction

   Reference_____

28. In a fully sprinklered Type IIA building, the maximum allowable area of protected exterior wall openings is _____ where the fire separation distance is 12 feet.

   a. 15 percent
   b. 25 percent
   c. 45 percent
   d. 60 percent

   Reference_____

Study Session 5

29. A parapet is not required on the exterior wall of a nonsprinklered building located a minimum of _____ feet from an interior lot line.

    a. 5
    b. 10
    c. 15
    d. 20

    Reference_____

30. Unprotected openings are not permitted in the exterior wall of a Group H-3 warehouse where the fire separation distance is less than _____ feet.

    a. 15
    b. 20
    c. 25
    d. 30

    Reference_____

31. Materials required to be noncombustible shall be tested in accordance with _____.

    a. ASTM E 119
    b. ASTM E 136
    c. ASTM E 84
    d. ASTM C 636

    Reference_____

32. The area of openings in an exterior wall of an open parking garage is permitted to be unlimited where there is a minimum fire separation distance of _____ feet.

    a. 5
    b. 10
    c. 20
    d. 30

    Reference_____

33. A parapet is not required on an exterior wall where the wall terminates at a roof with a minimum fire-resistance rating of _____.

    a. 45 minutes
    b. 1 hour
    c. 1 1/2 hour
    d. 2 hours

    Reference_____

34. A balcony projecting beyond the exterior wall on a Type IIA building is permitted to be constructed of fire-retardant-treated wood where the building is a maximum of _____ stories in height above grade plane.

 a. 2
 b. 3
 c. 4
 d. 5

 Reference _____

35. In other than Group H occupancies, unlimited unprotected openings are permitted in a first-story exterior wall that faces a street, provided the minimum fire separation distance is greater than _____ feet.

 a. 5
 b. 10
 c. 15
 d. 20

 Reference _____

Study Session 6

2012 IBC Sections 706 – 712
Fire and Smoke Protection Features II

OBJECTIVE: To gain an understanding of the fire-resistance-rated building components such as fire walls, fire barriers, fire partitions, smoke barriers, smoke partitions, and horizontal assemblies and vertical openings.

REFERENCE: Sections 706 through 712, 2012 *International Building Code*

KEY POINTS:
- What is the purpose of a fire wall?
- Where is a party wall located? How is it to be constructed?
- How must a fire wall perform structurally?
- What is a double fire wall? Under what conditions is such a condition permitted?
- Which types of materials are permitted in a fire wall? How is the minimum fire-resistance rating of a fire wall determined?
- How must the horizontal continuity of a fire wall be accomplished? Vertical continuity?
- Which options are possible where a fire wall serves a stepped building?
- How shall the penetration of combustible framing members entering into a masonry or concrete fire wall be addressed?
- What are the limitations on openings in a fire wall?
- Where are fire barriers utilized?
- How must a fire barrier be constructed? What restrictions are placed on openings?
- What is a fire area? What is its purpose? How is the minimum fire-resistance rating for fire barrier assemblies separating fire areas determined?
- How is a joint made at the intersection of a fire barrier and the underside of a fire-resistance-rated floor or roof deck above to be regulated? At the intersection of a fire barrier and a nonfire-resistance-rated roof assembly?
- Where are fire partitions required to be installed?
- What minimum fire-resistance rating is required for fire partitions? What rating is required in a sprinklered hotel or apartment building?
- To what extent must a fire partition extend above a ceiling?
- What is the function of a smoke barrier?
- What fire-resistance rating is mandated for smoke barriers?

KEY POINTS:
(Cont'd)
- How must openings in a smoke barrier be protected?
- Where are smoke barriers required?
- How is a smoke partition to be constructed? How are door openings, penetrations, ducts and air transfer openings regulated?
- How is the required rating of a horizontal assembly determined?
- What is the minimum fire-resistance rating for floor assemblies separating dwelling units in apartment buildings? Sleeping units in Group R-1 occupancies?
- What weight of ceiling panel requires no additional devices for the prevention of lateral displacement?
- When may the ceiling membrane in a fire-resistance-rated horizontal assembly be omitted? Floor membrane?
- How are skylights regulated in fire-resistance-rated roof construction?
- What various methods are permitted to address a joint in or between floor assemblies without a required fire-resistance rating?
- What are the various methods that have been established to address vertical openings between stories in a building?

Topic: Definition and Scope
Reference: IBC 706.1, 202
Category: Fire and Smoke Protection Features
Subject: Fire Walls

Code Text: *A fire wall is a fire-resistance-rated wall having protected openings, which restricts the spread of fire and extends continuously from the foundation to or through the roof, with sufficient structural stability under fire conditions to allow collapse of construction on either side without collapse of the wall. Each portion of a building separated by one or more fire walls that comply with the provisions of Section 706 shall be considered a separate building.*

Discussion and Commentary: By placing one or more fire walls in a large-area building, multiple smaller-area buildings are created. Each of these smaller spaces can then be considered a unique building for code purposes. Under various conditions, fire walls can serve to reduce a structure's type of construction, eliminate a required sprinkler system or increase the permitted size of a facility under one roof.

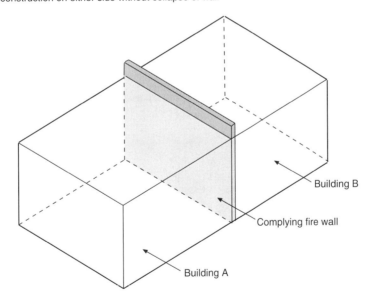

In a situation where a fire wall separates distinct occupancy groups that are required to be separated by a fire barrier wall, the most restrictive requirements of each separation apply. This includes both the wall's continuity and the required fire-resistance rating.

Study Session 6

Topic: Construction
Reference: IBC 706.3, 706.4

Category: Fire and Smoke Protected Features
Subject: Fire Walls

Code Text: *Fire walls shall be of any approved noncombustible materials. See exception for Type V construction. Fire walls shall have a fire-resistance rating of not less than that required by Table 706.4.*

Discussion and Commentary: A fire wall is designed to act in a manner similar to an exterior wall, as a barrier to prevent a fire in one building from spreading to the other building. Accordingly, construction of the fire wall must be commensurate with the exterior wall requirements for the construction type. In addition, the fire-resistance rating of the wall must be considerable in order to provide the necessary level of protection based on the anticipated fire loading that is due to the uses of the separate buildings. The required ratings vary based on occupancy and, to some degree, type of construction.

TABLE 706.4
FIRE WALL FIRE-RESISTANCE RATINGS

GROUP	FIRE-RESISTANCE RATING (hours)
A, B, E, H-4, I, R-1, R-2, U	3[a]
F-1, H-3[b], H-5, M, S-1	3
H-1, H-2	4[b]
F-2, S-2, R-3, R-4	2

a. In Type II or V construction, walls shall be permitted to have a 2-hour fire-resistance rating.
b. For Group H-1, H-2 or H-3 buildings, also see Sections 415.4 and 415.5.

A + B ≤ 25% of X

Each opening limited to 120 square feet unless both buildings are sprinklered

Fire-protection rating based on Tables 706.4 and 715.4

For SI: 1 square foot = 0.093m^2

Per Section 706.8, the total width of all openings in a fire wall is limited to 25 percent of the length of the wall in each story. There is no limit on the amount of total wall area containing openings; however, each opening is limited to 156 square feet in nonsprinklered buildings.

Topic: Horizontal Continuity
Reference: IBC 706.5
Category: Fire and Smoke Protection Features
Subject: Fire Walls

Code Text: *Fire walls shall be continuous from exterior wall to exterior wall and shall extend at least 18 inches (457 mm) beyond the exterior surface of exterior walls.* See three exceptions for various methods of terminating the fire wall at the interior surface of the exterior sheathing or finish materials.

Discussion and Commentary: Historically, the code has addressed the hazards of fire exposure at the fire wall from a vertical perspective, at the roof. There is also concern of a similar hazard from the horizontal perspective, at the intersection of the fire wall and the exterior wall. The 18-inch extension is intended to abate the potential for fire to travel from one building to the other around the fire wall. The 18-inch extension must extend the full height of the fire wall.

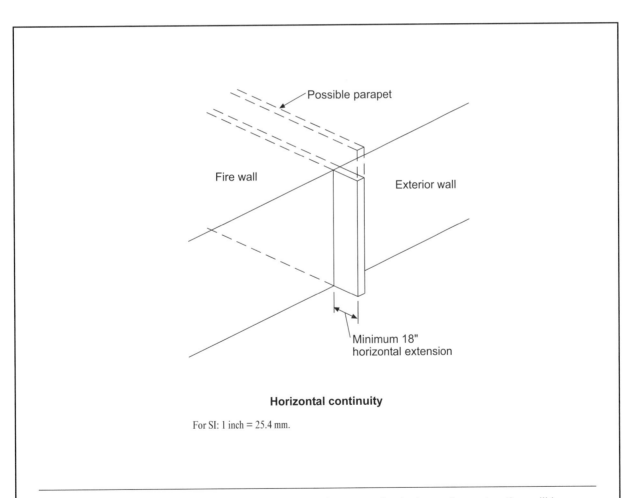

Horizontal continuity

For SI: 1 inch = 25.4 mm.

The three exceptions acknowledge the effect certain types of exterior wall construction will have on fire breaching the exterior of the building and exposing the adjacent building. These methods of protection are similar to those used at the roof construction where a parapet is not provided.

Topic: Vertical Continuity
Reference: IBC 706.6

Category: Fire and Smoke Protected Features
Subject: Fire Walls

Code Text: *Fire walls shall extend from the foundation to a termination point at least 30 inches (762 mm) above both adjacent roofs.* See exceptions for buildings with different roof levels, those with noncombustible roof construction, and those constructed under special provisions.

Discussion and Commentary: To ensure the *separate building* concept, a fire wall must be continuous vertically with no horizontal offsets from the foundation, through the roof to a point at least 30 inches above. Various exceptions to the parapet requirement allow the fire wall to terminate at the bottom of the roof deck or sheathing. According to many of the exceptions, the roof covering must be minimum Class B, and no openings in the roof are permitted within 4 feet of the fire wall.

Stepped buildings

For SI: 1 inch = 25.4 mm, 1 foot = 304.8 mm.

A stepped building, where the fire wall separates buildings having different roof levels, may require additional fire resistance to a point 15 feet above the lower roof. An alternative method provides for minimum 1-hour horizontal protection of the lower roof assembly.

Topic: Definition and Scope
Reference: IBC 707.1, 202
Category: Fire and Smoke Protection Features
Subject: Fire Barriers

Code Text: *A fire barrier is a fire-resistance-rated wall assembly of materials designed to restrict the spread of fire in which continuity is maintained. Fire barriers installed as required elsewhere in* the International Building Code *or the* International Fire Code *shall comply with* Section 707.

Discussion and Commentary: The term *fire barrier* is specific in the IBC and is used to describe a unique type of vertical fire separation element. Many of the building elements required to be constructed as fire barriers are listed in Section 707.3, including shaft enclosures, enclosures for interior exit stairways and exit access stairways, exit passageways, horizontal exits, atriums, incidental uses, control areas, separation of mixed occupancies and separation of fire areas.

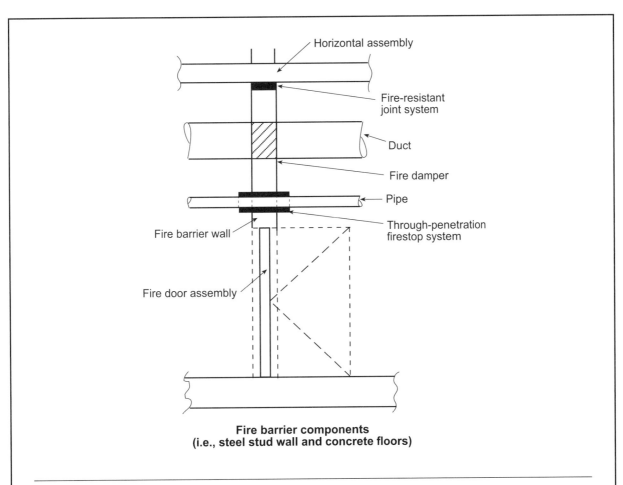

Fire barrier components
(i.e., steel stud wall and concrete floors)

Fire barriers may also be mandated for specific conditions not specifically mentioned in Section 707. Throughout the IBC, as well as the other *International Codes*, fire barriers are identified as the element used to provide the necessary fire separation for compartmentation of building spaces.

Study Session 6

Topic: Fire Areas
Reference: IBC 707.3.10, Table 707.3.10
Category: Fire and Smoke Protection Features
Subject: Fire Barriers

Code Text: *The fire barriers or horizontal assemblies, or both, separating a single occupancy into different fire areas shall have a fire-resistance rating of not less than that indicated in Table 707.3.10. The fire barriers or horizontal assemblies, or both, separating fire areas of mixed occupancies shall have a fire-resistance rating of not less than the highest value indicated in Table 707.3.10 for the occupancies under consideration.*

Discussion and Commentary: The code recognizes that in many buildings there are two methods to limit the spread of fire, either 1) the use of an automatic sprinkler system or 2) the creation of fire-resistive compartments that contain a fire's movement (fire areas). Section 903.2 identifies those occupancies where compartmentation is an acceptable alternative to a sprinkler system. Table 707.3.10 then mandates the minimum level of fire resistance of the fire barriers utilized to separate the building into two or more compartments (fire areas). As a result, the use of Table 707.3.10 is only applicable in buildings not protected by an automatic sprinkler system.

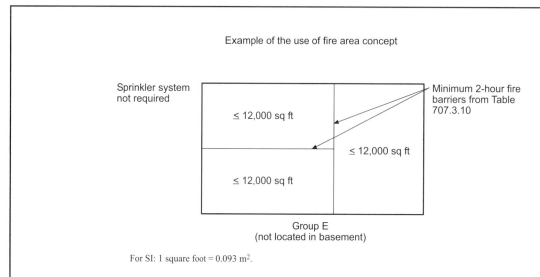

For SI: 1 square foot = 0.093 m².

TABLE 707.3.10
FIRE-RESISTANCE RATING REQUIREMENTS FOR FIRE BARRIER ASSEMBLIES OR HORIZONTAL ASSEMBLIES BETWEEN FIRE AREAS

OCCUPANCY GROUP	FIRE-RESISTANCE RATING (hours)
H-1, H-2	4
F-1, H-3, S-1	3
A, B, E, F-2, H-4, H-5, I, M, R, S-2	2
U	1

A fire area is considered the aggregate floor area enclosed and bounded by fire walls, fire barriers, exterior walls or horizontal assemblies of a building. The floor area under a canopy or similar horizontal projection is also included in the fire area determination.

Topic: Continuity
Reference: IBC 707.5, 707.5.1
Category: Fire and Smoke Protection Features
Subject: Fire Barriers

Code Text: *Fire barrier walls shall extend from the top of the floor/ceiling assembly below to the underside of the floor or roof sheathing, slab or deck above and shall be securely attached thereto. Such fire barriers shall be continuous through concealed spaces, such as the space above a suspended ceiling. The supporting construction for fire barrier walls shall be protected to afford the required fire-resistance rating of the fire barrier supported. See exceptions for fire barriers 1) separating tank storage per Section 415.6.2.1, 2) terminating at the lid of a shaft enclosure, 3) enclosing incidental uses and 4) enclosing interior exit stairways and exit access stairways.*

Discussion and Commentary: Where a wall is required to serve as a fire barrier, it must be tight from floor deck to floor or roof deck in order to provide a full separation. A fire barrier is often used in conjunction with a horizontal assembly in a multistory building to provide a complete separation.

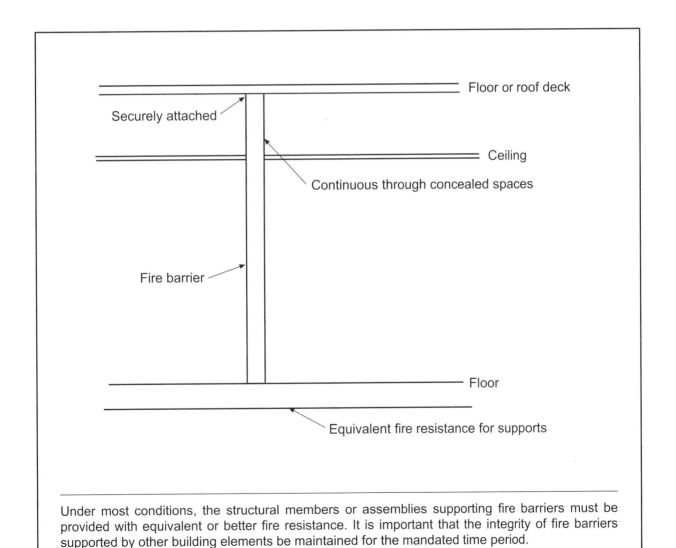

Under most conditions, the structural members or assemblies supporting fire barriers must be provided with equivalent or better fire resistance. It is important that the integrity of fire barriers supported by other building elements be maintained for the mandated time period.

Topic: Openings
Reference: IBC 707.6

Category: Fire and Smoke Protection Features
Subject: Fire Barriers

Code Text: *Openings in a fire barrier shall be protected in accordance with Section 716. Openings shall be limited to a maximum aggregate width of 25 percent of the length of the wall, and the maximum area of any single opening shall not exceed 156 square feet (15 m^2). Openings in enclosures for exit access stairways and ramps, interior exit stairways and exit passageways shall also comply with Sections 1022.3 and 1023.5, respectively.* See exceptions for 1) sprinklered adjoining floor areas, 2) fire doors serving an interior exit stairway or exit access stairway, 3) openings tested per ASTM E 119 or UL 263, 4) fire windows in atrium separation walls and 5) fire doors separating an interior exit stairway or exit access stairway from an exit passageway.

Discussion and Commentary: As openings in a fire barrier create a potential breach in the integrity of the fire-resistive separation, a limit is placed on the amount of permitted openings. The limitation allows for design flexibility without compromising the necessary level of fire separation. The aggregate area of such openings is not limited; however, each opening is limited to 156 square feet.

Opening in Fire Barriers

For SI: 1 square foot = 0.093 m².

The limiting size of an opening in a fire barrier wall is based on the maximum sizes established in fire door listings. Currently, the listing limitation on rolling steel fire doors is 156 square feet.

Topic: Definition and Scope
Reference: IBC 708.1, 202

Category: Fire and Smoke Protection Features
Subject: Fire Partitions

Code Text: *A fire partition is a vertical assembly of materials designed to restrict the spread of fire in which openings are protected. The following wall assemblies shall comply with Section 708: 1) walls separating dwelling units in the same building, 2) walls separating sleeping units in the same building, 3) walls separating tenant spaces in covered mall buildings as required by Section 402.4.2.1, 4) corridor walls as required by Section 1018.1 and 5) elevator lobby separation as required by Section 713.14.1. Fire partitions shall have a fire-resistance rating of not less than 1 hour. See exceptions for ratings reductions for corridor walls and dwelling unit/sleeping unit separations.*

Discussion and Commentary: Typically required to have a fire-resistance-rating of one-hour, fire partitions provide a moderate level of separation that is necessary under certain conditions. Although fire partitions have limited applications, they are important elements in the specific uses and areas in which they are mandated.

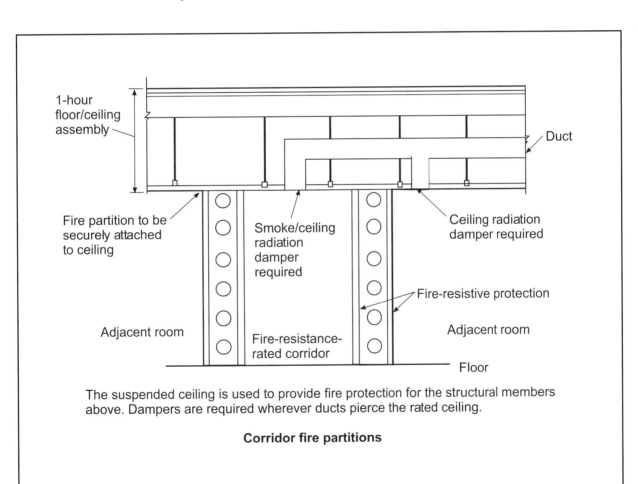

The suspended ceiling is used to provide fire protection for the structural members above. Dampers are required wherever ducts pierce the rated ceiling.

Corridor fire partitions

In sprinklered buildings of Types IIB, IIIB and VB construction, the 1-hour fire-resistance rating for dwelling unit and guestroom separations may be reduced to ½ hour. For a typical wood-stud wall system, this separation could be satisfied with $1/2$-inch gypsum board on each side.

Topic: Continuity
Reference: IBC 708.4

Category: Fire and Smoke Protection Features
Subject: Fire Partitions

Code Text: *Fire partitions shall extend from the top of the foundation or floor/ceiling assembly below to the underside of the floor or roof sheathing, slab or deck above or to the fire-resistance-rated floor/ceiling or roof/ceiling assembly above, and shall be securely attached thereto. The supporting construction shall be protected to afford the required fire-resistance rating of the wall supported, except for walls separating tenant spaces in covered and open mall buildings, walls separating dwelling units, walls separating sleeping units and corridor walls, in buildings of Type IIB, IIIB and VB construction.* See multiple exceptions.

Discussion and Commentary: The method of continuity is a primary difference between fire barriers and fire partitions. Fire partitions need not extend through a concealed space, such as the one above a suspended ceiling, provided that the ceiling is a portion of a fire-resistance-rated floor/ceiling or roof/ceiling assembly.

A variety of additional methods are described for creating a fire-resistance-rated corridor. A common construction technique is the "tunnel" corridor, where the corridor ceiling is constructed in a manner consistent with that of the corridor walls.

Topic: Definition and Continuity
Reference: IBC 709.4, 202
Category: Fire and Smoke Protection Features
Subject: Smoke Barriers

Code Text: *A smoke barrier is a continuous membrane, either vertical or horizontal, such as a wall, floor, or ceiling assembly, that is designed and constructed to restrict the movement of smoke. Smoke barriers shall form an effective membrane continuous from outside wall to outside wall and from the top of the foundation or floor/ceiling assembly below to the underside of the floor or roof sheathing, deck or slab above, including continuity through concealed spaces, such as those found above suspended ceilings, and including interstitial structural and mechanical spaces.* See exceptions for 1) fire/smoke-resistant ceilings, 2) elevator lobbies and 3) areas of refuge.

Discussion and Commentary: Where the primary concern of the code is the containment of smoke, the use of a smoke barrier is mandated. The locations for smoke barriers are found in various provisions in the IBC, including Section 407.5 for Group I-2 occupancies and Section 1007.6.2 for areas of refuge.

Required Use of Smoke Barriers

- Compartmentation of underground buildings (Sec. 405.4.2)
- Compartmentation of Group I-2 (Sec. 407.5)
- Compartmentation of Group I-3 (Sec. 408.6)
- Smoke control systems (Sec. 402.7.2, 404.5 and 909.5)
- Areas of refuge (Sec. 1007.6.2)

A smoke barrier must have a minimum 1-hour fire-resistance rating. Openings, penetrations, joints, ducts and transfer openings must also be protected to minimize the passage of smoke through the barrier. Opening protectives must have a minimum 20-minute fire-protection rating.

Topic: General Provisions
Reference: IBC 710
Category: Fire and Smoke Protection Features
Subject: Smoke Partitions

Code Text: *Smoke partitions installed as required elsewhere in the IBC shall comply with Section 710. The walls shall be of materials permitted by the building type of construction. Unless required elsewhere in the IBC, smoke partitions are not required to have a fire-resistance rating. Smoke partitions shall extend from the top of the foundation or floor below to the underside of the floor or roof sheathing, deck or slab above or to the underside of the ceiling above where the ceiling membrane is constructed to limit the transfer of smoke.*

Discussion and Commentary: A smoke partition is designed for a singular purpose, to limit the movement of smoke from one area to another. Therefore, windows in smoke partitions must be sealed, penetrations and joints must be adequately filled, and smoke dampers used to protect air transfer openings.

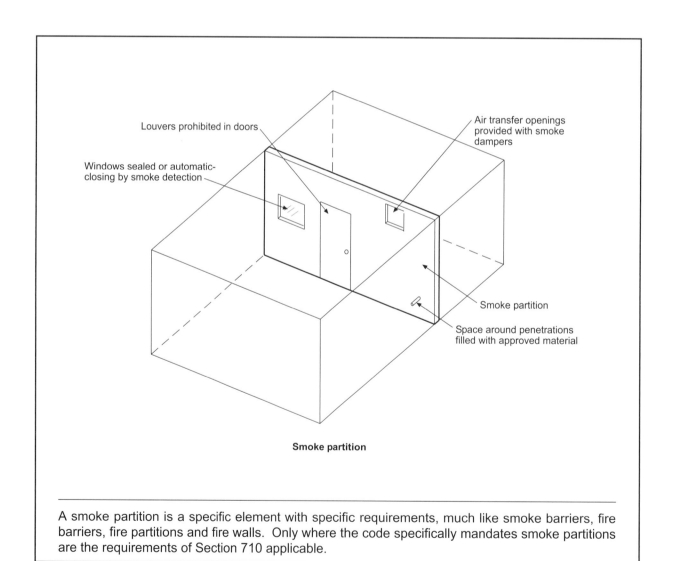

Smoke partition

A smoke partition is a specific element with specific requirements, much like smoke barriers, fire barriers, fire partitions and fire walls. Only where the code specifically mandates smoke partitions are the requirements of Section 710 applicable.

Topic: Fire-Resistance Rating
Reference: IBC 711.3

Category: Fire and Smoke Protection Features
Subject: Horizontal Assemblies

Code Text: *The fire-resistance rating of floor and roof assemblies shall not be less than that required by the building type of construction. Where the floor assembly separates mixed occupancies, the assembly shall have a fire-resistance rating of not less than that required by Section 508.4 based on the occupancies being separated. Where the floor assembly separates a single occupancy into different fire areas, the assembly shall have a fire-resistance rating of not less than that required by Section 707.3.10.*

Discussion and Commentary: Table 601 regulates the minimum fire-resistance ratings for floor construction based on the building's type of construction. This minimum level of fire-resistance must always be maintained. Other provisions of the code must also be considered where a horizontal separation is needed within a multistory building, such as for control areas or occupancy separations.

Omission of ceiling and flooring in horizontal assemblies per Section 711.3.3

Other than permitted openings, penetrations or joints, horizontal assemblies must be continuous in order to isolate totally one floor from another. An allowance is permitted for fire-resistance-rated roof construction, where skylights and other penetrations may be unprotected.

Study Session 6

Topic: Smoke Barriers
Reference: IBC 711.9

Category: Fire and Smoke Protection Features
Subject: Horizontal Assemblies

Code Text: *Where horizontal assemblies are required to resist the movement of smoke by other sections of the IBC in accordance with the definition of smoke barrier, penetrations and joints in such horizontal assemblies shall be protected as required for smoke barriers in accordance with Sections 714.5 and 715.6. Regardless of the number of stories connected by elevator shaft enclosures, doors located in elevator shaft enclosures that penetrate the horizontal assembly shall be protected by enclosed elevator lobbies complying with Section 713.14.1. Horizontal assemblies shall not be allowed to have unprotected vertical openings.*

Discussion and Commentary: Where smoke barriers are mandated in multistory buildings, all horizontal assemblies must be constructed in a manner to limit the vertical movement of smoke from story to story. This includes the protection of any joints and penetrations. Where the building is provided with an elevator, shaft enclosure protection must be extended in the form of elevator lobbies so as to restrict vertical smoke movement by means of the elevator shaft enclosure.

Horizontal smoke barrier

The primary use of smoke barriers is to create smoke compartments as required in Group I-2 occupancies. Where such occupancies occur in multistory buildings, the smoke compartments are limited vertically at each story as well as within the stories themselves.

Topic: Applications
Reference: IBC 712.1
Category: Fire and Smoke Protection Features
Subject: Vertical Openings

Code Text: The provisions of Section 712 shall apply to the vertical opening applications listed in Sections 712.1.1 through 712.1.18.

Discussion and Commentary: In multistory buildings, the upward transmission of fire, smoke and toxic gases through openings in the floor/ceiling assemblies continues to be a hazard of the highest degree. For many years, a shaft enclosure was considered the appropriate method to protect openings within a floor/ceiling assembly. Over time, other methods of protection have been identified as acceptable alternatives to the shaft enclosure approach. The IBC now places the emphasis on the presence of vertical openings while identifying the use of shaft enclosures as one of many protective measures that can be utilized to address the concern related to the vertical spread of fire, smoke and toxic gases

Shaft Enclosures	712.1.1
Individual Dwelling Units	712.1.2
Escalator Openings	712.1.3
Penetrations	712.1.4
Ducts	712.1.5
Atriums	712.1.6
Masonry Chimneys	712.1.7
Two-Story Openings	712.1.8
Parking Garages	712.1.9
Mezzanines	712.1.10
Joints	712.1.11
Unenclosed Stairs and Ramps	712.1.12
Floor Fire Doors	712.1.13
Group I-3 Occupancies	712.1.14
Elevators in Parking Garages	712.1.15
Duct Systems in Parking Garages	712.1.16
Nonfire-Resistance-Rated Joints	712.1.17
Openings Otherwise Permitted	712.1.18

In addition to a general reference to any other code provisions that might address the protection of vertical openings between stories, seventeen specific conditions—including shaft enclosures—are listed that identify various methods of compliance.

Study Session 6
IBC Sections 706 – 712

1. Fire walls must be constructed of noncombustible materials unless separating buildings of Type _____ construction.

    a. I b. III

    c. IV d. V

    Reference_____

2. A fire wall separating Type VA buildings housing Group M occupancies must have a minimum fire-resistance rating of _____ hour(s).

    a. 1 b. 2

    c. 3 d. 4

    Reference_____

3. In a Type IIB building containing a Group B occupancy, what is the minimum required fire-resistance rating for a fire wall?

    a. 1 hour b. 2 hour

    c. 3 hour d. 4 hour

    Reference_____

4. As a general provision, what minimum distance must a fire wall extend horizontally beyond the exterior surface of exterior walls?

   a. 18 inches
   b. 20 inches
   c. 30 inches
   d. 4 feet

   Reference_____

5. In general, what minimum distance above the roof must a fire wall extend?

   a. 18 inches
   b. 30 inches
   c. 3 feet
   d. 4 feet

   Reference_____

6. Where a fire wall occurs at a location where the roof heights differ, the fire wall may terminate at the underside of the lower roof slab, provided _____.

   a. the lower roof is of noncombustible construction
   b. the first 10 feet of the lower roof assembly has a minimum 1-hour rating
   c. the exterior wall of the higher portion is 1-hour to a height of 10 feet
   d. openings in the lower roof have a minimum 45-minute fire-protective rating

   Reference_____

7. Embedded ends of combustible members entering masonry or concrete fire walls shall be separated a minimum distance of _____ inch(es).

   a. 1
   b. 2
   c. 4
   d. 6

   Reference_____

8. The aggregate width of openings in a fire wall is limited to a maximum of _____ of the length of the wall.

   a. 10 percent
   b. 25 percent
   c. $33^1/_3$ percent
   d. 50 percent

   Reference_____

Study Session 6

9. Fire barriers are required for the separation of all of the following building elements, except _____ .

   a. shaft enclosures   b. exit passageways
   c. incidental uses   d. fire-resistance-rated corridors

   Reference_____

10. In nonrated buildings, construction supporting fire barriers need not be protected by equivalent fire resistance where the fire barriers are used for _____ .

    a. 1-hour occupancy separations
    b. 1-hour vertical exit enclosures
    c. 1-hour incidental use separations
    d. 2-hour horizontal exits

    Reference_____

11. In a nonsprinklered building, any single opening in a fire barrier is limited to a maximum area of _____ square feet.

    a. 20   b. 100
    c. 120   d. 156

    Reference_____

12. Openings in a fire barrier are permitted to be unlimited in area and aggregate width where the fire barrier is utilized as a(n) _____ .

    a. interior exit stairway enclosure
    b. horizontal exit wall
    c. incidental use separation
    d. control area separation

    Reference_____

13. Where vertical openings are created for an escalator system in a fully sprinklered hotel, special opening protection methods are established, other than automatic shutters, where the openings connect a maximum of _____ stories.

    a. two   b. three
    c. four   d. six

    Reference_____

14. A floor opening is permitted between a maximum of two stories without a shaft enclosure, under limited conditions, in all but which one of the following occupancies?

    a. Group A-2
    b. Group H-2
    c. Group I-2
    d. Group S-2

    Reference_____

15. Where double fire walls are constructed, they shall be in compliance with which of the following referenced standards?

    a. NFPA 286
    b. NFPA 221
    c. UL 1040
    d. UL 2079

    Reference_____

16. A fire barrier used to separate building areas connected by a horizontal exit shall have a minimum fire-resistance rating of _____.

    a. 1 hour
    b. 1½ hours
    c. 2 hours
    d. 3 hours

    Reference_____

17. A fire partition is not the appropriate wall assembly for walls separating _____.

    a. sleeping units in a Group R-1 hotel
    b. tenant spaces in a covered mall building
    c. control areas in a manufacturing occupancy
    d. dwelling units in an apartment building

    Reference_____

18. What is the minimum required fire-resistance rating for a fire partition separating sleeping units in a Group I-1 occupancy housed in a Type IIA building?

    a. 20 minutes
    b. 30 minutes
    c. 1 hour
    d. no rating is required

    Reference_____

Study Session 6

19. Where a fire-resistance-rated corridor ceiling is constructed as required for the corridor walls, the walls shall not terminate before reaching _____.

    a. the lower membrane of the ceiling assembly

    b. the upper membrane of the ceiling assembly

    c. the underside of the floor or roof deck above

    d. an approved fire-resistant joint system

    Reference_____

20. In other than a Group I-3 occupancy, what is the minimum required fire-resistance rating for a smoke barrier?

    a. no rating is required   b. 30 minutes

    c. 45 minutes   d. 1 hour

    Reference_____

21. In which one of the following types of construction must a smoke barrier be supported by fire-resistance-rated construction having at least an equivalent rating to the smoke barrier supported?

    a. Type IIB   b. Type IIIA

    c. Type IIIB   d. Type VB

    Reference_____

22. Automatic shutters utilized to protect escalator openings between stories shall have a minimum fire-resistance rating of _____.

    a. 45 minutes   b. 1 hour

    c. $1\frac{1}{2}$ hours   d. 2 hours

    Reference_____

23. Wire or other approved devices shall be installed above lay-in ceiling panels in a fire-resistance-rated floor/ceiling assembly to prevent vertical displacement where the weight of the panels is not adequate to resist a minimum upward force of _____.

    a. 1 psf   b. 2 psf

    c. 5 psf   d. 15 psf

    Reference_____

24. Under which of the following conditions may the ceiling membrane of a fire-resistance-rated horizontal assembly be omitted?

    a. where usable attic space occurs above

    b. where the assembly has a minimum fire-resistance rating of 2 hours

    c. where the floor construction is limited to combustible construction

    d. where an unusable crawl space occurs below

    Reference_____

25. Unprotected skylights are permitted through a fire-resistance-rated roof deck, provided _____.

    a. the skylights were tested as a part of the roof assembly

    b. the roof construction has not less than a 1-hour fire-resistance rating

    c. the structural integrity of the roof construction is maintained

    d. they are limited to 10 percent of the total roof area

    Reference_____

26. Where a fire barrier is utilized to separate a Group M occupancy into multiple fire areas, the minimum fire-resistance rating of the fire barrier shall be _____ hour(s).

    a. 1   b. 2

    c. 3   d. 4

    Reference_____

27. The voids created at the intersection of a fire barrier and a nonfire-resistance-rated roof assembly _____.

    a. are not regulated by the code

    b. shall be filled by an approved material or system

    c. shall be sealed with a noncombustible material

    d. must be protected with a fire-resistive joint system

    Reference_____

28. What hourly rating is mandated for the construction of smoke partitions?

    a. 0, no rating is required   b. ½ hour
    c. ¾ hour   d. 1 hour

    Reference_____

29. Windows in smoke partitions shall have a minimum fire-protection rating of _____.

    a. 0, no rating is required   b. 20 minutes
    c. 45 minutes   d. 1 hour

    Reference_____

30. In a Type IIB hotel that is sprinklered in accordance with NFPA 13, what is the minimum fire-resistance rating for the fire partitions and horizontal assemblies that separate sleeping units?

    a. 0, no rating is required   b. ½ hour
    c. 1 hour   d. 2 hours

    Reference_____

31. In general, fire walls shall extend to the outer edge of horizontal projecting elements that are located within _____ feet of the fire wall.

    a. 4   b. 5
    c. 8   d. 10

    Reference_____

32. Smoke compartments in Group I-3 occupancies may be constructed of nonrated steel smoke barriers provided the steel is a minimum of _____ inch in thickness.

    a. 0.08   b. 0.10
    c. 0.12   d. 0.15

    Reference_____

33. In other than Group _____ occupancies, the vertical openings created by an atrium condition are regulated under the provisions of IBC Section 404.

    a. A   b. H
    c. I   d. R

    Reference _____

34. Unless the installation of a damper will interfere with the operation of a smoke control system, air transfer openings in smoke partitions shall be provided with what type of damper?

    a. smoke damper only
    b. fire damper only
    c. combination fire/smoke damper
    d. no damper is required

    Reference _____

35. In a Group I-2 occupancy where a pair of opposite-swinging doors without a center mullion is installed in a smoke barrier across a corridor, the doors are permitted maximum undercuts of _____ inch.

    a. $1/4$ b. $3/8$
    c. $1/2$ d. $3/4$

    Reference _____

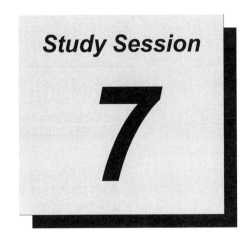

Study Session

7

2012 IBC Sections 713 – 720
Fire and Smoke Protection Features III

OBJECTIVE: To gain an understanding of shaft enclosures, fireblocking and draftstopping, as well as the methods of protecting fire-resistance-rated building components where they contain doors, windows, ducts, air transfer openings and penetrations.

REFERENCE: Sections 713 through 720, 2012 *International Building Code*

KEY POINTS:
- What is the purpose of a shaft enclosure? Where are such enclosures required?
- What type of fire-resistance-rated assemblies are used to construct shaft enclosures?
- How does the height of a building affect the fire-resistance ratings of shaft enclosures? How does the fire-resistance rating of the floor construction apply?
- Which methods are mandated for construction of a shaft enclosure?
- Which types of openings are permitted to penetrate a shaft enclosure? How must they be
- protected?
- How must shafts be enclosed at the top? At the bottom?
- Which special provisions govern refuse, recycling and laundry chutes?
- Under which conditions must elevator lobbies be provided?
- Which two types of penetrations are regulated by the code?
- What are the appropriate installation details where sleeves are used in penetrating a fire-resistance-rated assembly?
- What is a through penetration? Membrane penetration?
- Penetrating items of steel, ferrous or copper may be protected in what manner?
- What is an F rating and a T rating? What minimum fire-resistance rating is required of a through-penetration firestop system?
- How can the membrane penetration by a steel outlet box be addressed? A listed electrical box? A fire sprinkler?
- Which limitations are placed on noncombustible penetrating items connecting to combustible items?
- How are fire-resistance-rated horizontal assemblies penetrated? Nonfire-resistance-rated horizontal assemblies?

KEY POINTS:
(Cont'd)

- What is the purpose of a fire-resistance-rated joint system? When is such a system required?
- How should the intersection of an exterior curtain wall and a fire-resistance-rated floor assembly be accomplished? An exterior curtain wall and a nonrated floor assembly?
- How is the required fire-protection rating of an opening protective determined?
- What are the characteristics of a fire-resistance-rated corridor door or smoke barrier door?
- In what locations must a fire door be rated to a maximum transmission temperature end point?
- What are the labeling requirements for fire doors? Smoke and draft control doors? Fire-protection-rated glazing?
- Where are self-closing door assemblies mandated? Automatic-closing assemblies?
- What is the maximum fire-protection rating assigned to fire-protection-rated glazing?
- What are the different types of dampers? How should they be actuated?
- How must fire and smoke dampers be identified and accessed?
- Where are fire dampers required? Smoke dampers? Ceiling radiation dampers?
- In what specific locations are fire dampers not required for penetrations of shaft enclosures? Where smoke dampers are not required?
- Why is draftstopping and fireblocking unnecessary in noncombustible construction?
- Where is fireblocking required to be installed? Draftstopping?
- Which materials are acceptable as fireblocks? Draftstops?

Topic: Definition and Scope
Reference: IBC 713.1, 202

Category: Fire and Smoke Protection Features
Subject: Shaft Enclosures

Code Text: *A shaft is an enclosed space extending through one or more stories of a building, connecting vertical openings in successive floors, or floors and roof. A shaft enclosure is the walls or construction forming the boundaries of a shaft. The provisions of Section 713 shall apply to shafts required to protect openings and penetrations through floor/ceiling and roof/ceiling assemblies.*

Discussion and Commentary: It is not uncommon in multistory buildings to have openings that are provided to accommodate elevators, mechanical equipment or similar devices, and to transmit light or ventilation air. Because of the potential for the rapid spread of fire, smoke and gases vertically through buildings, such vertical openings must be protected with fire-resistance-rated shaft enclosures or otherwise addressed in accordance with Section 712.

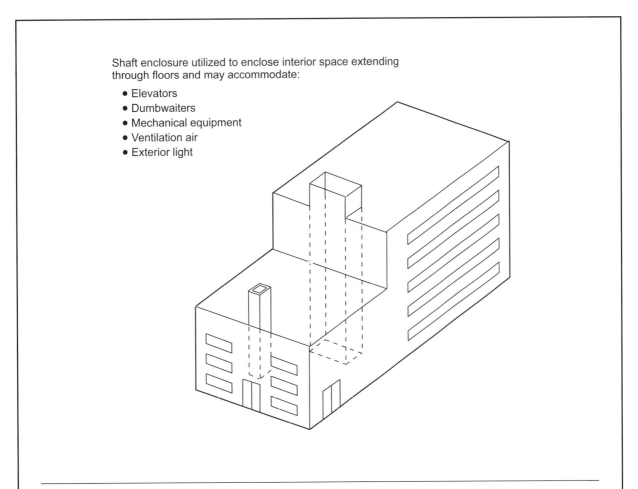

Shaft enclosure utilized to enclose interior space extending through floors and may accommodate:
- Elevators
- Dumbwaiters
- Mechanical equipment
- Ventilation air
- Exterior light

The fire-resistance rating required for a shaft enclosure is based on the building height, with 2 hours being required where four stories or more are connected. Where less than four stories are connected, 1 hour is required. The enclosure rating cannot be less than that of any floor penetrated.

Topic: Continuity
Reference: IBC 713.2, 713.7
Category: Fire-Resistance-Rated Construction
Subject: Shaft Enclosures

Code Text: *Shaft enclosures shall be constructed as fire barriers in accordance with Section 707 or horizontal assemblies constructed in accordance with Section 711, or both. Openings in a shaft enclosure shall be protected in accordance with Section 716 as required for fire barriers. Doors shall be self- or automatic-closing by smoke detection in accordance with Section 716.5.9.3. Openings other than those necessary for the purpose of the shaft shall not be permitted in shaft enclosures.*

Discussion and Commentary: The general provisions dictate that a shaft be completely enclosed with fire-resistance-rated construction. However, there are conditions that modify this rule. The protection of exterior shaft walls is often unnecessary. Additionally, for those shafts that do not extend to the bottom of the building, the code provides three methods of maintaining the integrity of the shaft enclosure.

Refuse, recycling and laundry chutes

Where a shaft enclosure terminates short of the roof, the top enclosure must be constructed to a fire-resistance degree consistent with that of the top floor penetrated. In no case may the enclosure at the top be rated less than that for the remainder of the shaft enclosure.

Topic: Elevator Lobbies
Reference: IBC 713.14.1
Category: Fire and Smoke Protection Features
Subject: Shaft Enclosures

Code Text: *An enclosed elevator lobby shall be provided at each floor where an elevator shaft enclosure connects more than three stories. The lobby shall separate the elevator shaft enclosure doors from each floor by fire partitions.* See exceptions for 1) the level of exist discharge, 2) elevators not required to be located in a shaft, 3) additional doors provided at the hoistway opening, 4) most sprinklered buildings, 5) the use of smoke partitions in fully sprinklered buildings, 6) elevator hoistways that are pressurized and 7) open parking garages.

Discussion and Commentary: The purpose of an elevator lobby under this provision is to reduce the potential for smoke to travel from the floor of fire origin to any other floor of the building by way of an elevator shaft enclosure. An allowance is provided for those low-rise buildings where the elevator shaft connects only two or three stories.

Regardless of the exceptions to Section 713.14.1, the requirements for accessible means of egress in Section 1007.4 may create the need for an elevator lobby to serve as a required area of refuge.

Topic: Definitions and Scope
Reference: IBC 714.1, 202

Category: Fire and Smoke Protection Features
Subject: Penetrations

Code Text: *A through penetration is a breach in both sides of a floor, floor-ceiling or wall assembly to accommodate an item passing through the breaches. A membrane penetration is a breach in one side of a floor-ceiling, roof-ceiling or wall assembly to accommodate an item installed into or passing through the breach. The provisions of Section 714 shall govern the materials and methods of construction used to protect through penetrations and membrane penetrations of horizontal assemblies and fire-resistance-rated wall assemblies.*

Discussion and Commentary: Fire-resistance-rated walls and horizontal assemblies are usually penetrated, both fully and partially, with piping, conduit, outlet boxes, cable, vents and similar penetrating items. The IBC regulates both the materials and the methods of penetration based on the specific conditions that exist. Where sleeves are used, they must be fastened securely in place, and all open space within and around the sleeve must be appropriately protected.

Penetrations of nonfire-resistance-rated horizontal assemblies are regulated by Section 714.4.2. Although some horizontal assemblies may not require a fire-resistance rating, the code intends that some degree of separation (compartmentalization) be provided from one story to another.

Topic: Fire-Resistance-Rated Walls
Reference: IBC 714.3
Category: Fire and Smoke Protection Features
Subject: Penetrations

Code Text: *Penetrations into or through fire walls, fire barriers, smoke barrier walls, and fire partitions shall comply with Sections 714.3.1 through 714.3.3. Penetrations in smoke barrier walls shall also comply with Section 714.5.*

Discussion and Commentary: In general, penetrations into or through fire-resistance-rated walls must be either protected with an approved through-penetration firestop system or installed as a tested component of an approved fire-resistance-rated assembly. These methods are considered proprietary, with each penetration being regulated by the specifics of the installation. Two generic methods are identified as exceptions to the general requirements; however, both methods are based on the penetration only of steel, ferrous or copper pipes or steel conduits. Under such conditions, the annular space around the penetrating items shall be filled with an appropriate material.

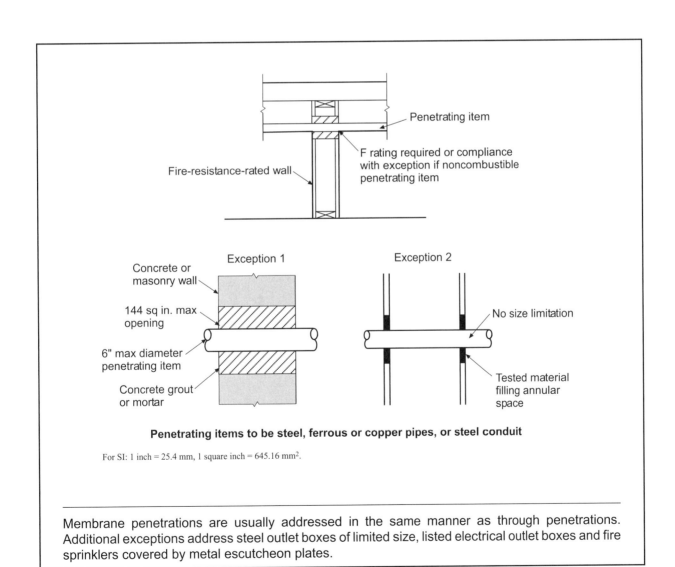

Penetrating items to be steel, ferrous or copper pipes, or steel conduit

For SI: 1 inch = 25.4 mm, 1 square inch = 645.16 mm².

Membrane penetrations are usually addressed in the same manner as through penetrations. Additional exceptions address steel outlet boxes of limited size, listed electrical outlet boxes and fire sprinklers covered by metal escutcheon plates.

Study Session 7 183

Topic: Fire-Resistance-Rated Walls
Reference: IBC 714.3.3, 717.1.1
Category: Fire and Smoke Protection Features
Subject: Penetrations

Code Text: *Noncombustible penetrating items shall not connect to combustible items beyond the point of firestopping unless it can be demonstrated that the fire-resistance integrity of the wall is maintained. Ducts that penetrate fire-resistance-rated assemblies and are not required by Section 717 to have dampers shall comply with the requirements of Sections 714.2 through 714.3.3.*

Discussion and Commentary: Duct penetrations of fire-resistance-rated wall assemblies are typically protected with fire dampers in accordance with Section 717.5. However, in those locations where dampers are not required, it is still necessary to address the structural integrity of the fire-resistive-rated wall where it is penetrated. Thus, the space between the duct and the wall must be protected in a manner consistent with that used for pipes, conduits and similar items.

To prevent the creation of an open penetration, the transition from a noncombustible penetrating item to a combustible item is prohibited beyond the point of firestopping. Such a condition is only permitted when its suitability has been demonstrated through testing.

Topic: Horizontal Assemblies
Reference: IBC 714.4
Category: Fire and Smoke Protection Features
Subject: Penetrations

Code Text: *Penetrations of a floor, floor/ceiling assembly or the ceiling membrane of a roof/ceiling assembly not required to be enclosed in a shaft by Section 712.1 shall be protected in accordance with Sections 714.4.1 through 714.4.2.2. Penetrations of the fire-resistance rated floor, floor/ceiling assembly or the ceiling membrane of a roof/ceiling assembly shall comply with Sections 714.4.1.1 through 714.4.1.4.*

Discussion and Commentary: Where horizontal construction is penetrated by a duct, pipe, tube, wire, conduit, cable, vent or similar item, the primary requirements are based on Section 713 for shaft enclosures. However, Section 712.1.4 permits the use of Section 714 for both through penetrations and membrane penetrations. The provisions for horizontal assemblies are very similar to those for walls, with special allowances for steel, copper or ferrous penetrating items. In addition to the typical penetrations, the code requires that recessed fixtures in fire-resistance-rated floor/ceiling assemblies be installed so that the required fire resistance is not reduced.

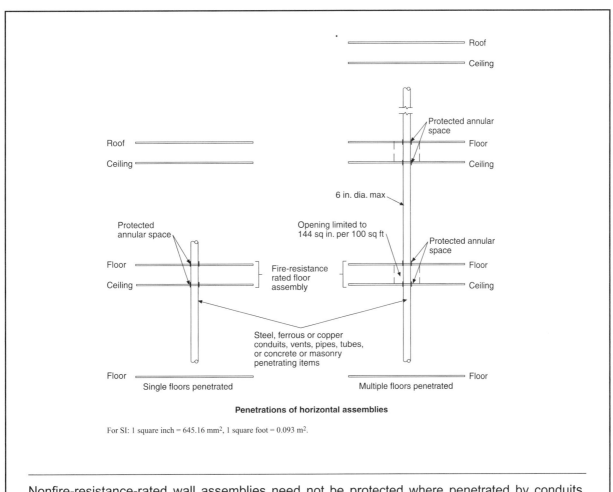

Penetrations of horizontal assemblies

For SI: 1 square inch = 645.16 mm², 1 square foot = 0.093 m².

Nonfire-resistance-rated wall assemblies need not be protected where penetrated by conduits, piping and similar penetrating items. However, such protection is mandated for the penetration of nonfire-resistance-rated horizontal assemblies to limit vertical fire spread.

Topic: General Provisions
Reference: IBC 715.1, 715.2

Category: Fire and Smoke Protection Features
Subject: Fire-Resistant Joint Systems

Code Text: *Joints installed in or between fire-resistance-rated walls, floor or floor/ceiling assemblies and roofs or roof/ceiling assemblies shall be protected by an approved fire-resistant joint system designed to resist the passage of fire for a time period not less than the required fire-resistance rating of the wall, floor or roof in or between which it is installed. See exception for nine locations where joint systems not required. Fire-resistant joint systems shall be securely installed in accordance with the listing criteria in or on the joint for its entire length so as not to dislodge, loosen or otherwise impair its ability to accommodate expected building movements and to resist the passage of fire and hot gases.*

Discussion and Commentary: Joints are created where the structural design of a building necessitates a separation between building components in order to accommodate anticipated structural displacements caused by thermal expansion and contraction, seismic activity, wind or other loads. The integrity of the fire-resistant separation must be maintained where such joints occur.

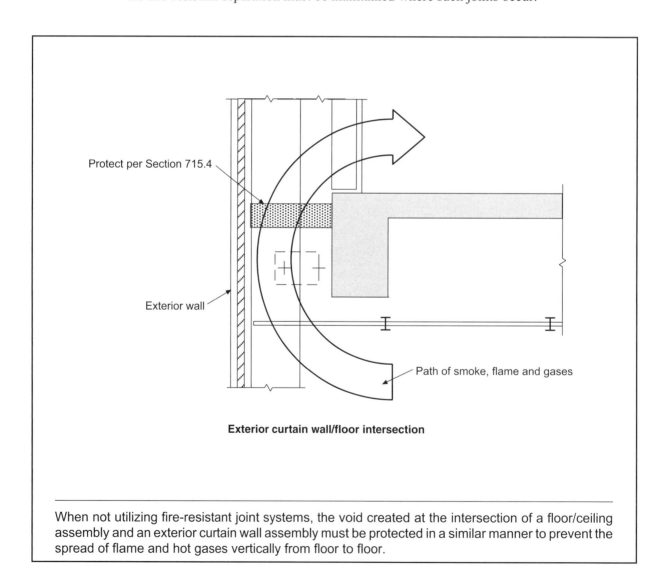

Exterior curtain wall/floor intersection

When not utilizing fire-resistant joint systems, the void created at the intersection of a floor/ceiling assembly and an exterior curtain wall assembly must be protected in a similar manner to prevent the spread of flame and hot gases vertically from floor to floor.

Topic: Fire Door Assemblies **Category:** Fire and Smoke Protection Features
Reference: IBC 716.5 **Subject:** Opening Protectives

Code Text: *Approved fire door and fire shutter assemblies shall be constructed of any material or assembly of component materials that conforms to the test requirements of Section 716.5.1 (side-hinged or pivoted swinging doors), 716.5.2 (other types of doors) or 716.5.3 (doors in corridors and smoke barriers) and the fire-protection rating indicated in Table 716.5.* See exceptions for tin-clad fire doors and floor fire doors.

Discussion and Commentary: The level of protection required for a fire door is commensurate with that required for the wall or partition in which it is installed. The minimum fire protection rating varies based on the wall's required rating as well as the type and use of the wall assembly under consideration. In addition to establishing the minimum fire-protection rating required for fire door assemblies, Table 716.5 also provides information on door vision panels, sidelights and transoms.

TABLE 716.5
OPENING FIRE PROTECTION ASSEMBLIES, RATINGS AND MARKINGS

TYPE OF ASSEMBLY	REQUIRED WALL ASSEMBLY RATING (hours)	MINIMUM FIRE DOOR AND FIRE SHUTTER ASSEMBLY RATING (hours)	DOOR VISION PANEL SIZE	FIRE RATED GLAZING MARKING DOOR VISION PANEL[e]	MINIMUM SIDELIGHT/ TRANSOM ASSEMBLY RATING (hours)		FIRE-RATED GLAZING MARKING SIDELITE/TRANSOM PANEL	
					Fire protection	Fire resistance	Fire protection	Fire resistance
Fire walls and fire barriers having a required fire-resistance rating greater than 1 hour	4	3	Not Permitted	Not Permitted	Not Permitted	4	Not Permitted	W-240
	3	3[a]	Not Permitted	Not Permitted	Not Permitted	3	Not Permitted	W-180
	2	1½	100 sq. in.[c]	≤100 sq.in. = D-H-90 >100 sq.in.= D-H-W-90	Not Permitted	2	Not Permitted	W-120
	1½	1½	100 sq. in.[c]	≤100 sq.in. = D-H-90 >100 sq.in.= D-H-W-90	Not Permitted	1½	Not Permitted	W-90
Shaft, exit enclosures and exit passageway walls	2	1½	100 sq. in.[c,d]	≤100 sq.in. = D-H-90 > 100 sq.in.= D-H-T-or D-H-T-W-90	Not Permitted	2	Not Permitted	W-120
Fire barriers having a required fire-resistance rating of 1 hour: Enclosures for shafts, exit access stairways, exit access ramps, interior exit stairways, interior exit ramps and exit passageway walls	1	1	100 sq. in.[c,d]	≤100 sq.in. = D-H-60 >100 sq.in.= D-H-T-60 or D-H-T-W-60	Not Permitted	1	Not Permitted	W-60
					Fire protection			
Other fire barriers	1	¾	Maximum size tested	D-H-NT-45	¾		D-H-NT-45	
Fire partitions: Corridor walls	1	⅓[b]	Maximum size tested	D-20	¾[b]		D-H-OH-45	
	0.5	⅓[b]	Maximum size tested	D-20	⅓		D-H-OH-20	
Other fire partitions	1	¾	Maximum size tested	D-H-45	¾		D-H-45	
	0.5	⅓	Maximum size tested	D-H-20	⅓		D-H-20	

(continued)

A fire door assembly installed in a fire-rated corridor wall or smoke barrier must have a minimum fire protection rating of 20 minutes. In addition, the door must pass an air leakage test to verify that it provides the necessary level of smoke protection.

Topic: Labeling Requirements
Reference: IBC 716.5.7.1

Category: Fire and Smoke Protection Features
Subject: Fire Door Assemblies

Code Text: *Fire doors shall be labeled showing the name of the manufacturer or other identification readily traceable back to the manufacturer, the name or trademark of the third-party inspection agency, the fire-protection rating, and where required for fire doors in interior exit stairways and ramps and exit passageways by Section 716.5.5, the maximum transmitted temperature end point. Smoke and draft control doors complying with UL 1784 shall be labeled as such and shall also comply with Section 716.5.7.3. Labels shall be approved and permanently affixed. The label shall be applied at the factory or location where fabrication and assembly are performed.*

Discussion and Commentary: To be certain that the proper protective assembly is installed in the proper location, it is critical that the assembly be listed and labeled. Field alteration of a fire door assembly is not permitted, because the assembly is usually only listed for use in the condition it was in when it left the factory.

Some fire door assemblies are too large to be tested in available furnaces. Therefore, the code recognizes a certificate of inspection as proof that the oversized doors comply with the requirements for materials, design and construction for a comparable fire door.

Topic: Fire Door Glazing Identification
Reference: IBC 716.5.8.3.1
Category: Fire and Smoke Protection Features
Subject: Opening Protectives

Code Text: *For fire-protection-rated glazing, the label shall bear the following four-part identification: "D – H or NH – T or NT – XXX." "D" indicates that the glazing shall be used in fire door assemblies and that the glazing meets the fire protection requirements of NFPA 252. "H" shall indicate that the glazing meets the hose stream requirements of NFPA 252. "NH" shall indicate that the glazing does not meet the hose stream requirements of the test. "T" shall indicate that the glazing meets the temperature requirements of Section 716.5.5.1. "NT" shall indicate that the glazing does not meet the temperature requirements of Section 716.5.5.1. The placeholder "XXX" shall specify the fire-protection rating period, in minutes.*

Discussion and Commentary: Glazing utilized in fire door assemblies can be easily identified for verification of its appropriate application. Such glazing must also be provided with the proper identification indicating its compliance as safety glazing in conformance with Section 2406.4.

Similar identification methods for glazing are found in Table 716.3 for fire-resistance-rated glazing utilized as wall assemblies and Table 716.6 for fire-protection-rated glazing used in fire window assemblies.

Topic: Door Closing
Reference: IBC 716.5.9, 716.5.9.1

Category: Fire and Smoke Protection Features
Subject: Fire Door Assemblies

Code Text: *Fire doors shall be self- or automatic-closing in accordance with Section 716. See exceptions for fire doors in common walls between Group R-1 guestrooms, and elevator car and associated hoistway doors. Unless otherwise specifically permitted, single fire doors and both leaves of pairs of side-hinged swinging fire doors shall be provided with an active latch bolt that will secure the door when it is closed.*

Discussion and Commentary: Fire doors must close and latch to be effective during a fire. The expectation is that the doors will normally be in a closed position and that the self-closing device will cause the door to close after use. Where specifically mandated by the code, automatic-closing devices must be installed. Such devices are intended for doors normally held in an open position.

Automatic-closing doors shall be actuated by smoke detection at: (Sec. 716.5.9.3)

- Doors installed across a corridor;
- Doors protecting openings in exits or corridors required to be fire-resistance-rated;
- Doors in walls of incidental uses required to resist the passage of smoke (Sec. 509.4);
- Doors installed in smoke barriers (Sec. 709.5);
- Doors installed in fire partitions (Sec. 708.6);
- Doors installed in fire walls (Sec. 706.8);
- Doors installed in shaft enclosures (Sec. 713.7);
- Doors installed in refuse and laundry access chutes and termination rooms (Sec. 713.13);
- Doors installed in compartmentation walls of underground buildings (Sec. 405.4.2);
- Doors installed in elevator lobby walls of underground buildings (Sec. 405.4.3); and
- Doors installed in smoke partitions (Sec. 710.5.2.3).

Automatic-closing fire door assemblies are required only where specifically addressed, such as in Section 709.5 for cross-corridor doors in Group I-2 occupancies. Where automatic-closing fire doors are provided, including nonrequired locations, they must typically be smoke activated.

Topic: Definitions
Reference: IBC 717, 202

Category: Fire and Smoke Protection Features
Subject: Fire and Smoke Dampers

Code Text: *A fire damper is a listed device, installed in ducts and air transfer openings designed to close automatically upon detection of heat and resist the passage of flame. A smoke damper is a listed device installed in ducts and air transfer openings that is designed to resist the passage of smoke. A ceiling radiation damper is a listed device installed in a ceiling membrane of a fire-resistance-rated floor/ceiling or roof/ceiling assembly to limit automatically the radiative heat transfer through an air inlet/outlet opening.*

Discussion and Commentary: The IBC identifies several types of dampers, each of which performs a specific function. The required type of damper is based on the function of the building element that is being penetrated by the duct or air transfer opening.

Figure courtesy
Sheet Metal and Air Conditioning Contractors National Association

Note: This illustration is not intended to exclusively endorse or indicate preference for a combination fire and smoke damper. Two separate dampers that satisfy the requirements for the respective functions may also be used for fire and smoke control.

Combination fire and smoke damper

Where both a fire and a smoke damper are mandated, the use of a combination damper is permitted. This type of listed device is designed to close automatically upon detecting heat and to resist the passage of air and smoke.

Topic: Damper Testing and Ratings
Reference: IBC 717.3.1
Category: Fire-Resistance-Rated Construction
Subject: Ducts and Air Transfer Openings

Code Text: *Dampers shall be listed and labeled in accordance with the standards in Section 717.3. Fire dampers shall comply with the requirements of UL 555. Smoke dampers shall comply with the requirements of UL 555S. Combination fire/smoke dampers shall comply with the requirements of both UL 555 and UL 555S. Ceiling radiation dampers shall comply with the requirements of UL 555C. Fire dampers shall have the minimum fire-protection rating specified in Table 717.3.2.1 for the type of penetration. Smoke damper leakage ratings shall be Class I or II. Elevated temperature ratings shall not be less than 250°F (121°C).*

Discussion and Commentary: Consistent with other openings that penetrate a fire-resistance-rated assembly, fire and smoke dampers must be provided where it is necessary to maintain the integrity of the assembly. The minimum damper rating is based on the rating of the assembly penetrated.

A smoke damper must close upon actuation of a listed smoke detector or smoke detection system. The IBC identifies five methods, one or more of which may be applicable, for the detector location and/or actuation.

Topic: Where Required
Reference: IBC 717.5
Category: Fire and Smoke Protection Features
Subject: Fire and Smoke Dampers

Code Text: *Fire dampers, smoke dampers and combination fire/smoke dampers shall be provided at the locations prescribed in Sections 717.5.1 through 717.5.7 and 717.6. Where an assembly is required to have both fire dampers and smoke dampers, combination fire/smoke dampers or a fire damper and a smoke damper shall be required.*

Discussion and Commentary: Only those specific building elements identified in the code need to be protected by fire and/or smoke dampers where penetrated by ducts or air transfer openings. As a general rule, fire dampers protect such openings in fire walls, fire barriers, shaft enclosures and fire partitions. Fire dampers may also be installed in some of those locations where a shaft enclosure is otherwise required. Smoke dampers are generally required for openings in shaft enclosures, smoke- and draft-control corridor enclosures, and smoke barriers.

There will be times when a fire-resistance-rated assembly is penetrated by a duct or transfer opening that is not required to be protected by a fire or smoke damper. In such situations, the condition will be regulated and protected as a penetration in accordance with Section 714.

Topic: Nonrated Floor Assemblies
Reference: IBC 717.6.3
Category: Fire and Smoke Protection Features
Subject: Ducts and Air Transfer Openings

Code Text: *Duct systems constructed of approved materials in accordance with the International Mechanical Code that penetrate nonfire-resistance-rated floor assemblies shall be protected by any of the following methods.* See three methods addressing 1) shaft enclosures, 2) ducts connecting only two stories, and 3) ducts connecting a maximum of three stories.

Discussion and Commentary: It is important when addressing the vertical spread of fire, smoke and gases—even in buildings where the floor assemblies are not required to have a fire-resistance rating—that some level of compartmentation is provided between stories.

Where Item 3 is utilized for a duct that connects three stories, a minimum 1$\frac{1}{2}$-hour fire damper is required at each floor line even though the floor or floor/ceiling assembly is not required to have a fire-resistance rating, because the damper is an alternative to a 1-hour shaft enclosure.

Topic: Fireblocking
Reference: IBC 718.2, 202
Category: Fire and Smoke Protection Features
Subject: Concealed Spaces

Code Text: *Fireblocking consists of materials installed to resist the free passage of flame to other areas of the building through concealed spaces. In combustible construction, fireblocking shall be installed to cut off concealed draft openings (both vertical and horizontal) and shall form an effective barrier between floors, between a top story and a roof or attic space. Fireblocking shall be installed in the locations specified in Sections 718.2.2 through 718.2.7.*

Discussion and Commentary: Experience has shown that the greatest fire damage to conventional light-framed wood buildings occurs when the fire travels unimpeded through concealed draft openings. Virtually any concealed air space within a building will provide an open channel through which high-temperature air and gases can spread. Fireblocking is invaluable to the control of fire prior to active fire suppression activities.

Platform framing　　　Balloon framing

In noncombustible construction, building materials located in concealed areas of the building construction do not contribute to the spread of fire. Therefore, fireblocking and draftstopping are required only in buildings of combustible construction.

Topic: Required Locations
Reference: IBC 718.2.2, 718.2.3

Category: Fire and Smoke Protection Features
Subject: Fireblocking

Code Text: *Fireblocking shall be provided in concealed spaces of stud walls and partitions, including furred spaces, and parallel rows of studs or staggered studs, as follows: 1) vertically at the ceiling and floor levels, and 2) horizontally at intervals not exceeding 10 feet (3048 mm). Fireblocking shall be provided at interconnections between concealed vertical . . . and horizontal spaces . . . such as occur at soffits, drop ceilings, cove ceilings and similar locations. See additional provisions for fireblocking at stairways; openings around vents, ducts and chimneys; concealed spaces of exterior architectural trim; and concealed sleeper spaces in floors.*

Discussion and Commentary: The platform framing techniques that are typically used in light-frame wood construction provide adequate fireblocking between stories in the stud walls. However, furred spaces and openings for penetrating elements such as vents should be addressed carefully as avenues for fire transmission between stories or along a wall.

Drawings courtesy
American Forest and Paper Association

In general, fireblocking materials must consist of lumber or wood structural panels of the thicknesses specified, gypsum board, cement fiber board, batts or blankets of mineral wool or glass fiber, or any other approved materials securely fastened in place.

Topic: Definition and Materials
Reference: IBC 718.3.1, 202
Category: Fire and Smoke Protection Features
Subject: Draftstopping

Code Text: *A draftstop is a material, device or construction installed to restrict the movement of air within open spaces of concealed areas of building components such as crawl spaces, floor/ceiling assemblies, roof/ceiling assemblies and attics. Draftstopping materials shall not be less than 0.5-inch (12.7 mm) gypsum board, 0.375-inch (9.5 mm) wood structural panel, 0.375-inch (9.5 mm) particleboard, 1-inch (25 mm) nominal lumber, cement fiberboard, batts or blankets of mineral wool or glass fiber, or other approved materials adequately supported. The integrity of draftstops shall be maintained.*

Discussion and Commentary: Draftstopping, like fireblocking, is required only in combustible construction. Although the role of draftstopping is important, it is less critical than that of fireblocking. Therefore, the protective materials used in draftstopping construction are permitted to be less substantial.

*Required to divide space into areas no greater than 1,000 sq ft

Drawings courtesy
American Forest and Paper Association

For SI: 1 square foot = 0.093 m².

The provisions for draftstops are categorized for two general occupancy categories: residential and all uses other than residential. Both floor/ceiling assemblies and attics are addressed for each category. Many of the requirements are eliminated in fully sprinklered buildings.

Study Session 7
IBC Sections 713 – 720

1. A shaft enclosure shall have a minimum 2-hour fire-resistance rating where connecting a minimum of _____ stories.

   a. two b. three
   c. four d. six

   Reference_____

2. Where the annular space at the penetration of horizontal assemblies is filled with an approved material, noncombustible piping may connect a maximum of _____ stories, provided the horizontal assemblies require no fire-resistance rating.

   a. 2 b. 3
   c. 4 d. 5

   Reference_____

3. In which of the following locations is a fire-resistant joint system required to protect all joints?

   a. floors within malls b. horizontal exit walls
   c. mezzanine floors d. roofs where openings are permitted

   Reference_____

4. Fire-resistant joint systems are not required for control joints having a maximum width of _____ inch when tested in accordance with ASTM E 119 or UL 263.

   a. 0.25
   b. 0.375
   c. 0.5
   d. 0.625

   Reference _____

5. In a nonsprinklered building, a fire door assembly in an exit passageway shall have a maximum transmitted temperature end point of _____ above ambient at the end of 30 minutes of standard fire test exposure.

   a. 250°F
   b. 450°F
   c. 600°F
   d. 650°F

   Reference _____

6. An opening into a laundry chute access room shall be protected by opening protectives having a minimum fire protection rating of _____ .

   a. 20 minutes
   b. 45 minutes
   c. 1 hour
   d. 1 1/2 hours

   Reference _____

7. A fire door assembly in a 1-hour fire barrier used in an interior exit stairway enclosure shall have a minimum fire protection rating of _____ hour.

   a. 1/3
   b. 1/2
   c. 3/4
   d. 1

   Reference _____

8. Fire door assemblies in a 2-hour exterior wall shall have a minimum fire protection rating of _____ .

   a. 20 minutes
   b. 45 minutes
   c. 1 hour
   d. 90 minutes

   Reference _____

9. Penetrations in _____ shall be tested for air leakage in accordance with the requirements of UL 1479.

   a. smoke partitions   b. smoke barriers
   c. fire barriers   d. fire walls

   Reference _____

10. Where an elevator lobby is required in a nonsprinklered building, it shall be separated from the floor level through the use of _____.

    a. smoke barriers   b. fire barriers
    c. fire partitions   d. smoke partitions

    Reference _____

11. In which one of the following locations may self-closing or automatic-closing devices be omitted from required fire doors?

    a. Group E corridor doors
    b. doors in common walls separating Group R-1 sleeping units
    c. doors between vertical exit enclosures and exit passageways
    d. Group I-2 smoke barrier doors

    Reference _____

12. For a fire door that is automatic closing by smoke detection, the maximum delay before the fire door starts to close is _____ seconds after the smoke detector is actuated.

    a. 3   b. 5
    c. 10   d. 15

    Reference _____

13. Where automatic-closing fire doors are installed, which one of the following locations does not require closing upon actuation of smoke detectors?

    a. cross-corridor doors in an office building
    b. doors in a fire barrier separating control areas
    c. a pair of doors installed in a fire wall
    d. a single door installed in a smoke partition

    Reference _____

14. Access openings for a refuse, recycling or laundry chute shall be located in rooms enclosed by minimum _____ fire barriers.

    a. 1   b. 1½

    c. 2   d. 3

    Reference_____

15. What is the minimum required fire-protection rating for a fire window assembly located in a 2-hour fire barrier?

    a. ¾ hour   b. 1 hour

    c. 1½ hours   d. fire windows are not permitted

    Reference_____

16. The total area of interior fire window assemblies located in fire partitions and fire barriers is limited to a maximum of _____ percent of the area of the common wall within any room.

    a. 10   b. 20

    c. 25   d. 50

    Reference_____

17. What is the minimum required fire damper rating for a damper located in a 1-hour fire barrier?

    a. 20 minutes   b. 45 minutes

    c. 1 hour   d. 1½ hour

    Reference_____

18. Where a spot-type detector is used to actuate a smoke damper installed within an air transfer opening in a smoke barrier wall, the detector shall be located within _____ of the damper.

    a. 12 inches vertically   b. 30 inches vertically

    c. 3 feet horizontally   d. 5 feet horizontally

    Reference_____

Study Session 7

19. Where a duct passes through a fire wall used as a horizontal exit, which of the following dampers is/are required?

    a. fire damper only

    b. smoke damper only

    c. both a fire damper and a smoke damper

    d. neither a fire damper nor a smoke damper

    Reference_____

20. In general, which of the following dampers is/are required where an air transfer opening penetrates a shaft enclosure?

    a. fire damper only

    b. smoke damper only

    c. both fire and smoke dampers

    d. neither a fire damper nor a smoke damper

    Reference_____

21. Which of the following materials is not specifically listed by the code as a fireblocking material?

    a. 2-inch nominal lumber

    b. two thicknesses of 1-inch nominal lumber with broken lap joints

    c. one thickness of 0.719-inch wood structural panel with joints backed

    d. one thickness of $1/2$-inch particleboard with joints backed

    Reference_____

22. The maximum spacing of fireblocking within concealed spaces of exterior architectural elements erected with combustible framing shall be _____ feet.

    a. 10   b. 20

    c. 60   d. 100

    Reference_____

23. Which of the following materials is not specifically listed by the code as an acceptable draftstopping material?

    a. $^3/_8$-inch gypsum board
    b. $^3/_8$-inch wood structural panels
    c. $^3/_8$-inch particleboard
    d. glass fiber batts

    Reference_____

24. In a nonsprinklered Type V office building, the maximum attic area permitted between draftstops is _____.

    a. 1,000 square feet
    b. 3,000 square feet
    c. 100 feet in any direction
    d. unlimited

    Reference_____

25. Other than for cellulose loose-fill insulation that is not spray applied, the maximum flame spread index of insulating materials concealed within a Type II building shall be _____.

    a. 25
    b. 50
    c. 75
    d. unlimited

    Reference_____

26. A through-penetration firestop system, where used to protect a through penetration in a fire barrier wall, shall at minimum have a(n) _____ rating of not less than the rating of the wall penetrated.

    a. F
    b. L
    c. T
    d. F and T

    Reference_____

27. A firestop system may not be required for the protection of a membrane penetration by an unlisted steel electrical box in a 1-hour fire partition, provided the box has a maximum individual size of _____ square inches.

    a. 16
    b. 25
    c. 100
    d. 144

    Reference_____

28. What is the minimum required fire protection rating for a fire window located in a ½-hour fire-resistance-rated fire partition?

    a. 0, no rating is required   b. 20 minutes

    c. 30 minutes   d. 45 minutes

    Reference_____

29. Where wood sleepers are installed for the installation of wood flooring in a church sanctuary, the maximum size of any open spaces under the flooring shall be _____ square feet unless the entire underfloor space is filled with an approved material.

    a. 10   b. 20

    c. 100   d. 144

    Reference_____

30. Where plaster is used for fire-resistance purposes, it shall be provided with an additional layer of approved lath where its thickness exceeds _____ inch.

    a. ³/₈   b. ½

    c. ³/₄   d. 1

    Reference_____

31. The operating temperature of a ceiling radiation damper actuation device shall be _____ above the normal temperature within the duct system, but not less than _____ .

    a. 40°F, 160°F   b. 50°F, 160°F

    c. 50°F, 175°F   c. 55°F, 175°F

    Reference_____

32. Draftstopping shall be installed in combustible floor/ceiling spaces of a nonsprinklered Group E occupancy such that the horizontal floor areas are a maximum of _____ square feet.

    a. 100   b. 400

    c. 1,000   d. 3,000

    Reference_____

33. A required smoke damper shall have a minimum leakage rating of Class _____.
   a. I
   b. II
   c. III
   d. I or II

   Reference _____

34. Fireblocking shall be installed within concealed spaces of exterior wall coverings erected with combustible framing so that any open spaces will be a maximum of _____ square feet in area.
   a. 100
   b. 200
   c. 1,000
   d. 3,000

   Reference _____

35. For fire-resistance purposes, $^1/_2$ inch of unsanded gypsum plaster is deemed to be equivalent to _____ inch of portland cement sand plaster.
   a. $^1/_4$
   b. $^3/_8$
   c. $^3/_4$
   d. 1

   Reference _____

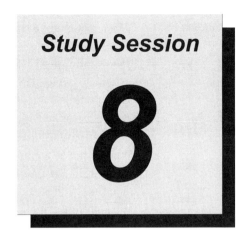

2012 IBC Chapter 9
Fire Protection Systems

OBJECTIVE: To obtain an understanding of the design and installation of fire protection systems, including automatic sprinkler systems, standpipe systems, fire alarm and detection systems, smoke control systems, and smoke and heat vents.

REFERENCE: Chapter 9, 2012 *International Building Code*

KEY POINTS:
- What are the quality standards for fire protection systems?
- How shall fire alarm systems be monitored?
- What are the different classifications for standpipe systems? For other types of systems?
- For Group A-1, A-3 and A-4 occupancies, what size fire area requires the installation of an automatic sprinkler system? How many occupants? What location in the building?
- When must a restaurant or café be sprinklered?
- Under what conditions must an ambulatory care facility be provided with an automatic sprinkler system?
- Which conditions would allow Group E occupancies to be nonsprinklered?
- A sprinkler system is required in Groups F-1, M and S-1 occupancies when the floor area within any fire area exceeds what size?
- Which high-hazard occupancies require installation of a sprinkler system? Institutional occupancies?
- Which residential occupancies require the installation of an automatic sprinkler system? In which portions of the building must the sprinkler system be located?
- Under what conditions must sprinkler protection be provided for exterior balconies and ground-floor patios of dwelling units?
- For the purpose of fire department access, what is considered a "windowless" building?
- In "windowless" buildings, how are basements regulated as compared to above-grade levels?
- When must buildings be sprinklered due to their height?
- Which locations or uses are exempt from the sprinkler requirements?
- What are the different classes of standpipe systems? How do they differ?
- When are standpipes required?

207

KEY POINTS:
(Cont'd)

- When a standpipe system is mandated, where must the hose connections be located?
- In what locations must portable fire extinguishers be installed? How is the size and distribution of portable fire extinguishers to be determined?
- In which occupancies are manual fire alarm systems required? Automatic fire alarm systems?
- Where are smoke alarms mandated?
- Where fire alarms are required, where must visible appliances be installed?
- What is the scope and purpose of a smoke-control system?
- What are the three different methods of mechanical smoke control? Which of the three is the primary means of controlling smoke?
- What are the critical elements for the acceptance of a smoke-control system?
- In which occupancies are smoke and heat vents required?
- When smoke and heat vents are provided, what are their minimum size and spacing requirements?
- What is the minimum distance between curtain boards? What is the maximum permitted size of the curtained area?
- Under which conditions can a mechanical smoke exhaust system be substituted for smoke and heat vents?
- What are the requirements for a fire command center?
- Where must fire department connections be located? What degree of fire department access is required to the connections?
- How are fire pump rooms required to be protected?
- What identification is required for shaftways accessible to the fire department from the exterior? From the interior?
- What types of buildings must be provided with emergency responder radio coverage?

Topic: Definition and Scope
Reference: IBC 901, 202
Category: Fire Protection Systems
Subject: Protection Systems

Code Text: *A fire protection system* consists of *approved devices, equipment and systems or combinations of systems used to detect a fire, activate an alarm, extinguish or control a fire, control or manage smoke and products of a fire or any combination thereof. The provisions of* Chapter 9 *shall specify where fire protection systems are required and shall apply to the design, installation and operation of fire protection systems. Fire protection systems shall be installed, repaired, operated and maintained in accordance with* the International Building Code and the International Fire Code.

Discussion and Commentary: The code provides requirements for three distinct systems considered vital to a safe building environment. The first system is intended to control and limit fire spread and to provide building occupants and fire personnel with the means of fighting a fire. The second system provides for detection of a fire condition and a means of notification. The third system is intended to control smoke migration.

General requirements for fire protection systems:

- Systems to be installed, repaired, operated and maintained in accordance with the *International Building Code* and *International Fire Code*.
- Systems not required by the IBC are permitted to be installed for partial or complete protection, provided such systems meet the requirements of the IBC.
- Any system for which an exception to, or reduction in, the provisions of the IBC has been granted shall be considered a required system.
- No person shall remove or modify any system installed or maintained under the provisions of either code without approval of the building official.
- All systems shall be tested in accordance with the requirements of the IBC and IFC, in the presence of the building official and at the expense of the owner or owner's representative.
- It is unlawful to occupy portions of a structure until the required fire protection systems within that portion have been tested and approved.

Unless specifically excepted, approved supervising stations in accordance with NFPA 72 are mandated for automatic sprinkler systems, fire alarm systems and Group H occupancy emergency alarm, detection, and automatic fire-extinguishing systems.

Topic: Fire Areas
Reference: IBC 901.7, 202
Category: Fire Protection Systems
Subject: Automatic Sprinkler Systems

Code Text: *A fire area is the aggregate floor area enclosed and bounded by fire walls, fire barriers, exterior walls or horizontal assemblies of a building. Where buildings, or portions thereof, are divided into fire areas so as not to exceed the limits established for requiring a fire protection system in accordance with* Chapter 9, *such fire areas shall be separated by fire barriers or horizontal assemblies, or both, having a fire-resistance rating of not less than that determined in accordance with Section 707.3.10.*

Discussion and Commentary: The concept behind fire areas is that of compartmentalization. As a building is subdivided into smaller spaces through the use of fire-resistance-rated elements, the potential hazards tend to be confined to each compartment. Therefore, as the level of hazards decreases, the need for protection diminishes. In the IBC, this reduced level of protection is reflected in the fact that an automatic sprinkler system may not be required. The primary purpose for the creation of fire areas is to address the requirements of Section 903.2 for sprinkler protection.

TABLE 707.3.10
FIRE-RESISTANCE RATING REQUIREMENTS FOR FIRE BARRIER ASSEMBLIES OR HORIZONTAL ASSEMBLIES BETWEEN FIRE AREAS

OCCUPANCY GROUP	FIRE-RESISTANCE RATING (hours)
H-1, H-2	4
F-1, H-3, S-1	3
A, B, E, F-2, H-4, H-5, I, M, R, S-2	2
U	1

To determine the appropriate level of fire resistance for fire barriers used to create one or more fire areas, refer to Table 707.3.10. This table mandates the minimum hourly rating for fire barriers separating one or more occupancies into different fire areas.

Topic: Group A Occupancies
Reference: IBC 903.2.1
Category: Fire Protection Systems
Subject: Automatic Sprinkler Systems

Code Text: *An automatic sprinkler system shall be provided throughout buildings and portions thereof used as Group A occupancies as provided in Section 903.2.1. For Group A-1, A-2, A-3 and A-4 occupancies, the automatic sprinkler system shall be provided throughout the floor area where the Group A-1, A-2, A-3 or A-4 occupancy is located, and in all floors from the Group A occupancy to, and including, the nearest level of exit discharge serving the Group A occupancy. For Group A-5 occupancies, the automatic sprinkler system shall be provided in the spaces indicated in Section 903.2.1.5.*

Discussion and Commentary: Although most Group A occupancies lack the combustible loading that creates a high degree of fire severity, protection provided by an automatic sprinkler system is deemed necessary due to the hazards of having large numbers of people in concentrated areas. Based on varying thresholds, a sprinkler system may be required in any of the five assembly occupancies.

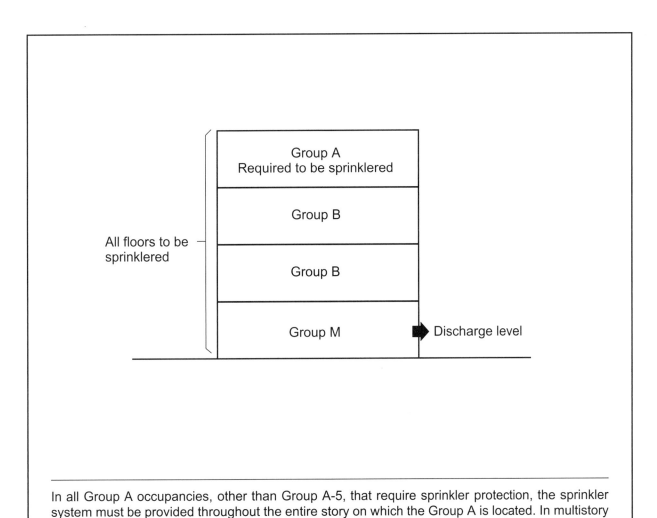

In all Group A occupancies, other than Group A-5, that require sprinkler protection, the sprinkler system must be provided throughout the entire story on which the Group A is located. In multistory buildings, the sprinkler system must also be provided on all stories between, and including, the Group A occupancy and the level of exit discharge.

Study Session 8

Topic: Group A Occupancies
Reference: IBC 903.2.1.1 – 903.2.1.5
Category: Fire Protection Systems
Subject: Automatic Sprinkler Systems

Code Text: *An automatic sprinkler system shall be provided for Group A-1 occupancies where one of the following conditions exists: 1) the fire area exceeds 12,000 square feet (1115 m²), 2) the fire area has an occupant load of 300 or more, or 3) the fire area is located on a floor other than a level of exit discharge serving such occupancy. The same criteria apply to Group A-3 and A-4 occupancies. An automatic sprinkler system shall be provided for Group A-2 occupancies where one of the following conditions exist: 1) the fire area exceeds 5,000 square feet (465 m²), 2) the fire area has an occupant load of 100 or more, or 3) the fire area is located on a floor other than a level of exit discharge serving such occupancies. An automatic sprinkler system shall be provided for Group A-5 occupancies in the following areas: concession stands, retail areas, press boxes and other accessory use areas in excess of 1,000 square feet (93 m²).*

Discussion and Commentary: The thresholds at which Group A occupancies are required to be sprinklered vary based upon fire records and the hazards associated with the different types of assembly uses.

Where a Group A-1 occupancy contains a multitheater complex, the installation of an automatic sprinkler system is required.

Topic: Group E Occupancies
Reference: IBC 903.2.3
Category: Fire Protection Systems
Subject: Automatic Sprinkler Systems

Code Text: *An automatic sprinkler system shall be provided for Group E occupancies as follows: 1) throughout all Group E fire areas greater than 12,000 square feet (1115 m²) in area, and 2) throughout every portion of educational buildings below the lowest level of exit discharge serving that portion of the building.* See exception for buildings with direct egress at ground level from each classroom.

Discussion and Commentary: As a group, educational occupancies tend to have a very good fire record. This stems from the ongoing supervision of activities in the building, as well as the rapid egress of students in response to emergencies. However, because of the amount of combustibles in Group E occupancies and the potentially high occupant load, it is typically necessary to provide sprinkler systems for those undivided floor areas exceeding 12,000 square feet or in basements of any size.

Option to automatic sprinkler protection for Group E occupancies

For SI: 1 square foot = 0.093 m².

Minimum 2-hour fire-resistance-rated fire barriers (Table 707.3.10) can be used to subdivide the building into small fire areas, thereby eliminating the sprinkler requirement. Direct egress at ground level from each classroom in the building is considered as an alternative to basement sprinkler protection.

Topic: Groups F-1, M and S-1 Occupancies   **Category:** Fire Protection Systems
Reference: IBC 903.2.4, 903.2.7, 903.2.9   **Subject:** Automatic Sprinkler Systems

Code Text: *An automatic sprinkler system shall be provided throughout all buildings contain a Group F-1 occupancy where one of the following conditions exists: 1) where a Group F-1 fire area exceeds 12,000 square feet (1115 m^2), 2) where a Group F-1 fire area is located more than three stories above grade plane, or 3) where the combined area of all Group F-1 fire areas on all floors, including any mezzanines, exceeds 24,000 square feet (2230 m^2). Same criteria applies to Group M and S-1 occupancies. A sprinkler system is also required in such occupancies where the manufacture, storage or display and sale of upholstered furniture or mattresses occurs in a significant amount.*

Discussion and Commentary: Because of the potential presence of high levels of combustible materials in factories, sales buildings and warehouses, the IBC limits the size and location of fire areas not protected by an automatic sprinkler system. If any fire area in the building exceeds the threshold, the sprinkler system must be provided throughout the entire building, not just in the fire area that exceeds the area or height limitations.

For SI: 1 square foot = 0.093 m².

Woodworking operations

In addition to the general provisions for sprinkler protection in manufacturing occupancies, IBC Section 903.2.4.1 mandates a more restrictive limitation for woodworking operations. Sprinkler protection for high-piled or rack storage is also addressed in the *International Fire Code*.

Topic: Groups H and I Occupancies
Reference: IBC 903.2.5.1, 903.2.6
Category: Fire Protection Systems
Subject: Automatic Sprinkler Systems

Code Text: *An automatic sprinkler system shall be installed in Group H occupancies. An automatic sprinkler system shall be provided throughout buildings with a Group I fire area.* See exceptions for Group I-1 and day-care facilities.

Discussion and Commentary: Hazardous occupancies require automatic sprinkler systems to protect not only the building's occupants and contents, but also the surrounding property. The sprinkler system only need be provided in the portion of the building classified as Group H. Buildings containing institutional uses must be protected throughout due to the lack of mobility of the occupants. The sprinkler system is intended to limit the size and the spread of a fire, thereby allowing extra time for moving occupants of the institutional building into an adjoining smoke compartment or through a horizontal exit.

[F] TABLE 903.2.5.2
GROUP H-5 SPRINKLER DESIGN CRITERIA

LOCATION	OCCUPANCY HAZARD CLASSIFICATION
Fabrication areas	Ordinary Hazard Group 2
Service corridors	Ordinary Hazard Group 2
Storage rooms without dispensing	Ordinary Hazard Group 2
Storage rooms with dispensing	Extra Hazard Group 2
Corridors	Ordinary Hazard Group 2

In a semiconductor fabrication facility classified as a Group H-5 occupancy, the sprinkler system must be installed throughout the entire building. For sprinkler design criteria, the code identifies the occupancy hazard classifications based on the various areas and locations.

Study Session 8

Topic: Group R Occupancies
Reference: IBC 903.2.8
Category: Fire Protection Systems
Subject: Automatic Sprinkler Systems

Code Text: *An automatic sprinkler system installed in accordance with Section 903.3 shall be provided throughout all buildings with a Group R fire area.*

Discussion and Commentary: Statistics bear out that the majority of fire deaths and injuries occur in residential occupancies. It has also been statistically shown that buildings provided with sprinkler systems perform quite well under fire conditions. This mandate for the installation of automatic sprinkler systems in all buildings containing any Group R occupancy is based upon the desire to reduce such fire deaths and injuries in all residential buildings regulated by the *International Building Code*. This provision, like most requirements found in Chapter 9, is also found in the *International Fire Code*.

The scope of the IBC, Section 101.2, defers certain residential occupancies to the construction regulations of the *International Residential Code*. As such, this sprinkler requirement applies only to those residential structures constructed under the requirements of the *International Building Code*.

Topic: Group S-2 Enclosed Parking Garage **Category:** Fire Protection Systems
Reference: IBC 903.2.10 **Subject:** Automatic Sprinkler Systems

Code Text: *An automatic sprinkler system shall be provided throughout buildings classified as enclosed parking garages in accordance with Section 406.6 as follows: 1) where the fire area of the enclosed parking garage exceeds 12,000 square feet (1115 m^2); or 2) where the enclosed parking garage is located beneath other groups.* See exception for enclosed parking garages located beneath Group R-3 occupancies.

Discussion and Commentary: Although parking garages are shown to have a very good fire record, the fire behavior in an enclosed parking garage is of greater concern than in an open parking garage. Because of the lack of exterior openings, smoke ventilation will be more difficult in the enclosed environment. Therefore, an automatic sprinkler system is required once the enclosed garage is sizeable in fire area. An additional concern occurs where the parking facility is in a mixed-occupancy building. If another occupancy group is located above the enclosed parking area, a sprinkler system is mandated regardless of the garage's fire area size.

Sprinkler system required throughout building where:

Fire area exceeds 12,000 sq ft

2-story Group S-2 enclosed parking garage

OR

Enclosed parking garage of any size located beneath another occupancy

Group B office building with Group S-2 enclosed parking garage below

Sprinkler protection of Group S-2 enclosed parking garages

Where the vehicles being stored consist of commercial trucks or buses, a more stringent fire area threshold is appropriate. The limitation of 5,000 square feet without sprinkler protection is typically applied to garages housing larger vehicles rather than pick-up trucks and similar-sized vehicles used for business activities.

Topic: Fire Department Access
Reference: IBC 903.2.11.1

Category: Fire Protection Systems
Subject: Automatic Sprinkler Systems

Code Text: *An automatic sprinkler system shall be installed throughout all stories, including basements of all buildings where the floor area exceeds 1,500 square feet (139.4 m^2) and where there are not complying exterior wall openings.*

Discussion and Commentary: The IBC considers those structures with inadequate exterior openings for fire department access and/or rescue to be "windowless buildings," which require the installation of an automatic sprinkler system. Two methods of providing appropriate openings are set forth; one method is for openings below grade, and the other is for openings entirely above adjoining ground level. In all cases, at least one side of the building must be provided with complying openings in each 50 lineal feet of exterior wall. Basements are more highly regulated than floors above grade.

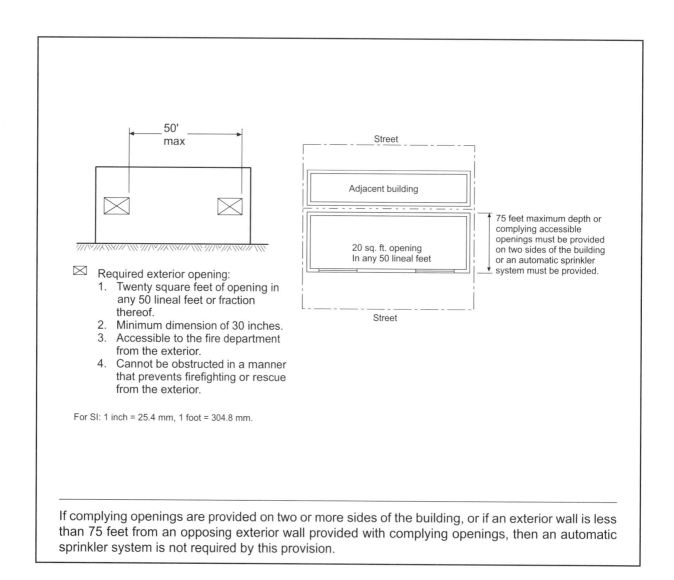

If complying openings are provided on two or more sides of the building, or if an exterior wall is less than 75 feet from an opposing exterior wall provided with complying openings, then an automatic sprinkler system is not required by this provision.

Topic: Buildings Over 55 Feet in Height
Reference: IBC 903.2.11.3
Category: Fire Protection Systems
Subject: Automatic Sprinkler Systems

Code Text: *An automatic sprinkler system shall be installed throughout buildings with a floor level having an occupant load of 30 or more that is located 55 feet (16 764 mm) or more above the lowest level of fire department vehicle access.* See exceptions that exempt airport control towers, open parking structures and Group F-2 occupancies.

Discussion and Commentary: Because of difficulties associated with manual suppression of a fire in buildings constructed a substantial height above the fire department's point of attack, an automatic sprinkler system is required throughout the building, regardless of occupancy. Note that buildings that qualify for a sprinkler system by this provision, often termed "mid-rise" buildings, are not necessarily high-rise buildings as defined in Section 202.

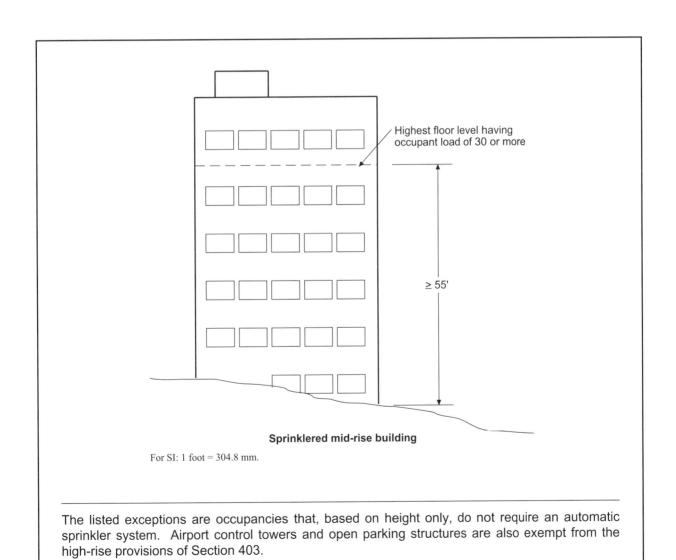

Sprinklered mid-rise building

For SI: 1 foot = 304.8 mm.

The listed exceptions are occupancies that, based on height only, do not require an automatic sprinkler system. Airport control towers and open parking structures are also exempt from the high-rise provisions of Section 403.

Study Session 8

Topic: Required Installations
Reference: IBC 905.3

Category: Fire Protection Systems
Subject: Standpipe Systems

Code Text: *Standpipe systems shall be installed where required by Sections 905.3.1 through 905.3.8. See exception for Group R-3 occupancies. Standpipe systems are allowed to be combined with automatic sprinkler systems.*

Discussion and Commentary: Installed exclusively for the fighting of fires, a standpipe system is a wet or dry system composed of piping, valves, outlets and related equipment designed to provide water at specified pressures. Standpipe systems are permitted to be combined with automatic sprinkler systems. Divided into Classes I, II and III, standpipe systems are generally required in structures of substantial height. Connections for Class I standpipes, which are solely for use by the fire department, shall be located in protected areas to allow for staging operations. Enclosures for interior exit stairways are typical locations for Class I connections.

REQUIRED STANDPIPE INSTALLATIONS

LOCATION OR USE	NONSPRINKLERED BUILDING	SPRINKLERED BUILDING
Building with floor level of the highest story located at more than 30 feet above lowest level of fire department vehicle access	Class III [1,2,3]	Class I
Building with floor level of the lowest story located at more than 30 feet below highest level of fire department vehicle access	Class III [1,2,3]	Class I
Group A occupancies with occupant load exceeding 1,000	Class I [4]	No requirement
Covered mall buildings	—	Class I
Stages more than 1,000 square feet	Class III	Class III [5]
Underground buildings	—	Class I

1. Class I standpipes permitted in basements equipped with automatic sprinkler system
2. Class I manual dry standpipes permitted in open parking garages subject to freezing temperatures, provided hose connections located as for Class II systems
3. Class I manual standpipes permitted in open parking garages where highest floor is less than 150 feet above the lowest level of fire department vehicle access
4. Not required in open-air seating spaces without enclosed spaces
5. Hose connections permitted to be supplied by sprinkler system

Fire hose cabinets in which hoses are attached to outlets on Class II standpipes (as well as the use of portable fire extinguishers) are provided as a means by which the building occupants can control the fire prior to either sprinkler activation or fire personnel arrival.

Topic: Where Required
Reference: IBC 906.1

Category: Fire Protection Systems
Subject: Portable Fire Extinguishers

Code Text: *Portable fire extinguishers shall be installed in the following locations. 1) in Group A, B, E, F, H, I, M, R-1, R-2, R-4 and S occupancies, 2) within 30 feet (9144 mm) of commercial cooking equipment, 3) in areas where flammable or combustible liquids are stored, used or dispensed, 4) on each floor of structures under construction, except Group R-3 occupancies, in accordance with Section 3315.1 of the IFC, 5) where required by the IFC sections indicated in Table 906.1, and 6) special-hazard areas, including but not limited to laboratories, computer rooms and generator rooms, where required by the fire code official.*

Discussion and Commentary: Portable fire extinguishers are typically required in all but Group R-3 and U occupancies to give occupants the means to suppress a fire in its incipient stage. The capability for manual fire suppression can contribute to the protection of the occupants by controlling the fire in its early stages.

In addition to portable fire extinguishers, many of the other fire protection components and systems found in the IBC are also replicated directly from the *International Fire Code* (IFC). IFC provisions addressing automatic sprinkler systems, standpipe systems, fire alarm systems, smoke and heat vents, fire pumps and emergency responder safety features are also inserted into the IBC to provide for greater convenience to the code user.

Study Session 8

Topic: Where Required
Reference: IBC 907.2

Category: Fire Protection Systems
Subject: Fire Alarm and Detection Systems

Code Text: *An approved fire alarm system installed in accordance with the provisions of the IBC and NFPA 72 shall be provided in new buildings and structures in accordance with Sections 907.2.1 through 907.2.23 and provide occupant notification in accordance with Section 907.5, unless other requirements are provided by another section of the IBC. A minimum of one manual fire alarm box shall be provided at an approved location to initiate a fire alarm signal for fire alarm systems employing automatic fire detectors or waterflow detection devices.*

Discussion and Commentary: For many of the occupancies identified by the IBC, it is necessary to provide some level of notification to the building occupants and/or a supervised location reserved for a fire emergency. The threshold at which an alarm and/or detection system is required varies according to the occupancy classification.

Audible alarm notification appliances are to be provided and shall create a distinctive sound that is used for no other purpose. Visual alarm notification appliances are also required, to varying degrees, in public and common areas, employee work areas, and Group I-1, R-1 and R-2 occupancies.

Topic: Manual Fire Alarm Boxes
Reference: IBC 907.4.2
Category: Fire Protection Systems
Subject: Fire Alarm and Detection Systems

Code Text: *Manual fire alarm boxes shall be located not more than 5 feet (1524 mm) from the entrance to each exit. Additional manual fire alarm boxes shall be located so that travel distance to the nearest box does not exceed 200 feet (60 960 mm). The height of the manual fire alarm boxes shall be a minimum of 42 inches (1067 mm) and a maximum of 48 inches (1219 mm), measured vertically, from the floor level to the activating handle or lever of the box. Manual fire alarm boxes shall be red in color.*

Discussion and Commentary: The required location of fire alarm boxes adjacent to exit doors provides an opportunity for the alarm to be transmitted in a timely manner. In multistory buildings, such locations also encourage the actuation of a manual fire alarm box on the fire floor prior to entering the stair enclosure, resulting in the alarm being received from the actual fire floor and not another floor along the path of egress.

Manual fire alarm boxes

For SI: 1 inch = 25.4 mm, 1 foot = 304.8 mm.

Where a manual fire alarm system is required, manual fire alarm boxes (pull stations) must be installed. However, in some occupancies the code permits the elimination of such boxes if water flow in an automatic sprinkler system installed throughout the building activates the notification appliances.

Topic: Scope
Reference: IBC 909.1

Category: Fire Protection Systems
Subject: Smoke Control Systems

Code Text: Section 909 *applies to mechanical or passive smoke control systems when they are required by some other provision of* the IBC. *The purpose of* Section 909 *is to provide a tenable environment for the evacuation or relocation of occupants. Smoke control systems regulated by* Section 909 *serve a different purpose than the smoke- and heat-venting provisions found in Section 910.*

Discussion and Commentary: The provisions Section 909 do not apply unless specifically mandated for a special use, such as an atrium. It is the intent that none of the requirements apply unless directed by other provisions of the code. Where a smoke control system is provided, it may be either passive or mechanical, or a combination of the two systems. A mechanical system is an engineered system that uses mechanical fans either to produce pressure differences across smoke barriers or to establish airflows to limit and direct smoke movement. A passive system is a system of smoke barriers arranged to limit the migration of smoke.

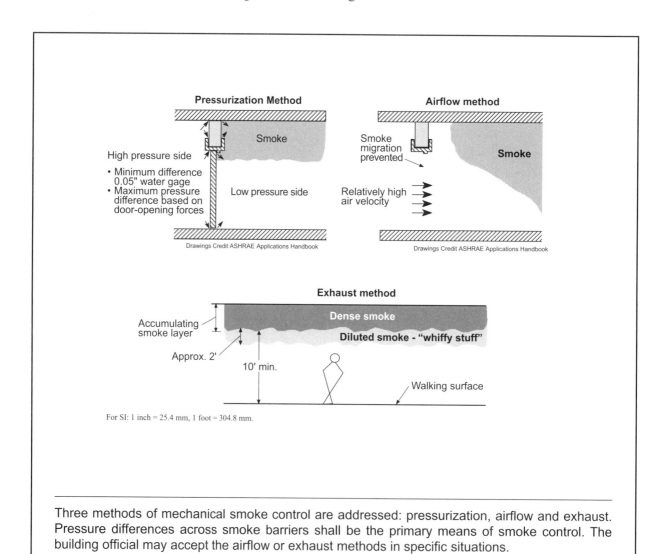

Three methods of mechanical smoke control are addressed: pressurization, airflow and exhaust. Pressure differences across smoke barriers shall be the primary means of smoke control. The building official may accept the airflow or exhaust methods in specific situations.

Topic: Where Required
Reference: IBC 910.2, 910.3.5
Category: Fire Protection Systems
Subject: Smoke and Heat Vents

Code Text: *Smoke and heat vents shall be installed in the roofs of one-story buildings or portions thereof 1) used as Group F-1 or S-1 occupancies having more than 50,000 square feet (4645 m^2) in undivided area* (see exception for aircraft repair hangars)*, and 2) containing high-piled combustible stock or rack storage in any occupancy group in accordance with Section 413 and the IFC. Where required by Table 910.3, draft curtains shall be installed on the underside of the roof in accordance with* Section 910.3.5.

Discussion and Commentary: Smoke and heat vents shall be installed in conjunction with draft curtains, which confine the smoke and hot gases so that they are not diluted. Draft curtains thus increase the effectiveness of automatic vents. Those areas of buildings that are equipped with early suppression fast response (ESFR) sprinklers are not required to be provided with automatic smoke and heat vents.

Note: In general, several small vents are more effective than a larger vent of equal area.

Roof vents and curtain boards

The code permits the building official to approve the use of a mechanical smoke exhaust system to ventilate the building as an alternative to smoke and heat vents. The exhaust fans in such a system are regulated for size, location, operation, wiring, control, supply air and interlocks.

Study Session 8
IBC Chapter 9

1. Automatic sprinkler systems shall be monitored by an approved supervising station where the minimum number of sprinklers is _____.

   a. 10 b. 20
   c. 50 d. 100

   Reference_____

2. Which one of the following classes of standpipe systems is intended primarily for the use of building occupants or the fire department during initial response?

   a. Class I b. Class II
   c. Class III d. Class IV

   Reference_____

3. Which one of the following types of standpipe systems requires water from a fire department pumper to be pumped into the system in order to supply the system demand?

   a. automatic dry b. automatic wet
   c. manual wet d. semi-automatic dry

   Reference_____

4. In a Group A-2 occupancy, an automatic sprinkler system shall be provided throughout any fire area having a minimum occupant load of _____.

   a. 50
   b. 60
   c. 100
   d. 3000

   Reference_____

5. A stadium press box in a Group A-5 occupancy having a maximum floor area of _____ square feet need not be sprinklered.

   a. 400
   b. 1,000
   c. 5,000
   d. 12,000

   Reference_____

6. A sprinkler system is not required for Group E fire areas having a maximum floor area of _____ square feet except where located below the exit discharge level.

   a. 2,500
   b. 5,000
   c. 12,000
   d. 20,000

   Reference_____

7. Where woodworking operations in a Group F-1 occupancy generate finely divided combustible waste, an automatic sprinkler system is required where such operations occupy a minimum size floor area of _____ square feet.

   a. 1,000
   b. 2,500
   c. 5,000
   d. 12,000

   Reference_____

8. In a single-story ambulatory care facility, an automatic sprinkler system is required where a minimum of _____ care recipients are incapable of self-preservation at any time.

   a. 4
   b. 6
   c. 10
   d. 20

   Reference_____

Study Session 8

9. An automatic sprinkler system shall be provided throughout a building with a Group M fire area located a minimum of _____ stories above grade plane.

    a. 2
    b. 3
    c. 4
    d. 6

    Reference_____

10. Buildings containing which of the following residential occupancies must be sprinklered under all conditions?

    a. Groups R-1 and R-2 only
    b. Groups R-1, R-2 and R-4 only
    c. Groups R-2 and R-4 only
    d. All Group R occupancies

    Reference_____

11. A building shall be fully sprinklered where the combined area of all Group S-1 fire areas exceeds _____ square feet.

    a. 2,500
    b. 5,000
    c. 12,000
    d. 24,000

    Reference_____

12. A single-story Group S-1 repair garage need not be fully sprinklered where the fire area contains a maximum of _____ square feet.

    a. 2,500
    b. 10,000
    c. 12,000
    d. 24,000

    Reference_____

13. A Group S-2 parking garage used to store commercial buses need not be sprinklered where the fire area has a maximum size of _____ square feet.

    a. 2,500
    b. 5,000
    c. 10,000
    d. 12,000

    Reference_____

14. Where any portion of a basement is located a minimum of _____ feet from complying exterior openings, the basement shall be provided with an automatic sprinkler system.

    a. 50   b. 75
    c. 100   d. 150

    Reference_____

15. Which of the following NFPA automatic sprinkler systems is designed for installation in one- and two-family dwellings and townhouses?

    a. 13   b. 13D
    c. 13R   d. 17A

    Reference_____

16. Which of the following spaces requiring sprinkler system protection is not required to use quick-response or residential sprinklers?

    a. light-hazard occupancies
    b. sleeping units in Group R-1
    c. care recipient sleeping units in Group I-2
    d. sales rooms of Group M

    Reference_____

17. A minimum of _____ shall be maintained between automatic sprinklers and the top of piles of combustible fibers.

    a. 12 inches   b. 18 inches
    c. 3 feet   d. 5 feet

    Reference_____

18. A secondary water supply, where required in high-rise buildings, shall have a minimum duration of _____ minutes.

    a. 15   b. 30
    c. 60   d. 90

    Reference_____

Study Session 8

19. The manual actuation device for a fire-extinguishing system for a commercial cooking system shall be located a minimum of _____ feet and a maximum of _____ feet from the kitchen exhaust system.

    a. 3, 6
    b. 5, 10
    c. 6, 12
    d. 10, 20

    Reference_____

20. Connections for Class II standpipe systems shall be located so that all portions of the building are within _____ feet of a nozzle attached to _____ feet of hose.

    a. 20, 50
    b. 30, 100
    c. 40, 125
    d. 40, 150

    Reference_____

21. In a nonsprinklered Group B occupancy, a manual fire alarm system shall be installed where there are a minimum of _____ occupants above or below the level of exit discharge.

    a. 101
    b. 201
    c. 301
    d. 501

    Reference_____

22. A manual fire alarm system is not required in a Group E occupancy with a maximum occupant load of _____ persons.

    a. 30
    b. 50
    c. 300
    d. 500

    Reference_____

23. Manual fire alarm boxes shall be located a maximum of _____ feet from the entrance to each exit.

    a. 5
    b. 10
    c. 12
    d. 20

    Reference_____

24. In a Group R-1 hotel providing 220 sleeping units, a minimum of _____ such units shall be provided with visible alarm notification devices.

    a. 3   b. 11
    c. 17   d. 22

    Reference_____

25. The maximum sound pressure level for audible alarm notification appliances shall be _____ dBA at the minimum hearing distance from the audible appliance.

    a. 60   b. 75
    c. 90   d. 110

    Reference_____

26. The manual actuation device for an automatic fire-extinguishing system serving a commercial cooking system shall be installed a minimum of _____ inches and a maximum of _____ inches above the floor..

    a. 38, 42   b. 34, 48
    c. 38, 42   d. 42, 48

    Reference_____

27. A Class III wet standpipe is not required for stages having a maximum size of _____ square feet.

    a. 100   b. 400
    c. 500   d. 1,000

    Reference_____

28. Where natural ventilation is utilized for venting a smokeproof enclosure, each vestibule shall be provided with a minimum _____ -square-foot opening in the exterior wall.

    a. 9   b. 16
    c. 24   d. 35

    Reference_____

Study Session 8

29. Other than for an aircraft repair hangar, a one-story Group S-1 occupancy shall be provided with smoke and heat vents where it exceeds _____ square feet in undivided area.

    a. 8,000   b. 10,000
    c. 15,000   d. 50,000

    Reference_____

30. Where required in a Group F-1 occupancy, smoke and heat vents shall be spaced at maximum intervals of _____ feet when measured center to center.

    a. 60   b. 75
    c. 100   d. 120

    Reference_____

31. Sprinkler protection is required for exterior balconies, decks and ground floor patios of dwelling units where the building is of _____ construction and a roof or deck exists above.

    a. combustible   b. Type III
    c. Type IV   d. Type V

    Reference_____

32. A manual fire alarm system shall be installed in all multistory Group F occupancies having a minimum combined occupant load of _____ above or below the lowest level of exit discharge.

    a. 50   b. 100
    c. 300   d. 500

    Reference_____

33. The vestibule space in a smokeproof enclosure shall have a minimum width of _____ inches and a minimum length in the direction of egress travel of _____ inches.

    a. 44, 60   b. 44, 72
    c. 60, 60   d. 60, 84

    Reference_____

34. Where automatic smoke and heat vents contain heat-sensitive glazing designed to shrink and drop out of the vent opening when exposed to fire, the vents shall open a maximum of _____ minute(s) after the established fire exposure.

   a. 1
   b. 2
   c. 5
   d. 10

   Reference _____

35. Where a fire command center is required by other provisions of the code, the room shall be a minimum of _____ square feet in area.

   a. 96
   b. 100
   c. 144
   d. 200

   Reference _____

Study Session 9

2012 IBC Sections 1001 – 1007, 1011 and 1013
Means of Egress I

OBJECTIVE: To obtain an understanding of the general system design requirements of a means of egress system, including the determination of occupant load, the required width and capacity of egress components, means of egress identification and illumination, accessible means of egress and the provisions regulating guards.

REFERENCE: Sections 1001 – 1007, 1011 and 1013, 2012 *International Building Code*

KEY POINTS:
- What is the definition of a means of egress system? What are its three distinct elements?
- What is the minimum ceiling height permitted along a means of egress?
- What limitations are placed on protruding objects extending below the minimum ceiling height? Projecting horizontally over a walking surface?
- For an elevation change along the egress path, at what point is a ramp required rather than a step or stairway?
- In areas without fixed seats, what is the correct method of determining occupant load?
- What method shall be used to calculate the occupant load in areas with fixed seating, such as benches, pews or booths?
- How may the design occupant load be increased over what is calculated?
- In which types of rooms or spaces must the maximum occupant load be posted?
- How is exiting addressed where stairways serve more than one floor?
- How is exiting from a mezzanine regulated?
- Are yards, patios and courts regulated in the same manner as interior areas?
- How shall the minimum width and capacity of different egress components be determined?
- What is the maximum allowable encroachment of a door into the required egress width?
- When must the means of egress be illuminated? What is the minimum required illumination at the floor level?
- Which locations must be provided with emergency power for egress illumination?
- For what duration is an emergency power system for means of egress illumination required to provide power?
- What is considered an accessible means of egress? How many are required in a building?

KEY POINTS:
(Cont'd)
- When is an area of refuge needed? How are they to be constructed?
- Under what conditions is a two-way communication system required? What are the minimum system requirements?
- How must an accessible means of egress be identified?
- What is the purpose of an exterior area for rescue assistance? How must this area be separated from the interior of the building it serves?
- Where are exit signs required? When must they be illuminated? What level of illumination is required? Which types of power sources are necessary?
- Under what conditions are low-level exit signs required? Where are such signs to be located?
- What is the definition of a guard?
- When are guards required? What is the minimum height requirement from the walking surface to the top of a guard?
- How must guards be constructed to limit passage through the protective barrier?
- How is window sill height regulated in Group R-2 and R-3 occupancies where openings are located more than 72 inches above grade? What are window fall prevention devices? Window opening control devices?

Topic: General
Reference: IBC 1001.1
Category: Means of Egress
Subject: Administration

Code Text: *Buildings or portions thereof shall be provided with a means of egress system as required by* Chapter 10. *The provisions of* Chapter 10 *shall control the design, construction and arrangement of means of egress components required to provide an approved means of egress from structures and portions thereof.*

Discussion and Commentary: The *International Building Code* regulates the design, construction and maintenance of an exiting system through two general categories—system design and egress components. Any building elements that are a part of the system must be reviewed for compliance with the criteria for the number, location, width or capacity, height, continuity and arrangement of egress components, and all other applicable provisions.

Definition of a means of egress

Building components along the path of egress travel might include aisle accessways, aisles, doors or doorways, intervening rooms, gates, corridors, ramps, exit access stairways, interior exit stairways, exit passageways, horizontal exits, exterior balconies, exterior exit stairways and egress courts.

Topic: Three-Part Egress System
Reference: IBC 1001.1, 202
Category: Means of Egress
Subject: Definitions

Code Text: *A means of egress is a continuous and unobstructed path of vertical and horizontal egress travel from any occupied portion of a building or structure to a public way. A means of egress consists of three separate and distinct parts: the exit access, the exit, and the exit discharge.*

Discussion and Commentary: The exit access begins at any occupied location within the building and does not end until it reaches the door to an interior exit stairway or ramp, a horizontal exit or exit passageway, an exterior exit stairway or ramp, or an exterior exit door at the level of exit discharge. Travel distance is regulated throughout the exit access, and the path of travel is seldom a fire-protected environment. At the exit discharge, which begins where the exit ends, egress remains regulated until the public way is reached.

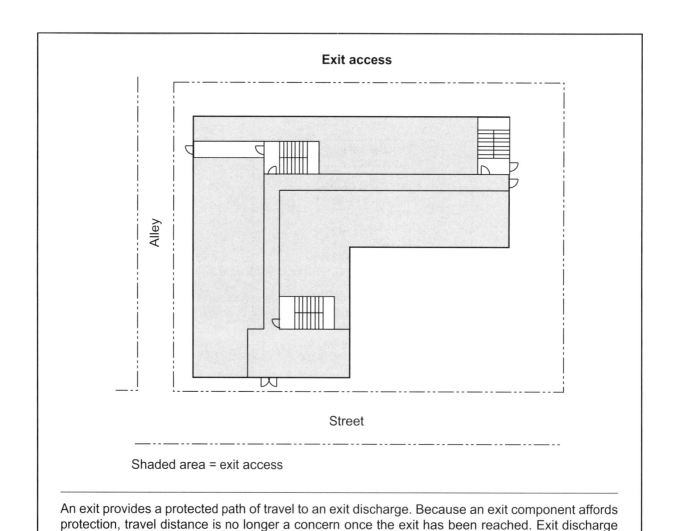

An exit provides a protected path of travel to an exit discharge. Because an exit component affords protection, travel distance is no longer a concern once the exit has been reached. Exit discharge travel distance to the public way is also unlimited.

Topic: Ceiling Height
Reference: IBC 1003.2
Category: Means of Egress
Subject: General Egress

Code Text: *The means of egress shall have a ceiling height of not less than 7 feet 6 inches (2286 mm). See exceptions for sloped ceilings, ceilings of dwelling units and sleeping units, allowable projections, stair and ramp headroom, door height, mezzanines and parking garages.*

Discussion and Commentary: In addition to providing a travel path of adequate width, the code requires that the clear height of the means of egress be maintained at least $7^{1}/_{2}$ feet above the walking surface. There are several exceptions to this general requirement that permit limited reductions in the mandated height. Under most conditions, the vertical clearance at a stairway or doorway may be reduced to 80 inches. Protruding objects, such as sprinklers and light fixtures, are also permitted to extend below the minimum required ceiling height for up to 50 percent of ceiling area of the means of egress, provided such objects maintain a headroom clearance of at least 80 inches. Special provisions are applicable to sloped ceilings.

Corridor, aisle, passageway or any walking surface along egress of path travel

For SI: 1 inch = 25.4 mm, 1 foot = 304.8 mm.

The minimum ceiling heights established for environmental concerns are addressed in Section 1208.2. Habitable spaces, such as bedrooms and living rooms in residential occupancies, occupiable spaces and corridors must be at least 7 feet 6 inches in height. In other areas, reduced headroom is permitted.

Study Session 9

Topic: Elevation Change
Reference: IBC 1003.5

Category: Means of Egress
Subject: General Egress

Code Text: *Where changes in elevation of less than 12 inches (305 mm) exist in the means of egress, sloped surfaces shall be used. Where the slope is greater than 1 unit vertical in 20 units horizontal (5-percent slope), ramps complying with Section 1010 shall be used. Where the difference in elevation is 6 inches (152 mm) or less, the ramp shall be equipped with either handrails or floor finish materials that contrast with adjacent floor finish materials. See exceptions for 1) a single 7-inch maximum step in Groups F, H, R-2, R-3, S and U; 2) a stair with one or two risers with a handrail provided; and 3) a step in aisles serving seating.*

Discussion and Commentary: Along the egress path, there is a concern about slight changes in elevation that are not readily apparent to persons seeking to exit under emergency conditions. Therefore, a single riser or a pair of shallow risers is not permitted. Steps used to achieve minor differences in elevation frequently go unnoticed, and as such, can cause accidents.

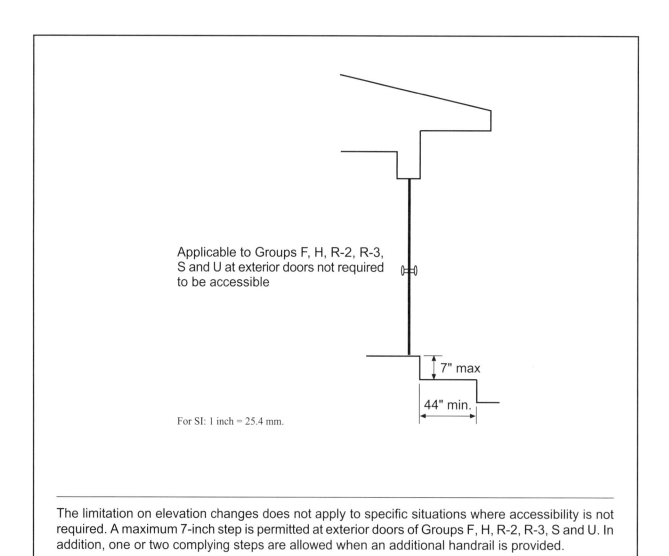

The limitation on elevation changes does not apply to specific situations where accessibility is not required. A maximum 7-inch step is permitted at exterior doors of Groups F, H, R-2, R-3, S and U. In addition, one or two complying steps are allowed when an additional handrail is provided.

Topic: Design Occupant Load
Reference: IBC 1004.1, 1004.1.2
Category: Means of Egress
Subject: Occupant Load

Code Text: *In determining means of egress requirements, the number of occupants for whom means of egress facilities shall be provided shall be determined in accordance with Section 1004. The number of occupants shall be computed at the rate of one occupant per unit of area as prescribed in Table 1004.1.2. For areas without fixed seating, the occupant load shall not be less than that number determined by dividing the floor area under consideration by the occupant load factor assigned to the function of the space as set forth in Table 1004.1.2. See exception where building official is authorized to reduce occupant load below that calculated.*

Discussion and Commentary: For occupant load determination, it must be assumed that under normal conditions all portions of a building are fully occupied at the same time. The density characteristics of the various uses identified in Table 1004.1.2 are considered "occupant load factors." For most occupancies, the gross floor area is to be considered. However, a few of the occupant load factors are based on net floor area, which allows the deduction of areas such as corridors, stairways, toilet rooms, equipment rooms and closets.

Where fixed seats are installed, Section 1004.4 specifies that the occupant load is determined simply by counting the number of seats. For benches and pews, the factor is one occupant per 18 inches of width. For booth seating, the factor is 24 inches per occupant.

Study Session 9

Topic: Outdoor Areas
Reference: IBC 1004.5
Category: Means of Egress
Subject: Occupant Load

Code Text: *Yards, patios, courts and similar outdoor areas accessible to and usable by the building occupants shall be provided with means of egress as required by* Chapter 10. *Where outdoor areas are to be used by persons in addition to the occupants of the building, and the path of egress travel from the outdoor areas passes through the building, means of egress requirements shall be based on the sum of the occupant loads of the building plus the outdoor areas.* See exceptions for service areas and dwellings.

Discussion and Commentary: Although not limited in application, the regulation of egress from outdoor areas often addresses the use of exterior spaces for dining and/or drinking in restaurants and similar establishments. In addition, the means of egress required from occupied roofs is determined based upon the requirements for outdoor areas. The building official is authorized to establish an occupant load for the outdoor space in accordance with its anticipated use and to apply all means of egress provisions that would be appropriate.

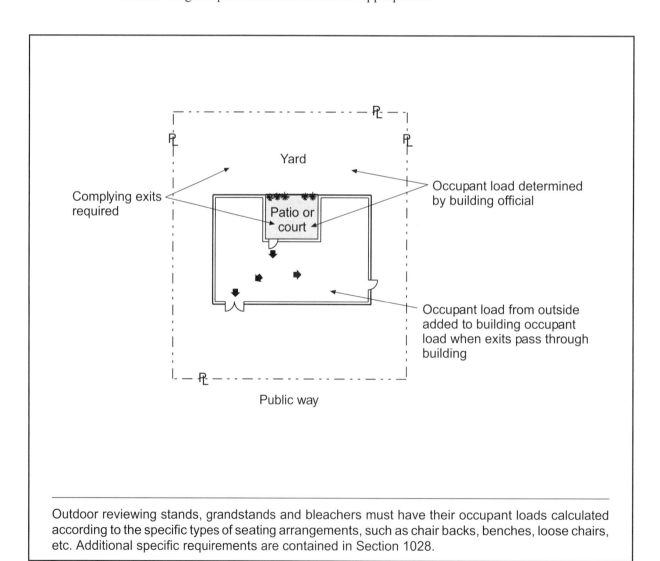

Outdoor reviewing stands, grandstands and bleachers must have their occupant loads calculated according to the specific types of seating arrangements, such as chair backs, benches, loose chairs, etc. Additional specific requirements are contained in Section 1028.

Topic: Width and Capacity
Reference: IBC 1005.2, 1005.3

Category: Means of Egress
Subject: Egress Sizing

Code Text: *The minimum width, in inches, of any means of egress components shall not be less than that specified for such component, elsewhere in* the IBC. *For other than Group H and I-2 occupancies, the capacity, in inches, of means of egress stairways shall be calculated by multiplying the occupant load served by such stairway by a means of egress capacity factor of 0.3 inches (7.62 mm) per occupant. The capacity, in inches, of means of egress components other than stairways shall be calculated by multiplying the occupant load served by such component by a means of egress capacity factor of 0.2 inches (5.08 mm) per occupant.* See exceptions that reduce the capacity factors to 0.2 inches and 0.15 inches, respectively, for buildings equipped throughout with an automatic sprinkler system and an emergency voice/alarm communication system.

Discussion and Commentary: In a given means of egress system, different components will afford different capacities. The most restrictive component will establish the capacity of the overall system. Doorways, aisles, stairways and corridors also have minimum established widths that must be provided.

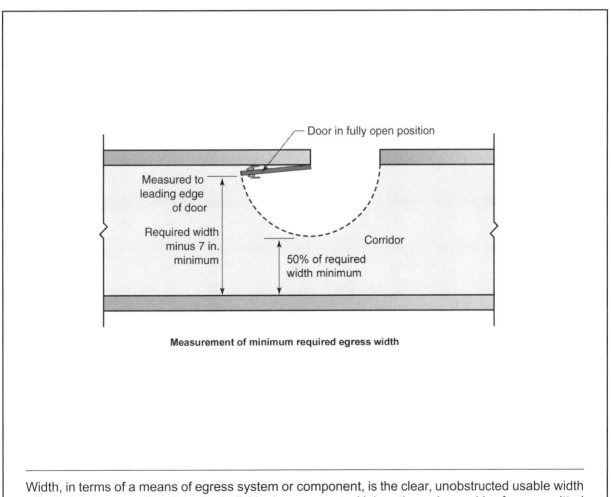

Measurement of minimum required egress width

Width, in terms of a means of egress system or component, is the clear, unobstructed usable width afforded along the exit path by the individual components. Unless the code provides for a permitted projection, the minimum required clear width may not be reduced throughout the travel path.

Topic: Distribution of Egress Capacity
Reference: IBC 1005.5

Category: Means of Egress
Subject: Egress Sizing

Code Text: *Where more than one exit, or access to more than one exit, is required, the means of egress shall be configured such that the loss of any one exit, or access to one exit, shall not reduce the available capacity to less than 50 percent of the required capacity.*

Discussion and Commentary: Where two complying means of egress are provided, the occupant load is to be distributed evenly between the two means of egress. However, where three or more means of egress are available, it is permissible to size one of the egress points for up to 50 percent of the occupant load, while distributing the remaining occupant load among the other means of egress. This distribution is not required to be equally applied; however, a dramatic imbalance of egress component capacities relative to occupant load distribution should be avoided.

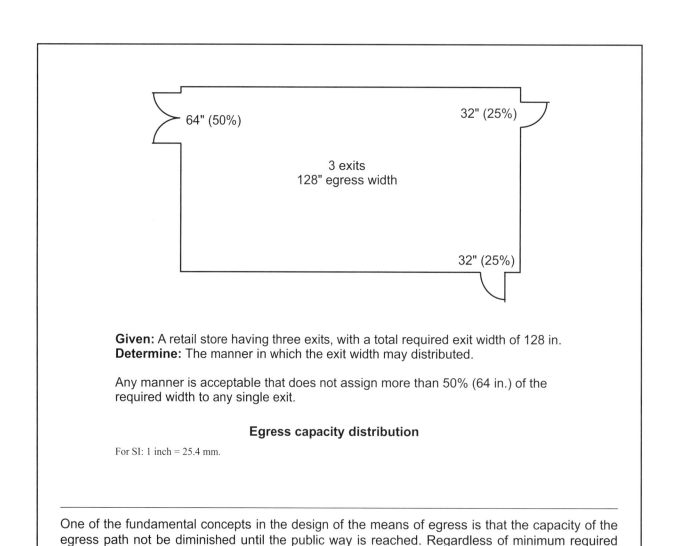

Given: A retail store having three exits, with a total required exit width of 128 in.
Determine: The manner in which the exit width may distributed.

Any manner is acceptable that does not assign more than 50% (64 in.) of the required width to any single exit.

Egress capacity distribution

For SI: 1 inch = 25.4 mm.

One of the fundamental concepts in the design of the means of egress is that the capacity of the egress path not be diminished until the public way is reached. Regardless of minimum required component width, the calculated width based on the occupant load served must be maintained.

Topic: Exiting from Multiple Levels **Category:** Means of Egress
Reference: IBC 1005.3.1, 1005.4, 1005.6 **Subject:** Egress Sizing

Code Text: *Where stairways serve more than one story, only the occupant load of each story considered individually shall be used in computing the required capacity of the stairway serving that story. The capacity of the means of egress required from any story of a building shall not be reduced along the path of egress travel until arrival at the public way. Where means of egress from stories above and below converge at an intermediate level, the capacity of the means of egress from the point of convergence shall not be less than the sum of the required capacities for the two adjacent stories.*

Discussion and Commentary: At other than mezzanines per Section 1004.1.1.2, it is not necessary to add occupants together for width calculations as they travel vertically from one floor level to the next. Only the capacity for each individual floor is utilized in establishing the minimum required stairway width at each flight. However, once a minimum required width has been established along the stair path, it cannot be reduced.

Given:
- A non sprinklered four-story office building with basement
- Occupant load exiting into stair enclosure at each level as indicated
- First floor occupants exit to exterior without entering the stair enclosure

Exit element	Occupant load served	Required width
Stair-point "A"	200	60"
Stair-point "B"	150	60"[1]
Stair-point "C"	240	72"
Stair-point "D"	150	45"
Stair-point "E"	390[2]	78"

[1] Required width from above must be maintained
[2] Door at point "E" serves occupants from "C" and "D" egress convergence per Section 1005.6

There is an allowance for adding the occupant loads of floors above and below an intermediate level together where they converge along the exit path. The aggregate occupant load of such converging floors is to be used in determining the minimum required exit width.

Topic: Emergency Power
Reference: IBC 1006.1, 1006.3

Category: Means of Egress
Subject: Illumination

Code Text: *The means of egress, including the exit discharge, shall be illuminated at all times the building space served by the means of egress is occupied.* See exceptions for 1) Group U occupancies; 2) aisle accessways in Group A; 3) dwelling and sleeping units in Groups R-1, R-2 and R-3; and 4) sleeping units of Group I. *The power supply for means of egress illumination shall normally be provided by the premises electrical supply. In the event of power supply failure, an emergency electrical system shall automatically illuminate aisles and unenclosed egress stairways in rooms and spaces which require two or more means of egress; and corridors, interior exit stairways and ramps, exit passageways, exterior egress components above grade, interior discharge elements, and exterior landings for exit discharge doorways in buildings required to have two or more exits.*

Discussion and Commentary: Often identified as *emergency lighting*, a completely separate source of power from the premise's wiring system is required when the life-safety risk in a building becomes sufficiently great. This threshold is recognized as the point at which the occupant load of the room, area or building is high enough so that two means of egress are required.

For the building occupant to be able to negotiate safely the means of egress system, the entire system must be illuminated any time the building is occupied. The illumination must provide an intensity of at least one foot-candle at the floor level.

Topic: General
Reference: IBC 1007.1, 1007.2
Category: Means of Egress
Subject: Accessible Means of Egress

Code Text: *Accessible spaces shall be provided with not less than one accessible means of egress. Where more than one means of egress is required by Sections 1015.1 or 1021.1 from any accessible space, each accessible portion of the space shall be served by not less than two accessible means of egress.* See exceptions for 1) alterations to existing buildings, 2) accessible mezzanines, and 3) assembly spaces with sloped floors. *Each required accessible means of egress shall be continuous to a public way and shall consist of one or more of the following components: accessible routes, interior exit stairways, interior exit access stairways, exterior exit stairways, elevators, platform lifts, horizontal exits, ramps, areas of refuge and exterior areas for assisted rescue.*

Discussion and Commentary: An accessible means of egress is *a continuous and unobstructed way of egress travel, from any accessible point in a building or facility to a public way.*

Exterior stairways and nonaccessible exit discharge areas may be served by exterior areas for assisted rescue. These specific exterior refuge areas must be adequately separated from the interior of the building by fire-resistance-rated construction and fire-protected openings.

Topic: Areas of Refuge
Reference: IBC 1007.6

Category: Means of Egress
Subject: Accessible Means of Egress

Code Text: *Every required area of refuge shall be accessible from the space it serves by an accessible means of egress. Every required area of refuge shall have direct access to a stairway complying with Section 1007.3 or an elevator complying with Section 1007.4.*

Discussion and Commentary: An area of refuge is defined as *an area where persons unable to use stairways can remain temporarily to await instructions or assistance during emergency evacuation.* An area of refuge needs to be separated from the remainder of the story by a smoke barrier or horizontal exit unless the refuge area is located within an enclosure for an exit access stairway or an interior exit stairway. A two-way communication system with appropriate instructions must be provided in each area of refuge and must also be identified by complying signs. There are several conditions under which areas of refuge are not required. The most commonly utilized exception to areas of refuge applies to buildings that are fully sprinklered.

For SI: 1 inch = 25.4 mm.

Required areas of refuge

Although three or more means of egress from an accessible space may be required, only two of the exitways must be accessible. However, where an area of refuge is used as part of the egress system, the maximum travel distance set forth in Section 1016.2 must be maintained.

Topic: Where Required
Reference: IBC 1007.8
Category: Accessible Means of Egress
Subject: Two-Way Communication Systems

Code Text: *A two-way communication system shall be provided at the elevator landing on each accessible floor that is one or more stories above or below the story of exit discharge comply with Sections 1007.8.1 and 1007.8.2.* See exceptions where 1) the two-way communication system is provided within complying areas of refuge, or 2) the floor level is provided with complying exit ramps.

Discussion and Commentary: In multistory buildings, two-way communication systems must be located at the elevator landing on each accessible floor level, with the exception of the level of exit discharge. The system is intended to offer a means of communication to disabled individuals who need assistance during an emergency situation. Such a system can be useful not only in the event of a fire but also in the case of a natural or technological disaster by providing emergency responders with the location of individuals who will require assistance to be safely evacuated from floor levels above or below the discharge level.

The provisions of Section 1007.6.3 addressing accessible means of egress require that all areas of refuge be provided with a two-way communication system. The specific requirements for the system are the same as those for the two-way communication systems mandated at elevator landings as set forth in Section 1007.8.

Topic: Where Required
Reference: IBC 1011.1

Category: Means of Egress
Subject: Exit Signs

Code Text: *Exit and exit access doors shall be marked by an approved exit sign readily visible from any direction of egress travel. The path of egress travel to exits and within exits shall be marked by readily visible exit signs to clearly indicate the direction of egress travel in cases where the exit or the path of egress travel is not immediately visible to the occupants.* See five exceptions for uses or conditions where exit signs are not required.

Discussion and Commentary: Exit signs are only mandated when the room or area under consideration is required to have multiple exits or exit access doors. Other locations are also specified where the presence of an exit sign is deemed unnecessary, such as clearly identifiable main exterior doors. Although the appropriate locations for exit signs should be identified during the plan review phase of a project, the true evaluation of their effectiveness should be done just prior to occupancy, when the correct location and orientation of the signs can be checked.

Additional exit signs may be necessary in lengthy exit access corridors and exit passageways to reinforce the direction of egress travel. It is possible that individuals subjected to extended travel would question the availability of an exit and attempt to locate an alternative egress path.

Topic: Low-level Signs in Group R-1
Reference: IBC 1011.2
Category: Means of Egress
Subject: Exit Signs

Code Text: *Where exit signs are required in Group R-1 occupancies by Section 1011.1, additional low-level exit signs shall be provided in all areas serving guest rooms in Group R-1 occupancies and shall comply with Section 1011.5. The bottom of the sign shall be not less than 10 inches (254 mm) nor more than 12 inches (305 mm) above the floor level. The sign shall be flush mounted to the door or wall. Where mounted on the wall, the edge of the sign shall be within 4 inches (102 mm) of the door frame on the latch side.*

Discussion and Commentary: In the means of egress system for Group R-1 occupancies, additional exit signs are mandated for those portions of the system serving the guest rooms. Occupants of such facilities are transient and typically not familiar with their surroundings. If a corridor or other egress component serving the guest rooms were to fill with smoke, the general exit signs located high in the space could quickly become obscured. The installation of additional signs at floor level provides for a secondary identification of the egress path.

Low-level exit signs must be either electrically powered, self-luminous or photoluminescent exit signs that are listed and labeled in accordance with UL 924 and installed in accordance with the manufacturer's instructions. Consistent with the requirements for all other exit signs, low-level signs shall be illuminated at all times.

Topic: Illumination and Power Source
Reference: IBC 1011.3, 1011.6.3

Category: Means of Egress
Subject: Exit Signs

Code Text: *Exit signs shall be internally or externally illuminated.* See exception for tactile signs. *Exit signs shall be illuminated at all times. To ensure continued illumination for a duration of not less than 90 minutes in case of primary power loss, the sign illumination means shall be connected to an emergency system provided from storage batteries, unit equipment or an on-site generator.*

Discussion and Commentary: To ensure visibility under all conditions, required exit signs must always be illuminated. For those signs that are internally illuminated, which make up the vast majority of exit signs, compliance with UL 924 is mandated. Such exit signs, which includes electrically-powered, self-luminous and photo luminescent signs, must be listed and labeled. In addition, they must be installed in accordance with the manufacturer's installation instructions.

Externally-illuminated exit signs are regulated through prescriptive requirements addressing the sign's graphics, illumination and power source. The word "EXIT" must be at least 6 inches in height, and at least 5 foot-candles of external illumination must be provided.

Topic: Location, Height and Openings
Reference: IBC 1013
Category: Means of Egress
Subject: Guards

Code Text: *Guards shall be located along open-sided walking surfaces, including mezzanines, equipment platforms, stairways, ramps and landings that are located more than 30 inches (762 mm) measured vertically to the floor or grade below at any point within 36 inches (914 mm) horizontally to the edge of the open side. See exceptions. Required guards shall be not less than 42 inches (1067 mm) high, measured vertically 1) from the adjacent walking surfaces, 2) on stairs, from the line connecting the leading edges of the tread nosings and 3) on ramps, from the ramp surface at the guard. See exceptions. Required guards shall not have openings which allow passage of a sphere 4 inches (102 mm) in diameter from the walking surface to the required guard height.* See exceptions.

Discussion and Commentary: Guards must be of adequate height and structural stability to prevent an individual from accidentally falling from the protected area. They must also be designed also to prevent small children from intentionally crawling through the barrier.

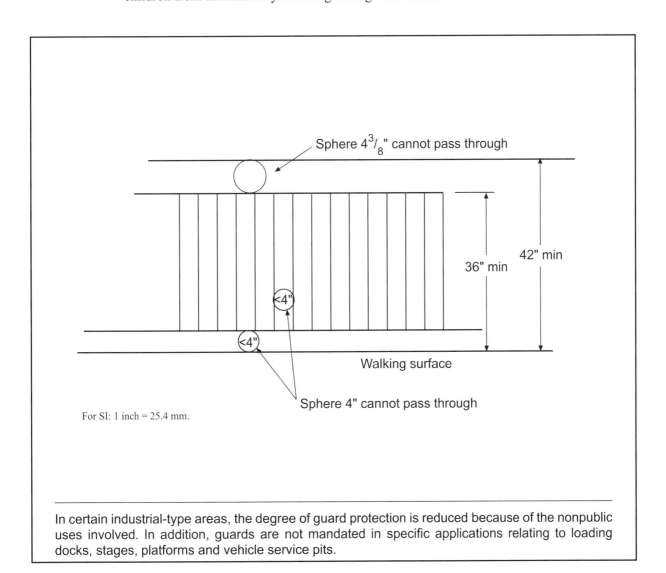

In certain industrial-type areas, the degree of guard protection is reduced because of the nonpublic uses involved. In addition, guards are not mandated in specific applications relating to loading docks, stages, platforms and vehicle service pits.

Topic: Rooftop Access and Equipment
Reference: IBC 1013.6

Category: Means of Egress
Subject: Guards

Code Text: *Guards shall be provided where appliances, equipment, fans, roof hatch openings or other components that require service are located within 10 feet (3048 mm) of a roof edge or open side of a walking surface and such edge or open side is located more than 30 inches (762 mm) above the floor, roof or grade below. The guard shall be constructed so as to prevent the passage of a sphere 21 inches in diameter (533 mm). The guard shall extend not less than 30 inches (762 mm) beyond each end of such appliance, equipment, fan or component.*

Discussion and Commentary: The requirement for guards primarily addresses the hazard created when service personnel are working on rooftop equipment. Where such activity occurs close to a roof edge, it is critical that guards be provided to prevent an accidental fall.

For SI: 1 inch = 25.4 mm, 1 foot = 304.8 mm

A guard is also mandated where a roof hatch is located near a roof edge. At times, these roof accesses are used during inclement weather, emergency situations or times of darkness. The area around roof hatch openings is also often utilized as a staging area or work area.

Topic: Window Sills
Reference: IBC 1013.8

Category: Exterior Walls
Subject: Exterior Windows and Doors

Code Text: *In Occupancy Groups R-2 and R-3, one- and two-family and multiple-family dwellings, where the opening of the sill portion of an operable window is located more than 72 inches (1829 mm) above the finished grade or other surface below, the lowest part of the clear opening of the window shall be at a height not less than 36 inches (915 mm) above the finished floor surface of the room in which the window is located. Operable sections of windows shall not permit openings that allow passage of a 4 inch (102 mm) diameter sphere where such openings are located within 36 inches (915 mm) of the finished floor. See exceptions for complying window fall prevention devices and opening control devices.*

Discussion and Commentary: The minimum sill height requirement for operable windows located a considerable height above the surface below is intended to reduce the number of falls by children from such windows. The minimum height is established at a point above the center of gravity of most children.

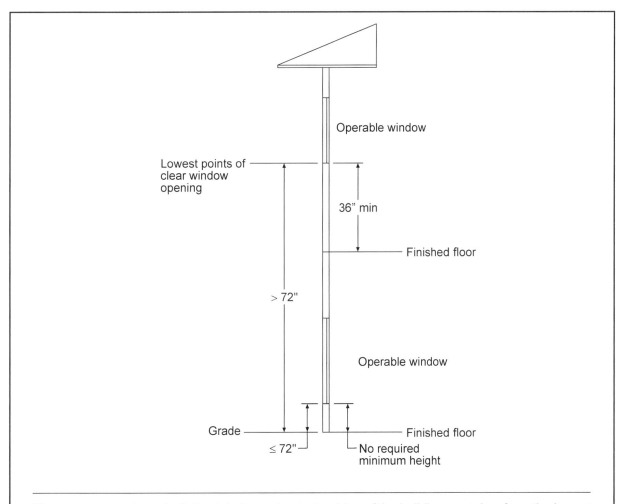

The measurements on both the interior and exterior sides of the building are taken from the lowest part of the clear window opening, providing for consistent application. Where the lower window panel is inoperable, the measurement is to be taken to the lowest point of the lowest operable panel.

Study Session 9

Study Session 9
IBC Sections 1001 – 1007, 1011 and 1013

1. A court or yard that provides access to a public way for one or more exits is considered a(n) _____ .

   a. exit access way   b. egress court
   c. public way   d. horizontal exit

   Reference_____

2. Which of the following elements is not a distinct and separate part of the means of egress?

   a. exit discharge   b. exit access
   c. exit   d. exit convergence

   Reference_____

3. That portion of the exit access that occupants must traverse before two separate and distinct paths of egress travel to two exits are available is considered a _____.

   a. means of egress   b. single egress path
   c. common path of egress travel   d. limited egress travel distance

   Reference_____

4. Panic hardware that is listed for use on fire door assemblies is considered to be _____ hardware.

   a. fire egress
   b. fire exit
   c. panic
   d. panic and fire

   Reference_____

5. An alternating tread device has a series of steps that are positioned a minimum of _____ degrees and maximum of _____ degrees from horizontal.

   a. 30, 45
   b. 45, 60
   c. 50, 70
   d. 60, 75

   Reference_____

6. In a dining room without fixed seating, the calculated occupant load is determined by dividing the floor area by a factor of one occupant per _____ square feet.

   a. 7 net
   b. 15 net
   c. 15 gross
   d. 20 net

   Reference_____

7. A 1,500-square-foot (net) woodworking shop classroom in a high school is considered to have a design occupant load of _____ persons.

   a. 25
   b. 30
   c. 75
   d. 100

   Reference_____

8. For areas having fixed seats and aisles, the occupant load for bench seating without armrests is based on one occupant for each _____ inches of seating length.

   a. 15
   b. 18
   c. 24
   d. 30

   Reference_____

9. In a fully-sprinklered Group M retail sales building having an occupant load of 3,200 occupants, the minimum total calculated means of egress width for egress elements other than stairways shall be _____ inches if an emergency voice/alarm communication system is provided.

   a. 960 b. 640
   c. 480 d. 320

   Reference_____

10. A stairway serving 160 occupants in a fully-sprinklered Group I-2 hospital shall be a minimum of _____ inches in width.

    a. 42 b. 44
    c. 48 d. 60

    Reference_____

11. Multiple means of egress shall be sized so that the loss of any one means of egress shall not reduce the available capacity to less than _____ of the required capacity.

    a. 10 percent b. 25 percent
    c. $33^1/_3$ percent d. 50 percent

    Reference_____

12. When fully open, a door is permitted to project into the required width of the path of egress travel a maximum of _____ .

    a. one-half the required width b. one-half the actual width
    c. $3^1/_2$ inches d. 7 inches

    Reference_____

13. Up to 50 percent of the ceiling area of a means of egress may have a minimum ceiling height of _____ where reduced by protruding objects.

    a. 78 inches b. 80 inches
    c. 84 inches d. 90 inches

    Reference_____

14. At a doorway, the minimum headroom clearance below any door closer or stop shall be _____ inches.

    a. 76   b. 78
    c. 80   d. 84

    Reference_____

15. Other than handrails serving stairs and ramps, the maximum projection into the travel path of a horizontal projection located between 27 and 80 inches above the walking surface shall be _____ inches.

    a. 1½   b. 3½
    c. 4   d. 4½

    Reference_____

16. At an exterior door not required to be accessible in a Group F-1 occupancy, what is the maximum permitted elevation change?

    a. ½ inch   b. 1 inch
    c. 7 inches   d. 8 inches

    Reference_____

17. Exit signs shall be located so that the maximum distance from any point in an exit access corridor or exit passageway to the nearest visible exit sign is _____ feet, or the listed viewing distance for the sign, whichever is less.

    a. 50   b. 75
    c. 100   d. 150

    Reference_____

18. Externally-illuminated exit signs shall have a minimum intensity at the face of the sign of _____ foot-candles.

    a. 1   b. 5
    c. 10   d. 12

    Reference_____

19. Emergency lighting facilities for means of egress illumination shall initially provide _____ along the path of egress at floor level.

    a. at least 1 foot-candle
    b. an average of 1 foot-candle
    c. at least 5 foot-candles
    d. an average of 0.2 foot-candle

    Reference_____

20. A guard need not be located along an open-sided walking surface located a maximum of _____ inches above the floor below.

    a. 15
    b. 30
    c. 36
    d. 42

    Reference_____

21. In a mercantile occupancy, a required guard shall form a protective barrier a minimum of _____ inches in height.

    a. 36
    b. 38
    c. 42
    d. 44

    Reference_____

22. In areas of a Group S-1 occupancy not open to the public, horizontal intermediate rails in a required guard shall be constructed so that a _____ sphere cannot pass through any opening.

    a. 4-inch
    b. 6-inch
    c. 12-inch
    d. 21-inch

    Reference_____

23. Unless a roof-top HVAC unit requiring occasional service and maintenance is located a minimum of _____ from the roof edge, a complying guard shall be provided.

    a. 10 feet
    b. 5 feet
    c. 3 feet
    d. 30 inches

    Reference_____

24. In a nonsprinklered building, a stairway utilized as an accessible means of egress shall be a minimum of _____ inches in clear width between handrails.

    a. 36   b. 44
    c. 48   d. 60

    Reference_____

25. An area of refuge serving 450 occupants shall be provided with a minimum of _____ wheelchair space(s).

    a. 1   b. 2
    c. 3   d. 5

    Reference_____

26. Where a barrier is installed below a protruding object having a vertical clearance of less than 80 inches, the leading edge of the barrier shall be located a maximum of _____ inches above the floor.

    a. 27   b. 30
    c. 36   d. 42

    Reference_____

27. In all cases, the occupant load in a room or building shall not be increased beyond a maximum of one occupant per _____ square feet.

    a. 3   b. 5
    c. 6   d. 7

    Reference_____

28. The means of egress in which of the following areas is required to be illuminated when the space is occupied?

    a. aisle accessways in Group A   b. sleeping units in Group I
    c. dwelling units in Group R-2   d. aisles in Groups F and S

    Reference_____

29. The exterior wall adjacent to an exterior area for assisted rescue landing does not require a fire-resistance rating where the landing is located a minimum of _____ feet horizontally from the wall.

   a. 5 b. 10
   c. 15 d. 20

   Reference_____

30. A tactile sign stating EXIT shall be provided adjacent to the door to all of the following means of egress components, except for _____.

   a. an exit stairway b. the exit discharge
   c. an exit access corridor d. an exit passageway

   Reference_____

31. Where seating booths are provided, the occupant load shall be based on one person for each _____ inches of booth length.

   a. 18 b. 20
   c. 21 d. 24

   Reference_____

32. In any position during the course of the door swing, a door is permitted to project into the required width of the path of egress travel a maximum of _____.

   a. one-half the required width b. one-half the actual width
   c. 4 inches d. 4$\frac{1}{2}$ inches

   Reference_____

33. Under general conditions, the means of egress illumination level shall not be less than _____ foot-candle(s) at the walking surface level.

   a. 0.1 b. 1
   c. 5 d. 10

   Reference_____

34. From a height of 36 inches to 42 inches in a required guard, a sphere with a maximum diameter of _____ inches shall not pass through.

    a. 4   b. $4^3/_8$

    c. 6   d. 8

Reference _____

35. Within an individual dwelling unit of a Group R-2 occupancy, openings in required guards on the open sides of stairs shall not allow the passage of a maximum _____-inch-diameter sphere.

    a. 4   b. $4^3/_8$

    c. 6   d. 8

Reference _____

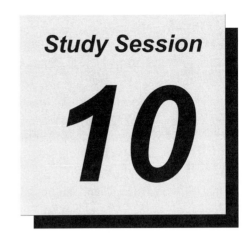

Study Session 10

2012 IBC Sections 1008 – 1010, and 1012
Means of Egress II

OBJECTIVE: To obtain an understanding of the general component requirements of a means of egress system, including those regulating doors, gates, stairways, ramps and turnstiles located along the egress path.

REFERENCE: Sections 1008 – 1010, and 1012, 2012 *International Building Code*

KEY POINTS:
- What is the minimum height and width of an egress door?
- What are the limitations on projections into the required clear door width?
- When must doors swing in the direction of egress travel?
- What is the maximum opening force permitted for an interior side-swinging door without a closer? Other side-swinging doors? Sliding and folding doors?
- What are "special" doors? How are they regulated differently than other doors?
- At a door, what change in elevation is permitted for a landing or floor surface?
- How must landings at doors be sized? What is the maximum allowable amount that doors may encroach into the required landing size?
- What is the maximum height of a threshold at a doorway?
- Which types of locks and latches are required on egress doors?
- When is the unlatching of an egress door permitted to take more than a single operation?
- Why do turnstiles create special egress concerns? What are the limitations on their use?
- When is panic hardware required?
- How are gates regulated differently than doors?
- How do interior exit stairways differ from exit access stairways? What code section regulates interior exit stairways? Exit access stairways?
- Under what conditions is an exit access stairway permitted to be unenclosed? Where enclosure is required, what degree of construction is mandated?
- How is the minimum required width of a stairway determined?
- At a stairway and its landings, what is the minimum headroom clearance?
- What is the minimum rise of a stair riser? Maximum rise? Minimum tread run?
- What degree of tolerance is permitted between the largest and the smallest tread run within a flight of stairs? Between the greatest and the smallest riser height?

265

KEY POINTS:
(Cont'd)
- How are stairway landings regulated for size?
- What is the maximum vertical rise permitted between stairway landings?
- Which special limitations apply to winders? Curved or spiral stairways? Alternate tread devices?
- When is a stairway required to provide access to a roof?
- What is the maximum permissible ramp slope? The maximum permissible rise for a ramp?
- How are ramp landings regulated for width? Length? Construction? Edge protection?
- When are handrails required for ramps?
- What must be the minimum height of a handrail located above the nosing of stairway treads and landings? Maximum height?
- How are handrails to be regulated for graspability?
- What are the exceptions to the general provision that handrails be continuous and without interruption?
- To what extent must handrails extend beyond the top and bottom risers of a stair flight?
- How much clear space is needed between a handrail and a wall or other surface?
- Under which conditions are intermediate handrails required?

Topic: Additional Doors and Identification **Category:** Means of Egress
Reference: IBC 1008.1 **Subject:** Doors, Gates and Turnstiles

Code Text: *Doors provided for egress purposes in numbers greater than required by the IBC shall meet the requirements of Section 1008. Means of egress doors shall be readily distinguishable from the adjacent construction and finishes such that the doors are easily recognizable as doors. Mirrors or similar reflecting materials shall not be used on means of egress doors. Means of egress doors shall not be concealed by curtains, drapes, decorations or similar materials.*

Discussion and Commentary: During a fire or other incident, occupants will attempt to exit through those doors that they believe will eventually lead to the exterior. Accordingly, any doors that would suggest an egress path must meet all of the door requirements. In addition, means of egress doors must be obvious and available for immediate use by the building occupants.

For SI: 1 inch = 25.4 mm, 1 foot = 304.8 mm.

In accordance with Section 1020.2, any building or structure used for human occupancy must have at least one exterior door opening that complies with the minimum width (32 inches) and height (80 inches) requirements of Section 1008.1.1.

Study Session 10

Topic: Sizes of Doors
Reference: IBC 1008.1.1
Category: Means of Egress
Subject: Doors, Gates and Turnstiles

Code Text: *The minimum width of each door opening shall be sufficient for the occupant load thereof and shall provide a clear width of not less than 32 inches (813 mm). Clear openings of doorways with swinging doors shall be measured between the face of the door and the stop, with the door open 90 degrees (1.57 rad). The height of door openings shall not be less than 80 inches (2032 mm). See multiple exceptions for both width and height.*

Discussion and Commentary: A clear width of 32 inches is required only to a height of 34 inches above the floor or ground. Beyond this point, projections up to 4 inches into the required width are permitted. Although a single doorway is expected to be used for the egress of one individual at a time, it must also be of adequate width for wheelchair users.

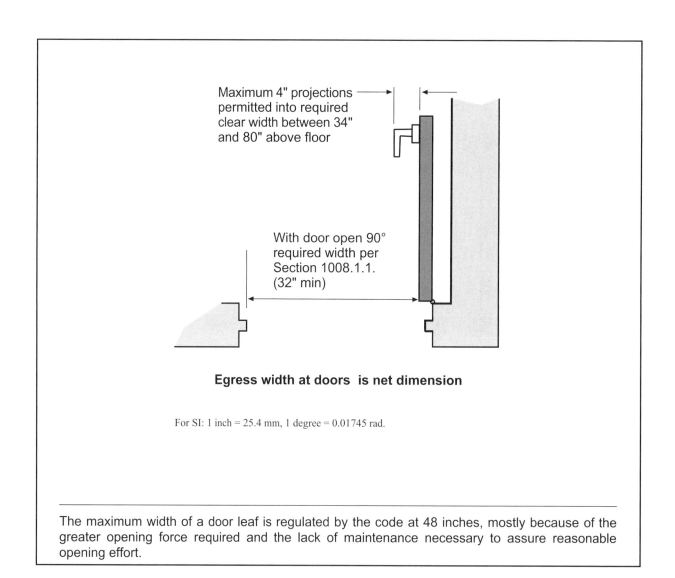

Egress width at doors is net dimension

For SI: 1 inch = 25.4 mm, 1 degree = 0.01745 rad.

The maximum width of a door leaf is regulated by the code at 48 inches, mostly because of the greater opening force required and the lack of maintenance necessary to assure reasonable opening effort.

Topic: Door Swing
Reference: IBC 1008.1.2
Category: Means of Egress
Subject: Doors, Gates and Turnstiles

Code Text: *Egress doors shall be of the pivoted or side-hinged swinging type.* See multiple exceptions addressing special conditions. *Doors shall swing in the direction of egress travel where serving a room or area containing an occupant load of 50 or more persons or a Group H occupancy.*

Discussion and Commentary: Numerous fire deaths in buildings have been attributed to improper exit doors, but no single incident is more infamous than the 1942 Coconut Grove fire in Boston. Inward-swinging exterior exit doors were a significant factor in the loss of 492 lives. As a result, doors serving sizable occupant loads or Group H occupancies must swing in the direction of exit flow. For assembly and educational occupancies, the use of panic hardware increases the likelihood that egress doors can be opened easily.

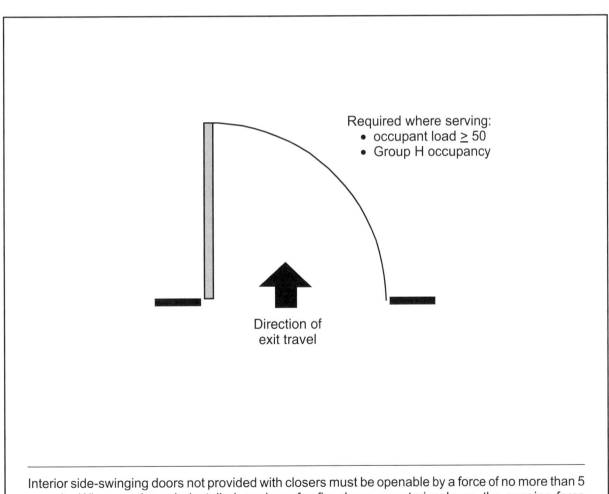

Interior side-swinging doors not provided with closers must be openable by a force of no more than 5 pounds. Where a closer is installed, such as for fire doors or exterior doors, the opening force limitation is increased to 15 pounds to ensure that the device can bring the door to a fully closed position.

Topic: Special Doors
Reference: IBC 1008.1.4

Category: Means of Egress
Subject: Doors, Gates and Turnstiles

Code Text: *Special doors and security grilles shall comply with the requirements of Sections 1008.1.4.1 through 1008.1.4.4.*

Discussion and Commentary: In general, doors in means of egress systems must be of the pivoted or side-hinged swinging type. Other doors, identified as special doors, are also addressed in the code. Such doors include revolving doors, power-operated doors, horizontal sliding doors and security grilles. These types of doors are specifically limited in their use because the difficult or unusual operation of such doors increases the likelihood of obstructed travel in an emergency.

The various types of special doors are permitted to be used for egress purposes when regulated by occupancy, occupant load, operation, opening force, power supply or other factors that contribute to the effectiveness and reliability of the egress door.

Topic: Floor Elevation
Reference: IBC 1008.1.5
Category: Means of Egress
Subject: Doors, Gates and Turnstiles

Code Text: *There shall be a floor or landing on each side of a door. Such floor or landing shall be at the same elevation on each side of the door. See exceptions.*

Discussion and Commentary: To avoid a surprise change in the elevation of a walking surface as it passes through a doorway, limitations have been placed on the height differential. The IBC generally requires that no elevation change occur at a door. In many cases, however, such a change in elevation takes place due to a variation in the type or thickness of floor finish materials. Where this occurs, a difference of no more than $^1/_2$ inch is permitted. Otherwise, only those exceptions that apply to certain dwelling units or to exterior doors not on an accessible route may have an elevation change at a doorway.

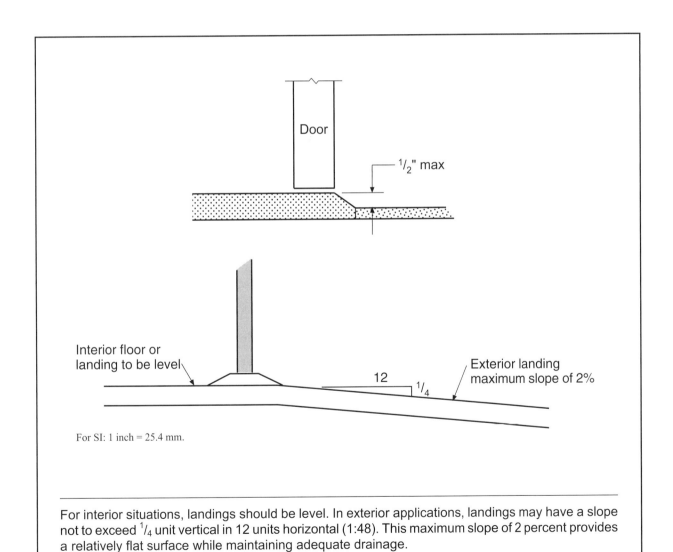

For interior situations, landings should be level. In exterior applications, landings may have a slope not to exceed $^1/_4$ unit vertical in 12 units horizontal (1:48). This maximum slope of 2 percent provides a relatively flat surface while maintaining adequate drainage.

Topic: Landings at Doors
Reference: IBC 1008.1.6
Category: Means of Egress
Subject: Doors, Gates and Turnstiles

Code Text: *Landings shall have a width not less than the width of the stairway or the door, whichever is greater. Doors in the fully open position shall not reduce a required dimension by more than 7 inches (178 mm). When a landing serves an occupant load of 50 or more, doors in any position shall not reduce the landing to less than one-half its required width.*

Discussion and Commentary: This provision, which allows a door to project into the path of exit travel on a stairway's landing, comprises two issues. The first issue is that a door is not a fixed obstruction; it swings across the landing when it is used by occupants of the building. The second issue is that a door in any position is allowed to obstruct only one-half of the required width of the landing. The expectation is that the additional width of the landing will be provided for the occupants using the stairway as the door swings toward its fully open or fully closed position.

In Group R-3 and U occupancies, and within individual units of Group R-2 occupancies, the length of a landing can be no less than 36 inches. In all other occupancy groups, the minimum required landing length is 44 inches, measured in the direction of travel.

Topic: Door Arrangement
Reference: IBC 1008.1.8

Category: Means of Egress
Subject: Doors, Gates and Turnstiles

Code Text: *Space between two doors in series shall be 48 inches (1219 mm) minimum plus the width of a door swinging into the space. Doors in series shall swing either in the same direction or away from the space between doors.* See exceptions for dwelling units and horizontal sliding power-operated doors.

Discussion and Commentary: Where two doors are installed in a manner to create a vestibule or similar space, they must be located so to allow building occupants effective and efficient movement through one door prior to continuing through the second door. This is especially true where the person opening the door has limited mobility and is required to make special effort in opening the door and passing through the doorway. The IBC recognizes these concerns by mandating an adequate spatial separation between doors provided in a series.

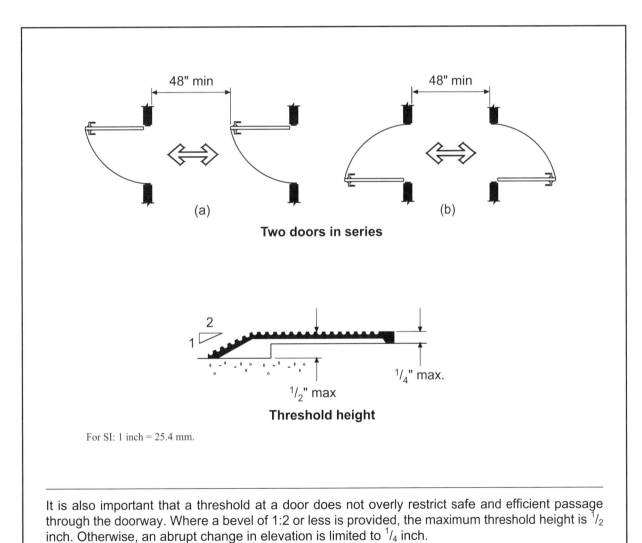

Two doors in series

Threshold height

For SI: 1 inch = 25.4 mm.

It is also important that a threshold at a door does not overly restrict safe and efficient passage through the doorway. Where a bevel of 1:2 or less is provided, the maximum threshold height is ¹/₂ inch. Otherwise, an abrupt change in elevation is limited to ¹/₄ inch.

Topic: Operations
Reference: IBC 1008.1.9
Category: Means of Egress
Subject: Doors, Gates and Turnstiles

Code Text: *Except as specifically permitted by Section 1008.1.9, egress doors shall be readily openable from the egress side without the use of a key or special knowledge or effort. Door handles, pulls, latches, locks and other operating devices shall be installed 34 inches (864 mm) minimum and 48 inches (1219 mm) maximum above the finished floor. Locks and latches shall be permitted to prevent operation of doors where any of five listed conditions exists. Manually operated flush bolts or surface bolts are not permitted. See five exceptions. The unlatching of any door or leaf shall not require more than one operation. See exceptions identifying four locations that allow for multiple operations.*

Discussion and Commentary: Every element along the path of exit travel through a means of egress system, particularly doors, must be under the control of, and operable by, the person seeking egress. The intent is that the hardware installed be of a type familiar to most users, readily recognizable and usable under any emergency conditions.

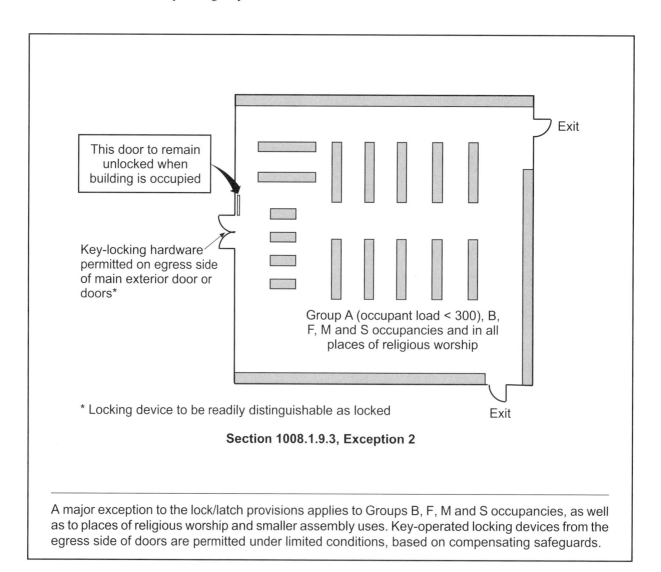

Section 1008.1.9.3, Exception 2

A major exception to the lock/latch provisions applies to Groups B, F, M and S occupancies, as well as to places of religious worship and smaller assembly uses. Key-operated locking devices from the egress side of doors are permitted under limited conditions, based on compensating safeguards.

Topic: Panic and Fire Exit Hardware
Reference: IBC 1008.1.10

Category: Means of Egress
Subject: Doors, Ramps and Turnstiles

Code Text: *Doors serving a Group H occupancy and doors serving rooms or spaces with an occupant load of 50 or more in a Group A or E occupancy shall not be provided with a latch or lock unless it is panic hardware or fire exit hardware.* See exception for main exit of Group A occupancy.

Discussion and Commentary: Panic hardware is a door-latching assembly incorporating a device that releases the latch when force is applied in the direction of exit travel. It is utilized in assembly occupancies because of the hazard that occurs when a large number of occupants reach an exit door at the same instance. In educational occupancies, the same concern is present, along with the need for children to be able to operate the latch of an exit door easily during an emergency. Panic hardware is mandated in all Group H occupancies, regardless of occupant load, on account of the increased hazard level anticipated within the building. Fire exit hardware is merely panic hardware listed for use on a fire door assembly.

To ensure that contact with the door actuates the releasing device, the code requires that the actuating portion extend for at least one half of the door width. Where balanced or pivoted doors are used, the device width is again limited to one-half of the door width for leverage purposes.

Study Session 10

Topic: Turnstiles
Reference: IBC 1008.3
Category: Means of Egress
Subject: Doors

Code Text: *Turnstiles or similar devices that restrict travel to one direction shall not be placed so as to obstruct any required means of egress.* See exception that allows installation of turnstiles serving as a portion of the means of egress if each turnstile serves a maximum occupant load of 50 persons and meets four additional provisions.

Discussion and Commentary: Turnstiles may serve as an egress component under very specific conditions. Where the listed conditions are met, each turnstile can be assigned a maximum exit capacity of 50 persons. However, permanent turnstiles cannot be considered as providing any of the required egress capacity when serving an occupant load over 300, no matter how many turnstiles are installed. In such cases, additional side-hinged swinging doors must be installed within 50 feet of the permanent turnstiles.

Each turnstile credited for up to 50-person capacity for egress where each turnstile:

- Will turn freely in direction of egress when power is lost, and upon manual release by employee in area
- Only given credit for 50% of required egress capacity (egress other than by turnstiles required)
- Limited to 39 inches in height
- Has minimum of 16$\frac{1}{2}$ inches clear width at and below height of 39 inches
- Has minimum of 22 inches clear width at height above 39 inches

Where the turnstile has a height exceeding 39 inches, the restriction to egress is much like that of a revolving door. Therefore, the egress limitations for revolving doors as set forth in Section 1008.1.4.1 apply to this higher type of turnstile.

| Topic: General Provisions | Category: Means of Egress |
| Reference: IBC 1009.1 – 1009.3 | Subject: Stairways |

Code Text: *Stairways serving occupied portions of a building shall comply with the requirements of Section 1009. Interior exit stairways shall lead directly to the exterior of the building or shall be extended to the exterior of the building with an exit passageway. Interior exit stairways shall be included, as necessary, to meet one or more means of egress design requirements, such as required number of exits or exit access travel distance. Floor openings between stories created by exit access stairways shall be enclosed.* See 10 exceptions identifying conditions under which exit access stairways are permitted to be unenclosed.

Discussion and Commentary: There are two types of means of egress stairways within a building, interior exit stairways and exit access stairways. Interior exit stairways, regulated by Section 1022, are considered *exits* and must always be enclosed with fire-resistance-rated construction. Exit access stairways are considered portions of the *exit access* and are permitted to be unenclosed under a variety of special conditions.

The provisions of Section 1009 regulating the design and construction of stairways are applicable to all stairways, including those that may be considered only "convenience" stairs and not considered a portion of any required means of egress.

Study Session 10

Topic: Stairway Width
Reference: IBC 1009.4, 1012.8
Category: Means of Egress
Subject: Stairways and Handrails

Code Text: *The width of stairways shall be determined as specified in Section 1005.3.1 (calculated based on occupant load) but such width shall not be less than 44 inches (1118 mm). See exceptions for small occupant loads, spiral stairways, aisle stairs, and where a stairway lift is installed. Projections into the required width at each side shall not exceed $4^1/_2$ inches (114 mm) at or below the handrail height. Projections into the required width shall not be limited above the minimum headroom height required in Section 1009.5.*

Discussion and Commentary: A stairway is considered one or more flights of stairs, each made up of one or more risers and any connecting landings. Any change of elevation along a travel path, unless accomplished by a ramp, must include a stair or stairway. Although stairways are generally required to be at least 44 inches in width, a 36-inch-wide stairway is permitted where serving an occupant load of 50 or less.

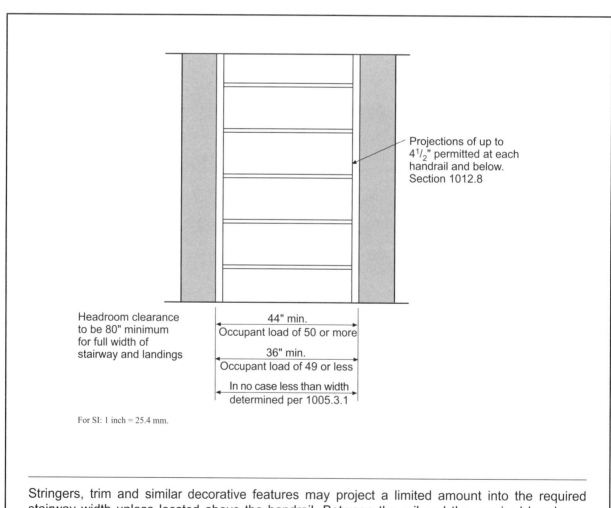

Stringers, trim and similar decorative features may project a limited amount into the required stairway width unless located above the handrail. Between the rail and the required headroom height of 80 inches, no projection into the required width is permitted.

Topic: Stair Treads and Risers
Reference: IBC 1009.7.2
Category: Means of Egress
Subject: Stairways and Handrails

Code Text: *Stair riser heights shall be 7 inches (178 mm) maximum and 4 inches (102 mm) minimum. Rectangular tread depths shall be 11 inches (279 mm) minimum.* See exceptions, including the allowance for greater riser heights ($7^3/_4$ inches) and shallower tread depths (10 inches) in Group R-3 and associated Group U occupancies, and within individual dwelling units of Group R-2.

Discussion and Commentary: The stairway 7-11 rule is the result of much research and discussion on stairway design. In addition to the proportional criteria that has been developed, the uniformity of the treads and risers in a flight of stairs is critical. The maximum variation between the highest and lowest risers and between the shallowest and deepest treads is limited to $^3/_8$ inch within any flight, which is intended as a permissible construction tolerance.

Treads and risers to be of uniform size and shape
($^3/_8$" tolerance permitted between least and greatest within flight)

For SI: 1 inch = 25.4 mm, 1 degree = 0.01745 rad.

Curved stairways, winders, spiral stairways, aisle stairs and alternating tread devices are unique configurations requiring special consideration. The use of these stairways is limited to varying degrees based on occupancy, occupant load, design and use as a required means of egress.

Study Session 10

Topic: Stairway Landings
Reference: IBC 1009.8
Category: Means of Egress
Subject: Stairways and Handrails

Code Text: *There shall be a floor or landing at the top and bottom of each stairway. The width of landings shall be not less than the width of stairways they serve. Every landing shall have a minimum width measured in the direction of travel equal to the width of the stairway. Where the stairway has a straight run the depth need not exceed 48 inches (1219 mm). See exception for aisle stairs.*

Discussion and Commentary: A landing that serves a stairway is required to have a length equal to or greater than the stairway width unless the stairway has a straight run. This measurement is based on the actual width of the stairway, not the required width. It is important to ensure that the capacity of the stairway is not reduced as occupants travel between stairway flights. A maximum length of 48 inches is acceptable for straight stairway travel, insofar as the capacity is not reduced for travel through the landing.

For SI: 1 inch = 25.4 mm.

Because of the difficulty many individuals encounter while negotiating stairs, the code requires a maximum vertical rise between landings of 12 feet. When placed at limited intervals, landings can be used as a resting place for the stair user and can also make stair travel less intimidating.

Topic: Handrail Locations
Reference: IBC 1009.15, 1012.9
Category: Means of Egress
Subject: Stairways and Handrails

Code Text: *Stairways shall have handrails on each side and shall comply with Section 1012. See five exceptions where a single handrail or no handrail is required. Stairways shall have intermediate handrails located in such a manner so that all portions of the stairway width required for egress capacity are within 30 inches (762 mm) of a handrail. On monumental stairs, handrails shall be located along the most direct path of egress travel.*

Discussion and Commentary: The handrail, a very important safety element of a stairway, must be located within relatively easy reach of every stair user. Therefore, in most applications, a rail must be provided on both sides of the stairway. In the case of extremely wide stairways, such as monumental stairs, the requirement for additional rails located throughout the width of the stairway is based on the required stairway width, not the actual width.

Various exceptions permit the use of a single handrail, and in some cases no rail, within a dwelling unit. In addition, aisle stairs with a center handrail, or those serving seating on only one side, are permitted to use a single handrail.

Topic: Slope, Rise, Width and Handrails **Category:** Means of Egress
Reference: IBC 1010 **Subject:** Ramps

Code Text: *Ramps used as part of a means of egress shall have a running slope not steeper than one unit vertical in 12 units horizontal (8-percent slope) See exception for assembly uses per Section 1028.11. The slope of other pedestrian ramps shall not be steeper than one unit vertical in eight units horizontal (12.5-percent slope). The rise for any ramp run shall be 30 inches (762 mm) maximum. The minimum width of a means of egress ramp shall not be less than that required for corridors by Section 1018.2. The clear width of a ramp and the clear width between handrails, if provided, shall be 36 inches (914 mm). Ramps with a rise greater than 6 inches (152 mm) shall have handrails on both sides. Handrails shall comply with Section 1012.*

Discussion and Commentary: Although many of the governing ramp provisions are designed for accessibility purposes, egress capabilities must also be considered. Handrails may project into the required ramp width up to $4^1/_2$ inches at each handrail at or below the handrail height, but in no case shall the clear width between handrails be less than 36 inches. A minimum headroom of 80 inches must be provided at all portions of the ramp.

For SI: 1 inch = 25.4 mm, 1 foot = 304.8 mm.

To provide adequate clearance at ramp landings, doors cannot reduce the clear landing width to less than 42 inches. A landing must be at least 60 inches in length and at least as wide as the widest ramp run adjoining the landing.

Topic: Edge Protection
Reference: IBC 1010.10
Category: Means of Egress
Subject: Ramps

Code Text: *Edge protection complying with Sections 1010.10.1* (curb, rail, wall or barrier) *or 1010.10.2* (extended floor or ground surface) *shall be provided on each side of ramp runs and at each side of ramp landings.* See exceptions for curb ramps, ramp landings, and ramps in fixed-seating assembly spaces. *A curb must be a minimum of 4 inches (102 mm) in height. Barriers must be constructed so that the barrier prevents the passage of a 4-inch-diameter (102 mm) sphere, where any portion of the sphere is within 4 inches (102 mm) of the floor or ground surface,* or *the floor or ground surface of the ramp run or landing shall extend 12 inches (305 mm) minimum beyond the inside face of a handrail complying with Section 1012.*

Discussion and Commentary: Edge protection at ramps and ramp landings is necessary to prevent the wheels of a wheelchair from leaving the ramp or landing surface, or becoming lodged between the edge of the ramp and any adjacent construction. The protection is also beneficial to those individuals who utilize various forms of walking aids, including canes and crutches.

For SI: 1 inch = 25.4 mm

Edge protection is different than that type of protection provided by a guard. The presence of a complying guard does not necessarily provide adequate edge protection, and the presence of adequate edge protection does not typically satisfy the requirements for a guard.

Topic: Handrail Dimensions
Reference: IBC 1012.2, 1012.3
Category: Means of Egress
Subject: Stairways and Handrails

Code Text: *Handrail height, measured above stair tread nosings, or finish surface of ramp slope, shall be uniform, not less than 34 inches (864 mm) and not more than 38 inches (965 mm). See lower height range for alternating tread devices and ship ladders. Handrails with a circular cross section shall have an outside diameter of at least $1^1/_4$ inches (32 mm) and not greater than 2 inches ((51 mm). Where the handrail is not circular, it shall have a perimeter dimension of at least 4 inches (102 mm) and not greater than $6^1/_4$ inches (160 mm) with a maximum cross-section dimension of $2^1/_4$ inches (57 mm) and a minimum cross section dimension of 1 inch (25 mm). See allowances for Type II rails and those handrail shapes providing equivalent graspability.*

Discussion and Commentary: Handrail height shall be measured from the nosing of the treads to the top of the rail. Handrails located above or below this height range are not easily reached by most individuals. The shape of the rail should allow for easy grasping by most users.

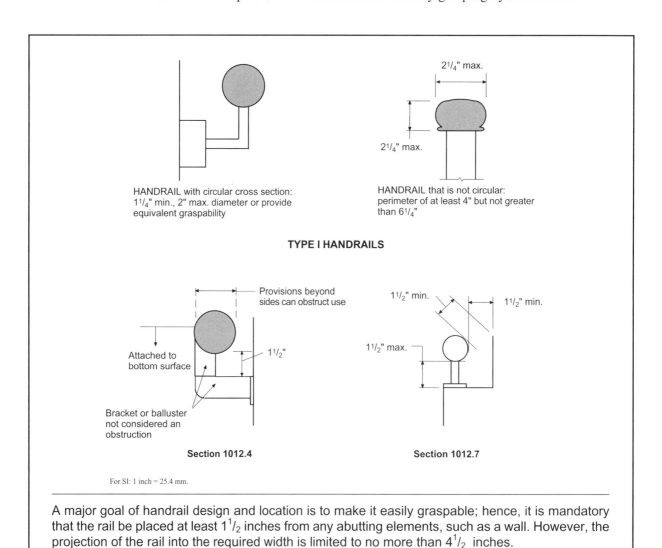

A major goal of handrail design and location is to make it easily graspable; hence, it is mandatory that the rail be placed at least $1^1/_2$ inches from any abutting elements, such as a wall. However, the projection of the rail into the required width is limited to no more than $4^1/_2$ inches.

Study Session 10
IBC Sections 1008 – 1010, and 1012

1. In general, a door opening shall provide a minimum clear width of _____ inches.

   a. 30   b. 32
   c. 34   d. 36

   Reference_____

2. In a Group I-2 occupancy, means of egress doors used for the movement of beds shall have a minimum width of _____ inches.

   a. 32   b. 36
   c. $41^1/_2$ d. 44

   Reference_____

3. The minimum required width for door openings does not apply to storage closets less than _____ square feet in floor area.

   a. 10   b. 25
   c. 50   d. 100

   Reference_____

4. At a door opening, the maximum permitted projection into the required clear width shall be _____ inch(es) at any point between 34 inches and 80 inches above the floor.

   a. 0, no projections are permitted

   b. ½

   c. 1

   d. 4

   Reference_____

5. In which of the following uses must an egress door be a side-hinged swinging door or manually operated horizontal sliding door where serving an occupant load of 10 or less?

   a. office   b. storage

   c. manufacturing   d. retail sales

   Reference_____

6. In other than a Group H occupancy, egress doors shall swing in the direction of egress travel where serving a minimum occupant load of _____ persons.

   a. 10   b. 30

   c. 50   d. 100

   Reference_____

7. For a side-swinging fire door provided with a self-closing device, the egress door shall swing to a full-open position when subjected to a maximum _____ force.

   a. 5-pound   b. 15-pound

   c. 30-pound   d. 50-pound

   Reference_____

8. A revolving door shall be provided with a side-hinged swinging door located in the same wall and within _____ feet of the revolving door.

   a. 5   b. 10

   c. 20   d. 50

   Reference_____

9. Entrance doors in a means of egress in a Group B office building may be equipped with an approved entrance and egress access control system under limited conditions, including that any manual unlocking device be located a maximum of _____ feet from the secured doors.

    a. 3   b. 4
    c. 5   d. 10

    Reference_____

10. Exterior landings at doors shall have a maximum slope of _____ unit vertical in 12 units horizontal.

    a. $1/8$ b. $1/4$
    c. $1/2$ d. 1

    Reference_____

11. Other than at sliding doors serving dwelling units, the maximum height for thresholds at doorways is _____ inch.

    a. $1/8$ b. $1/4$
    c. $1/2$ d. 1

    Reference_____

12. Approved, listed delayed egress locks are permitted under specific conditions in all but which one of the following occupancies?

    a. Group A   b. Group I
    c. Group F   d. Group S

    Reference_____

13. In general, door handles, pulls, latches, locks and other operating devices shall be installed a minimum of _____ inches and a maximum of _____ inches above the finished floor.

    a. 32, 48   b. 34, 42
    c. 34, 48   d. 36, 54

    Reference_____

Study Session 10

14. Where an exit access stairway connecting two stories is required to be enclosed, the enclosure shall have a minimum fire-resistance rating of _____ hour(s).

    a. 0 (no rating is required)   b. 1/2
    c. 1   d. 2

    Reference_____

15. Stairways, other than spiral stairways, shall have a minimum headroom clearance of _____ inches.

    a. 76   b. 78
    c. 80   d. 84

    Reference_____

16. Within an individual dwelling unit of a Group R-2 apartment building, a stair shall have a maximum riser height of _____ inches and a minimum tread depth of _____ inches.

    a. 7, 11   b. $7^3/_4$, 10
    c. 8, 9   d. $8^1/_4$, 9

    Reference_____

17. For a 60-inch-wide stairway having a straight run, any intermediate landing shall be a minimum of _____ inches in length.

    a. 36   b. 44
    c. 48   d. 60

    Reference_____

18. For winder treads, the minimum permitted tread depth within the clear width of the stair shall be _____ inches.

    a. 4   b. 6
    c. 8   d. 10

    Reference_____

19. Where serving a maximum of five occupants, a spiral stairway may be used as a means of egress component from a space having a maximum floor area of _____ square feet.

    a. 200 b. 250
    c. 500 d. 1,000

    Reference_____

20. Stairway handrails shall be located a minimum of _____ inches and a maximum of _____ inches above stair tread nosings.

    a. 30, 34 b. 30, 38
    c. 32, 34 d. 34, 38

    Reference_____

21. Where handrails are not continuous between stair flights, they shall continue to slope for _____ beyond the bottom riser.

    a. the depth of one tread
    b. a minimum of 12 inches
    c. 12 inches plus one tread depth
    d. one-half the length of the landing

    Reference_____

22. Projections into the required width at each stairway handrail shall be limited to a maximum of _____ inches at any point at and below the handrail height.

    a. 0, no projections permitted b. $1^1/_2$
    c. $3^1/_2$ d. $4^1/_2$

    Reference_____

23. The maximum permitted rise of any ramp shall be _____ inches.

    a. 30 b. 44
    c. 48 d. 60

    Reference_____

24. Only those ramps having a maximum rise of _____ inches are permitted without complying handrails.

    a. 6   b. 12

    c. 18   d. 24

Reference_____

25. Unless having a maximum height of _____ inches, turnstiles shall be regulated as for revolving doors.

    a. 34   b. 36

    c. 38   d. 39

Reference_____

26. Where a power-operated door must be opened manually, the maximum force to set the door in motion shall be _____ pounds.

    a. 5   b. 15

    c. 30   d. 50

Reference_____

27. The space required between two doors in a series shall be a minimum of _____ inches plus the width of a door swinging into the space.

    a. 30   b. 44

    c. 48   d. 60

Reference_____

28. A stairway serving an occupant load of 35 in an office suite shall have a minimum width of _____ inches.

    a. 30   b. 36

    c. 42   d. 44

Reference_____

29. Type I stairway handrails having a circular cross section shall have a minimum outside diameter of _____ inches and a maximum diameter of _____ inches.

    a. 1¼, 2
    b. 1¼, 2 ⁵⁄₈
    c. 1½, 2
    d. 1½, 2 ⁵⁄₈

    Reference_____

30. Where a railing is used as edge protection along the side of a ramp run, it must have a rail mounted to prevent the passage of a maximum _____-inch sphere, where any portion of the sphere is within 4 inches of the ground.

    a. 4
    b. 6
    c. 8
    d. 12

    Reference_____

31. Doors from electrical rooms with equipment rated at a minimum of _____ amperes and over 6 feet wide that contain overcurrent, switching or control devices must be provided with panic hardware.

    a. 600
    b. 1,000
    c. 1,200
    d. 1,500

    Reference_____

32. In general, the minimum required width for a stairway in a Group B occupancy shall be _____ inches.

    a. 36
    b. 42
    c. 44
    d. 48

    Reference_____

33. Where a ramp is not a part of an accessible route, the length of the landing shall be a minimum of _____ inches measured in the direction of travel.

    a. 44
    b. 48
    c. 54
    d. 60

    Reference_____

34. At ramps where handrails are not continuous between runs, the handrails shall extend horizontally a minimum of _____ inches beyond the top and bottom of the ramp.

    a. 6	b. 8
    c. 12	d. 18

    Reference _____

35. Stairways shall have intermediate handrails located so that all portions of the stairway width required for egress capacity are within _____ inches of a handrail.

    a. 30	b. 32
    c. 35	d. 44

    Reference _____

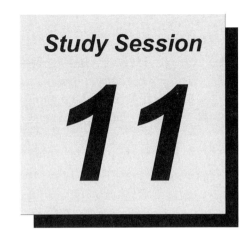

2012 IBC Sections 1014 – 1019
Means of Egress III

OBJECTIVE: To obtain an understanding of the system design requirements for the exit access, including number of exits, separation of egress doorways and maximum travel distances, as well as the requirements for the exit access components, including aisles, corridors and egress balconies.

REFERENCE: Section 1014 through 1019, 2012 *International Building Code*

KEY POINTS:
- What portion of the means of egress system is the exit access?
- Is egress permitted through an adjoining or intervening room? If so, under what conditions?
- How must egress be provided where more than one tenant occupies any single floor of a building or structure?
- What is a common path of egress travel?
- Why is the common path of egress travel so limiting?
- When are at least two exits or exit access doorways required from a room?
- At what point is access required to three or more exits or exit access doorways?
- Why are multiple exit paths required to be separated at a specified minimum distance from each other?
- At what minimum distance must two exits or exit access doorways be separated? Three or more exits or exit access doorways?
- What benefit is derived from exit separation in a sprinklered building?
- Why is travel distance regulated?
- How is the maximum travel distance measured? Where does it start? Where does it end?
- How is travel distance measured where travel involves exit access stairways?
- Which occupancies permit the least travel distance? The most?
- What are the travel distance limitations for low-hazard storage and manufacturing buildings? Which special conditions must be met?
- To what amount can the travel distance be increased on an exterior egress balcony?
- Where must complying aisles be provided?
- What is the minimum permitted aisle width in public areas of Groups B and M? In nonpublic areas?

KEY POINTS:
(Cont'd)
- What is an aisle accessway? What is the minimum aisle accessway width where the aisle serves merchandise pads?
- What is the purpose of a corridor? When must a corridor be of fire-resistance-rated construction?
- What is the minimum required width of a corridor?
- What is a dead-end condition? When are dead ends limited in length?
- How is the use of corridors for supply, return, exhaust or ventilation air regulated?
- How do the exiting provisions for egress balconies compare with those for corridors?
- When is a fire-resistance-rated separation mandated between the building and an egress balcony?

Topic: Egress Through Intervening Spaces **Category:** Means of Egress
Reference: IBC 1014.2, #1 **Subject:** Exit Access

Code Text: *Egress from a room or space shall not pass through adjoining or intervening rooms or areas, except where such adjoining rooms or areas are accessory to one or the other; are not a Group H occupancy; and provide a discernible path of egress travel to an exit.* See exception for Group H, S and F occupancies.

Discussion and Commentary: A workable means of egress system must be direct, obvious and unobstructed. Therefore, egress may only travel through an intervening room, space or area where the exit path is discernable. Travel must be such that it is readily apparent which direction the occupant must go to continue toward the exit. In addition, access through a high-hazard space is prohibited unless traveling from another high-hazard space. It is also expected that any intervening room used for egress will be related to the room from which egress begins.

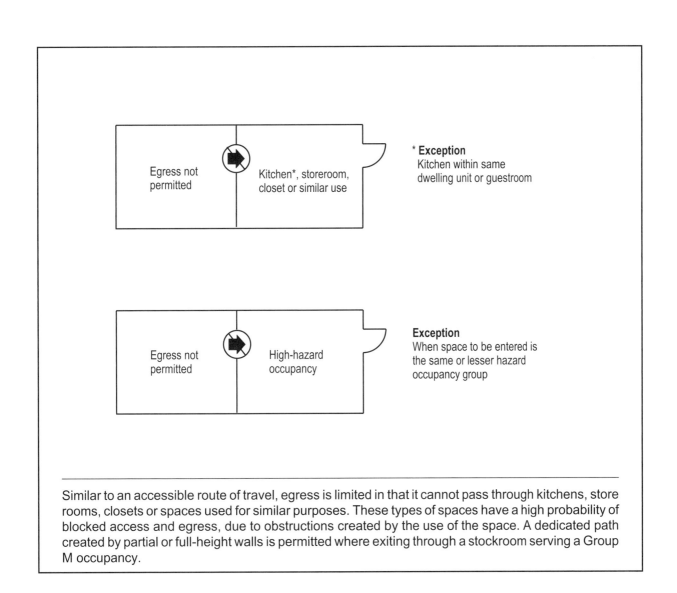

Similar to an accessible route of travel, egress is limited in that it cannot pass through kitchens, store rooms, closets or spaces used for similar purposes. These types of spaces have a high probability of blocked access and egress, due to obstructions created by the use of the space. A dedicated path created by partial or full-height walls is permitted where exiting through a stockroom serving a Group M occupancy.

Study Session 11

Topic: Common Path of Egress Travel
Reference: IBC 1014.3
Category: Means of Egress
Subject: Exit Access

Code Text: *The common path of egress travel shall not exceed the common path of egress travel distances in Table 1014.3.*

Discussion and Commentary: A common path of egress travel is defined as *that portion of exit access which the occupants are required to traverse before two separate and distinct paths of egress travel to two exits are available.* The concept of limiting the common path of egress travel addresses the concern that multiple egress options must be available to occupants where the expected egress travel distance becomes excessive. Although the overall travel distance in a building may be of considerable length, such travel is greatly limited where only one egress path is available. A limit of 75 feet of common path travel is applied to most occupancies and under most conditions. The limitation is reduced to 25 feet for Group H-1, H-2 and H-3 occupancies.

In business, factory and storage occupancies, the code permits up to 100 feet of common path of egress travel where the building is fully sprinklered. The 100-foot limitation also applies to nonsprinklered buildings housing Groups B, S or U with an occupant load of 30 or less.

Topic: Doorways Required
Reference: IBC 1015.1

Category: Means of Egress
Subject: Exit and Exit Access Doorways

Code Text: *Two exits or exit access doorways from any space shall be provided where 1) the occupant load of the space exceeds one of the values in Table 1015.1, or 2) the common path of egress travel exceeds one of the limitations of Section 1014.3, or (3) required by Sections 1015.3, 1015.4, 1015.5 or 1015.6. See exceptions for Group I-2 and R-2 occupancies.*

Discussion and Commentary: Two basic criteria establish the point at which it is necessary to provide at least two paths of egress travel from a portion of a building. The first is a function of both the occupancy involved and the occupant load of the space. The second is related to the maximum travel distance an occupant encounters prior to reaching a point where a choice of egress paths is available. Based on the occupancy involved, where the established occupant load or the common travel distance is exceeded, the space must have multiple egress doorways.

TABLE 1015.1
SPACES WITH ONE EXIT OR EXIT ACCESS DOORWAY

OCCUPANCY	MAXIMUM OCCUPANT LOAD
A, B, E, F, M, U	49
H-1, H-2, H-3	3
H-4, H-5, I-1, I-2, I-3, I-4, R	10
S	29

Requirements for access to three or four exits or exit access doorways from an individual space are found in Section 1015.1.1 and are based solely on occupant load. The same criteria applies for exits within a story as indicated by Section 1021.2.4.

Study Session 11

Topic: Doorway Arrangement
Reference: IBC 1015.2

Category: Means of Egress
Subject: Exit and Exit Access Doorways

Code Text: *Required exits shall be located in a manner that makes their availability obvious. Where two exits or exit access doorways are required from any portion of the exit access, the exit doors or exit access doorways shall be placed a distance apart equal to not less than one-half of the length of the maximum overall diagonal dimension of the building or area to be served measured in a straight line between exit doors or exit access doorways. Interlocking or scissor stairs shall be counted as one exit stairway.* See exceptions for sprinklered buildings and where a rated corridor connects two interior exit stairways.

Discussion and Commentary: One of the fundamental concepts of exiting is that a single fire incident should not render all means of egress unusable. In this regard, egress doorways are required to be located so as to minimize the probability of such an occurrence. The required separation in sprinklered buildings is reduced to a distance of one-third of the overall diagonal.

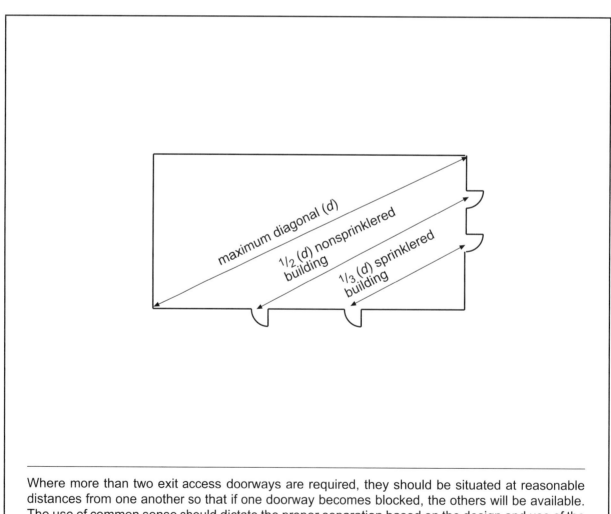

Where more than two exit access doorways are required, they should be situated at reasonable distances from one another so that if one doorway becomes blocked, the others will be available. The use of common sense should dictate the proper separation based on the design and use of the space or room.

Topic: Travel Distance Limitations
Reference: IBC 1016.2, 1016.3
Category: Means of Egress
Subject: Exit Access Travel Distance

Code Text: *Exit access travel distance shall not exceed the values given in Table 1016.2. Exit access travel distance shall be measured from the most remote point within a story along the natural and unobstructed path of horizontal and vertical egress travel to the entrance to an exit. See exceptions for open parking garages and outdoor facilities with open exit access components.*

Discussion and Commentary: Travel distance is considered the portion of egress travel between any occupiable location in a building and the nearest entrance to an exit. Because quick evacuation from a building is the foremost method of protecting the occupants in many fire incidents, the length of travel is limited until the occupant reaches one of the "protected components." Travel should be measured around any obstruction that is considered fixed or permanent, including low-height office partitions, retail shelving, storage racks, fixed seating, etc.

In most sprinklered buildings, the code permits a moderate increase in the permitted travel distance over that permitted in nonsprinklered buildings. An increase of 50 feet is typical of most occupancies; however, a travel distance increase of 100 feet is permitted for Group B occupancies protected by a sprinkler system.

Topic: Travel on Exit Access Stairways
Reference: IBC 1016.3.1
Category: Means of Egress
Subject: Exit Access Travel Distance

Code Text: *Travel distance on exit access stairways or ramps shall be included in the exit access travel distance measurement. The measurement along stairways shall be made on a plane parallel and tangent to the stair tread nosings in the center of the stair and landings.*

Discussion and Commentary: Travel distance may include travel on an exit access stairway or ramp. In such cases, travel up or down the stairway or ramp would need to be included in the travel distance determination. Exit access stairways are addressed under the provisions of Section 1009.3.

TABLE 1016.2
EXIT ACCESS TRAVEL DISTANCE[a]

OCCUPANCY	WITHOUT SPRINKLER SYSTEM (feet)	WITH SPRINKLER SYSTEM (feet)
A, E, F-1, M, R, S-1	200	250[b]
I-1	Not Permitted	250[c]
B	200	300[c]
F-2, S-2, U	300	400[c]
H-1	Not Permitted	75[c]
H-2	Not Permitted	100[c]
H-3	Not Permitted	150[c]
H-4	Not Permitted	175[c]
H-5	Not Permitted	200[c]
I-2, I-3, I-4	Not Permitted	200[c]

For SI: 1 foot = 304.8 mm.

a. See the following sections for modifications to *exit access* travel distance requirements:
   Section 402.8: For the distance limitation in *malls*.
   Section 404.9: For the distance limitation through an *atrium* space.
   Section 407.4: For the distance limitation in Group I-2.
   Sections 408.6.1 and 408.8.1: For the distance limitations in Group I-3.
   Section 411.4: For the distance limitation in special amusement buildings.
   Section 1015.4: For the distance limitation in refrigeration machinery rooms.
   Section 1015.5: For the distance limitation in refrigerated rooms and spaces.
   Section 1021.2: For buildings with one *exit*.
   Section 1028.7: For increased limitation in assembly seating.
   Section 1028.7: For increased limitation for assembly open-air seating.
   Section 3103.4: For temporary structures.
   Section 3104.9: For pedestrian walkways.

b. Buildings equipped throughout with an *automatic sprinkler system* in accordance with Section 903.3.1.1 or 903.3.1.2. See Section 903 for occupancies where *automatic sprinkler systems* are permitted in accordance with Section 903.3.1.2.

c. Buildings equipped throughout with an *automatic sprinkler system* in accordance with Section 903.3.1.1.

As an example, where an exit access stairway is provided as a sole means of egress from a mezzanine, the travel distance would be measured from the most remote point on the mezzanine, down the stairway and continue until reaching the entrance to the nearest exit.

Topic: Aisles
Reference: IBC 1017.1
Category: Means of Egress
Subject: Exit Access

Code Text: *Aisles or aisle accessways shall be provided from all occupied portions of the exit access which contain seats, tables, furnishings, displays and similar fixtures or equipment. The required width of aisles shall be unobstructed.* See exception for permissible encroachments. *Aisles and aisle accessways serving a room or space used for assembly purposes shall comply with Section 1028. In Group B and M occupancies, the minimum clear aisle width shall be determined by Section 1005.1 for the occupant load served, but shall not be less than 36 inches (914 mm).* See exception for nonpublic aisles.

Discussion and Commentary: Well-defined aisles must be provided throughout office spaces, retail stores and similar facilities. The mandated clear width varies based on the presence of obstructions, such as chairs, clothes racks or other items that can easily interrupt the egress flow to an exit.

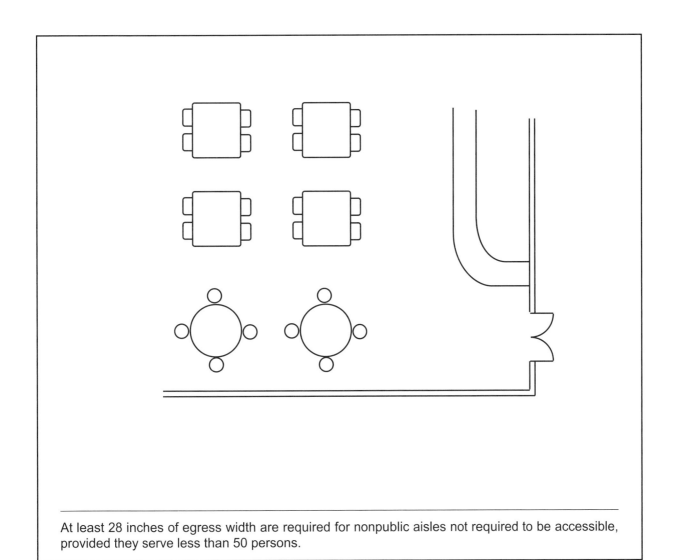

At least 28 inches of egress width are required for nonpublic aisles not required to be accessible, provided they serve less than 50 persons.

Topic: Aisle Accessways in Group M
Reference: IBC 1017.4
Category: Means of Egress
Subject: Exit Access

Code Text: *An aisle accessway shall be provided on at least one side of each element within the merchandise pad. The minimum clear width for an aisle accessway not required to be accessible shall be 30 inches (762 mm). The required clear width of the aisle accessway shall be measured perpendicular to the elements and merchandise within the merchandise pad. The 30-inch (762 mm) minimum clear width shall be maintained to provide a path to an adjacent aisle or aisle accessway.*

Discussion and Commentary: A merchandise pad is defined as *the merchandise display area that contains multiple counters, shelves, racks and other movable fixtures.* Bounded by aisles, permanent fixtures and walls, the merchandise pad also includes aisle accessways utilized to provide both access to an aisle and circulation throughout the pad area.

For SI: 1 inch = 25.4 mm

Within a merchandise pad, the common path of travel is limited to 75 feet in length. Where the occupant load of the area served by the common path exceeds 50 persons, the common path cannot exceed 30 feet in length from any point in the merchandise pad.

Topic: Corridor Construction
Reference: IBC 1018.1
Category: Means of Egress
Subject: Corridors

Code Text: *Corridors shall be fire-resistance rated in accordance with Table 1018.1. The corridor walls required to be fire-resistance rated shall comply with Section 708 for fire partitions.* See four exceptions where a rating is not required.

Discussion and Commentary: A fire-resistance-rated corridor is intended to protect occupants of the corridor during egress travel from an incident in an enclosed space bordering the corridor. The construction of the corridor provides a minimum level of protection from fire and smoke through the use of fire-resistance-rated walls and ceilings, as well as fire-protected openings. Smoke infiltration is limited also by smoke- and draft-control door assemblies and smoke dampers. Occupancy group, occupant load and presence of a fire sprinkler system are the major factors in determining whether or not a corridor must have a fire-resistance rating.

TABLE 1018.1
CORRIDOR FIRE-RESISTANCE RATING

OCCUPANCY	OCCUPANT LOAD SERVED BY CORRIDOR	REQUIRED FIRE-RESISTANCE RATING (hours)	
		Without sprinkler system	With sprinkler system[c]
H-1, H-2, H-3	All	Not Permitted	1
H-4, H-5	Greater than 30	Not Permitted	1
A, B, E, F, M, S, U	Greater than 30	1	0
R	Greater than 10	Not Permitted	0.5
I-2[a], I-4	All	Not Permitted	0
I-1, I-3	All	Not Permitted	1[b]

a. For requirements for occupancies in Group I-2, see Sections 407.2 and 407.3.
b. For a reduction in the fire-resistance rating for occupancies in Group I-3, see Section 408.8.
c. Buildings equipped throughout with an automatic sprinkler system in accordance with Section 903.3.1.1 or 903.3.1.2 where allowed.

Exceptions eliminate the need for a fire-resistance-rated corridor in certain Group E occupancies, in sleeping units or dwelling units of residential occupancies, in open parking garages and in Group B occupancies that are permitted a single means of egress by Section 1015.1.

Topic: Corridor Width
Reference: IBC 1018.2
Category: Means of Egress
Subject: Corridors

Code Text: *The minimum width of corridors specified in Table 1018.2 shall be as determined in Section 1005.1 (calculated width). Table 1018.2 identifies a minimum required width of 44 inches except for: 1) 24 inches for access to building systems or equipment, 2) 36 inches for occupant loads less than 50, 3) 36 inches within a dwelling unit, 4) 72 inches in a Group E corridor with an occupant load of 100 or more, 5) 72 inches in specified health care centers, and 6) 96 inches in Group I-2 areas where required for bed movement.*

Discussion and Commentary: To allow for adequate circulation throughout a building and, more importantly, for egress purposes, a minimum width requirement is established. In addition, complying routes of travel to accessible spaces must be provided. Only in areas used for access to electrical, mechanical or plumbing systems or equipment is the width permitted to be reduced to less than 36 inches. A minimum 36-inch width is mandated for corridors within dwelling units, or for those corridors serving an occupant load of 50 or less. Only specific projections such as doors are permitted to encroach a limited distance into the required corridor width.

See 1018.2 for additional width provisions

For SI: 1 inch = 25.4 mm, 1 foot = 304.8 mm.

Certain occupancies require additional corridor widths based on their specialized uses. Corridors serving 100 or more occupants in Group E educational occupancies must be at least 72 inches in width, and healthcare occupancies require increased widths for bed movement.

Topic: Dead Ends
Reference: IBC 1018.4
Category: Means of Egress
Subject: Corridors

Code Text: *Where more than one exit or exit access doorway is required, the exit access shall be arranged such that there are no dead ends in corridors more than 20 feet (6096 mm) in length. See exceptions for increased dead-end lengths.*

Discussion and Commentary: Limitations on dead-end corridors are established where two or more exit or exit access doorways are required. The intent is to limit the distance building occupants must travel before they determine that there is no way out and that they must retrace steps in order to locate an exit or exit access doorway. Where only a single means of egress is permitted, a dead-end condition is not limited in length; however, the provisions of Section 1014.3 for common paths of travel must be considered.

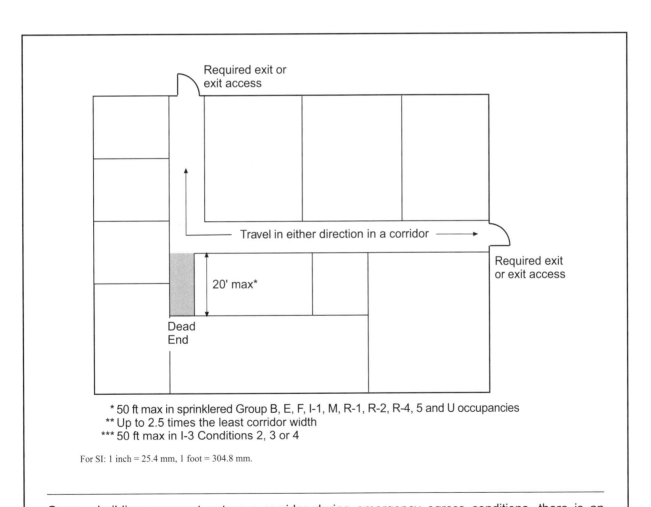

* 50 ft max in sprinklered Group B, E, F, I-1, M, R-1, R-2, R-4, 5 and U occupancies
** Up to 2.5 times the least corridor width
*** 50 ft max in I-3 Conditions 2, 3 or 4

For SI: 1 inch = 25.4 mm, 1 foot = 304.8 mm.

Once a building occupant enters a corridor during emergency egress conditions, there is an expectation that a direct and obvious exit path is available. Dead-end configurations should be minimal, if not eliminated, to expedite the exiting process.

Topic: Air Movement in Corridors
Reference: IBC 1018.5, 1018.5.1
Category: Means of Egress
Subject: Corridors

Code Text: *Corridors shall not serve as supply, return, exhaust, relief or ventilation air ducts or plenums.* See four exceptions addressing return air, make-up air and incidental air movement. *Use of the space between the corridor ceiling and the floor or roof structure above as a return air plenum is permitted for one or more of the following conditions:* See five conditions.

Discussion and Commentary: The use of corridors for the movement of air is strictly limited by the code. Because a corridor is intended to be a relatively safe environment for occupants exiting a building, it is not advisable to introduce air movement that might increase the potential for fire, smoke or toxic gases to enter the corridor. It is possible, under specific conditions, to use the space above a corridor ceiling as a return air plenum. For example, where the corridor is not required to be of fire-resistance-rated construction, or where the above-ceiling space is isolated from a rated corridor by fire-resistance-rated construction, the upper area may be used for return air.

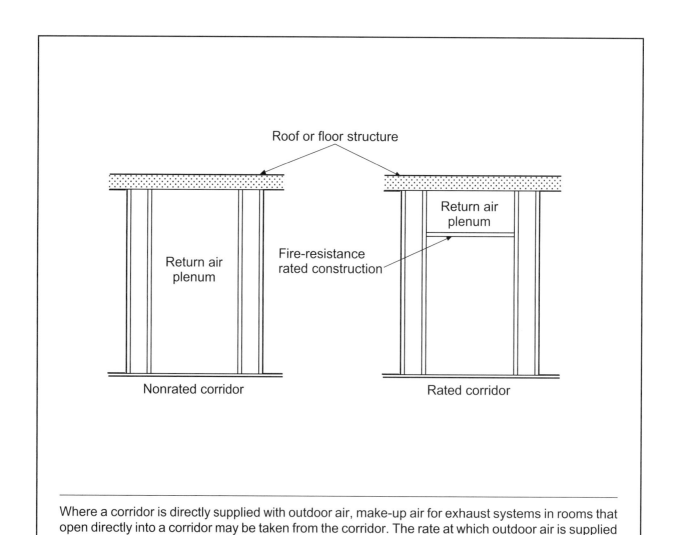

Where a corridor is directly supplied with outdoor air, make-up air for exhaust systems in rooms that open directly into a corridor may be taken from the corridor. The rate at which outdoor air is supplied to the corridor must exceed the rate of makeup air taken from the corridor.

Topic: Corridor Continuity
Reference: IBC 1018.6

Category: Means of Egress
Subject: Corridors

Code Text: *Fire-resistance-rated corridors shall be continuous from the point of entry to an exit, and shall not be interrupted by intervening rooms.* See exceptions for travel through foyers, lobbies and reception rooms.

Discussion and Commentary: Once an occupant enters a corridor required to be of fire-resistance-rated construction, he or she expects that travel to an exit will be direct. The level of protection within the corridor must not be reduced at any point along the egress path. Where an intervening room or space interrupts the corridor, it is quite likely that the exitway will be obstructed or confusing. Where corridor travel includes or terminates at a lobby, foyer or reception room, the condition is considered acceptable, insofar as such spaces are usually an extension of the circulation and egress path.

Where the path of travel occurs in a corridor not required to be fire-resistance-rated, such travel may then proceed through other intervening spaces, provided all other requirements of the code are met, such as those for common path of egress travel and travel distance.

Study Session 11

Topic: Egress Balconies
Reference: IBC 1019
Category: Means of Egress
Subject: Exit Access

Code Text: *Balconies used for egress purposes shall conform to the same requirements as corridors for width, headroom, dead ends and projections. Exterior egress balconies shall be separated from the interior of the building by walls and opening protectives as required for corridors.* See exception for elimination of separation.

Discussion and Commentary: Although the openness of exterior balconies provides some degree of protection from smoke and toxic gases created by a fire, travel along such balconies usually places the occupants at considerable risk. Therefore, the IBC regulates egress balcony travel in a manner consistent with unprotected travel inside the structure. An increase of 100 feet in maximum allowable travel distance is permitted by Section 1016.2.1 for egress balcony travel.

For SI: 1 foot = 304.8 mm

For an exit access element to be considered an egress balcony, it must be sufficiently open to the exterior to minimize the potential for smoke and toxic gases to accumulate. The code considers openings for at least 50 percent of the long side to be adequately open.

Study Session 11
IBC Sections 1014 – 1019

1. Where the common path of travel is within the permitted limits, a Group B occupancy may have a single means of egress where the maximum occupant load is _____.

   a. 10 b. 20
   c. 29 d. 49

   Reference_____

2. In which one of the following occupancies having an occupant load of 40 must at least two exits or exit access doorways always be provided?

   a. A-2 b. B
   c. F-1 d. S-1

   Reference_____

3. What is the minimum number of exits or exit access doorways required for a Group M sales floor having an occupant load of 1,200?

   a. 2 b. 3
   c. 4 d. 6

   Reference_____

4. Where two means of egress are required from a room in a fully sprinklered building, they shall be separated a minimum of _____ of the length of the maximum overall diagonal dimension of the area served.

   a. one-half   b. one-third
   c. one-fourth   d. one-sixth

   Reference_____

5. Which of the following conditions is not specifically required where egress from a room passes through an adjoining room?

   a. a discernable path of travel must be provided
   b. the adjoining room and the area served are accessory to one or the other
   c. no more than one means of egress can pass through the adjoining room
   d. the adjoining room cannot be classified as Group H

   Reference_____

6. In general, through which one of the following spaces is egress not specifically prohibited?

   a. private offices   b. storage rooms
   c. kitchens   d. closets

   Reference_____

7. The means of egress is permitted to pass from a small tenant space through a larger adjoining tenant space, provided the smaller tenant space occupies less than _____ of the area of the larger tenant space.

   a. 10   b. 15
   c. 25   d. $33^1/_3$

   Reference_____

8. Access to at least two exits or exit access doorways shall be provided in rooms or spaces of day-care facilities where care is provided for a minimum of _____ children that are $2^1/_2$ years of age or less.

   a. 6   b. 11
   c. 20   d. 29

   Reference_____

9. Three exits or exit access doorways shall be provided from any space with a minimum occupant load of _____ and a maximum occupant load of _____.

   a. 301, 500
   b. 301, 600
   c. 501, 800
   d. 501, 1000

   Reference_____

10. In a nonsprinklered Group A-3 occupancy, what is the maximum permitted exit access travel distance?

    a. 150 feet
    b. 200 feet
    c. 250 feet
    d. 300 feet

    Reference_____

11. Up to an additional 100 feet of travel distance is available where the last portion of exit access travel occurs _____.

    a. within a corridor
    b. within a 1-hour-rated corridor
    c. on an exterior egress balcony
    d. on an exterior exit stairway

    Reference_____

12. In a Group H-3 occupancy, the common path of egress travel is limited to a maximum of _____ feet.

    a. 25
    b. 75
    c. 100
    d. 150

    Reference_____

13. In an office tenant space having an occupant load of 24 persons, the maximum length of a common path of egress travel shall be _____ feet.

    a. 25
    b. 75
    c. 100
    d. 200

    Reference_____

14. In the public areas of a Group M occupancy, aisles having fixtures or equipment on both sides shall be a minimum of _____ inches in width.

    a. 28 b. 36
    c. 42 d. 44

    Reference_____

15. Where accessibility is not required, what is the minimum required width of an aisle serving 15 persons in a nonpublic area of a Group B occupancy?

    a. no minimum width required b. 28 inches
    c. 36 inches d. 44 inches

    Reference_____

16. In other than rooms or spaces used for assembly purposes and Group B and M occupancies, aisles shall be a minimum of _____ inches in clear width.

    a. 32 b. 36
    c. 42 d. 44

    Reference_____

17. A corridor having a required capacity of 125 persons in a Group E occupancy shall be a minimum of _____ inches in width.

    a. 36 b. 44
    c. 60 d. 72

    Reference_____

18. A corridor in a Group B occupancy serving an occupant load of 40 persons shall have a minimum width of _____ inches.

    a. 36 b. 44
    c. 48 d. 60

    Reference_____

19. In a nonsprinklered Group M retail sales building, a corridor serving a minimum occupant load of _____ persons shall be fire-resistance rated.

    a. any number of   b. 11
    c. 31   d. 51

    Reference_____

20. In a fully-sprinklered building, a corridor serving 100 persons in which of the following occupancies must be fire-resistance rated?

    a. Group A-2   b. Group I-2
    c. Group M   d. Group R-1

    Reference_____

21. In a Group I-2 occupancy, corridors requiring more than one exit or exit access doorway may have dead ends if limited to a maximum of _____ feet in length.

    a. 8   b. 20
    c. 30   d. 50

    Reference_____

22. In a fully-sprinklered Group E occupancy, the maximum permitted length of a dead-end condition in a corridor requiring at least two means of egress is _____.

    a. 20 feet   b. 50 feet
    c. twice the corridor width   d. four times the corridor width

    Reference_____

23. Utilization of corridors as return air plenums is permitted within tenant spaces having a maximum floor area of _____ square feet.

    a. 400   b. 1,000
    c. 3,000   d. 5,000

    Reference_____

Study Session 11

24. What is the minimum required width of an egress balcony serving an apartment building where the required capacity of the balcony is 55 persons?

    a. 36 inches b. 42 inches
    c. 44 inches d. 48 inches

    Reference_____

25. The long side of an egress balcony shall be a minimum of _____ percent open.

    a. 25 b. 40
    c. 50 d. 66²/₃

    Reference_____

26. In a Group A-2 dining room having a single exit access door, the common path of egress travel is limited to a maximum of _____ feet.

    a. 25 b. 75
    c. 100 d. 200

    Reference_____

27. At least two exits or exit access doors are required from a refrigeration machinery room where the room has a floor area exceeding _____ square feet.

    a. 0, two exits are always required
    b. 200
    c. 500
    d. 1,000

    Reference_____

28. All portions of refrigeration machinery rooms shall be within _____ feet of an exit or exit access doorway.

    a. 25 b. 75
    c. 100 d. 150

    Reference_____

29. What is the maximum permitted travel distance for a Group H-1 occupancy protected by an automatic sprinkler system?

    a. 50 feet
    b. 75 feet
    c. 100 feet
    d. 125 feet

    Reference_____

30. A 16-foot-wide dead-end corridor in a nonsprinklered Group E high school is prohibited where the dead end is a minimum of _____ feet in length.

    a. 20
    b. 30
    c. 32
    d. 40

    Reference_____

31. Where egress is permitted through a stockroom of a Group M occupancy, a minimum aisle width of _____ inches is required, defined by full or partial height walls.

    a. 30
    b. 36
    c. 44
    d. 48

    Reference_____

32. An aisle accessway in a merchandise pad of a Group M occupancy shall be a minimum of _____ inches in width on at least one side of each merchandising element.

    a. 30
    b. 36
    c. 44
    d. 48

    Reference_____

33. A minimum of two exit access doorways are required in boiler rooms with a floor area exceeding a minimum of _____ square feet and any fuel-fired equipment has an input capacity greater than 400,000 Btu.

    a. 100
    b. 400
    c. 500
    d. 1,000

    Reference_____

34. In a Group M occupancy, the common path of egress travel shall not exceed _____ feet from any point within a merchandise pad with an occupant load of 40.

    a. 20   b. 30
    c. 50   d. 75

    Reference _____

35. What is the minimum required corridor width in areas of Group I-2 occupancies where the movement of beds is required?

    a. 44 inches   b. 60 inches
    c. 72 inches   d. 96 inches

    Reference _____

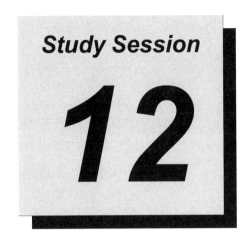

Study Session

12

2012 IBC Sections 1020 – 1029
Means of Egress IV

OBJECTIVE: To obtain an understanding of the provisions governing the exit and exit discharge portions of the means of egress, the special requirements applicable to egress from assembly occupancies, and the details for emergency escape and rescue openings.

REFERENCE: Sections 1020 through 1029, 2012 *International Building Code*

KEY POINTS:
- What is the definition of an exit? What elements of the building are considered exits?
- For which purposes are exits permitted to be used?
- How many exits from a building are required? From any story within a building?
- What degree of fire resistance is required for interior exit stairways and ramps?
- Where must an interior exit stairway or ramp terminate?
- How are openings in the enclosure for an interior exit stairway or ramp regulated? Elevators? Penetrations? Ventilation?
- How must exterior walls of an interior exit stairway or ramp be protected?
- How is travel in an interior exit stairway or ramp that extends beyond the discharge level addressed?
- Where are stairway floor number signs to be located?
- What is a smokeproof enclosure? When is a smokeproof enclosure required?
- What is an exit passageway?
- How is an exit passageway regulated for width?
- What level of fire-resistance-rated construction is mandated for an exit passageway?
- Which types of openings and penetrations are permitted in an exit passageway or a vertical exit enclosure? How are openings and penetrations to be protected?
- In what manner can an exit passageway be provided with ventilation?
- In what types of buildings are luminous egress path markings required? Where within the building are the markings to be provided?
- What is the function of a horizontal exit? How is it to be constructed?
- How is the capacity of a horizontal exit refuge area determined?
- What is the maximum height permitted for an exterior exit stairway?
- What is the minimum size exterior opening required for an exterior exit stairway?

KEY POINTS:
(Cont'd)
- Where does the exit discharge begin? Where does it end?
- What is an egress court?
- What is the minimum size of an egress court? When are egress court walls required to be of fire-resistance-rated construction with protected openings?
- What is a safe dispersal area? What conditions must apply where such an area is utilized?
- What is the minimum capacity of the main exit in an assembly occupancy?
- What is smoke-protected assembly seating?
- What minimum aisle widths are required in assembly occupancies without smoke protection? With smoke protection?
- How is the guard height regulated in assembly areas where the guards interfere with occupant sightlines?
- Where are emergency escape and rescue openings required?
- What is the minimum height and width of an emergency escape and rescue opening? Minimum clear opening size? Maximum sill height from floor?

Topic: Definition
Reference: IBC 1020, 202
Category: Means of Egress
Subject: Exits

Code Text: An exit is *that portion of a means of egress system between the exit access and the exit discharge or public way. Exit components include exterior exit doors at the level of exit discharge, interior exit stairways, interior exit ramps, exit passageways, exterior exit stairways and exterior exit ramps and horizontal exits.*

Discussion and Commentary: The path of travel through the exit access portion of the egress system must be designed to lead to one or more exits, which are locations where some degree of protection or safety from fire hazards is provided. Travel distance is no longer regulated once an exit is reached; therefore, travel within an exit component is virtually unlimited.

Because an exit must be maintained for egress, it cannot be used for any purpose that interferes with egress. In addition, once a mandated level of protection is provided for occupants reaching an exit, that level cannot be diminished prior to their reaching the exit discharge.

Topic: General Provisions
Reference: IBC 1020.1, 1020.2
Category: Means of Egress
Subject: Exits

Code Text: *An exit shall not be used for any purpose that interferes with its function as a means of egress. Once a given level of exit protection is achieved, such level of protection shall not be reduced until arrival at the exit discharge. Buildings or structures used for human occupancy shall have at least one exterior door that meets the requirements of Section 1008.1.1.*

Discussion and Commentary: Exits constitute those portions of the means of egress where the occupant first achieves a significant level of fire protection. An exit is expected to provide a mandated level of protection, and that level cannot be reduced until arrival at the exit discharge. The primary function of an exit component is to provide egress, and any other use cannot interfere with that function. Use of the exit for other purposes can often lead to the egress path being obstructed and possibly unusable. It is not uncommon to find the storage of equipment or furniture occurring within an exit enclosure, creating an unsafe condition.

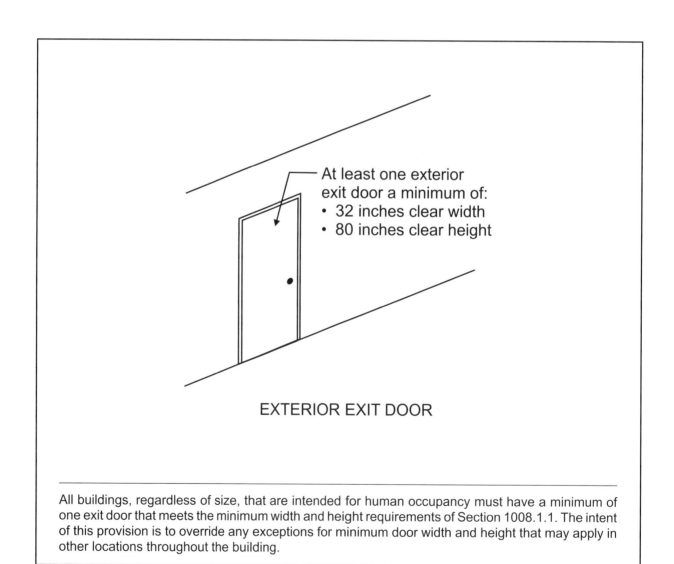

All buildings, regardless of size, that are intended for human occupancy must have a minimum of one exit door that meets the minimum width and height requirements of Section 1008.1.1. The intent of this provision is to override any exceptions for minimum door width and height that may apply in other locations throughout the building.

Topic: Exits from Stories
Reference: IBC 1021.2
Category: Means of Egress
Subject: Number of Exits

Code Text: *Two exits, or exit access stairways or ramps providing access to exits, from any story or occupied roof shall be provided where: 1) the occupant load or number of dwelling units exceeds one of the values in Table 1021.2(1) or 1021.2(2), or 2) the exit access travel distance exceeds that specified in Table 1021.2(1) or 1021.2(2), or 3) helistop landing areas located on buildings or structures shall be provided with two exits, or exit access stairways or ramps providing access to exits.* See six exceptions that allow for a single exit.

Discussion and Commentary: The general provisions call for at least two unique and separate exits from any story of a building. Buildings and stories with one exit are permitted where the configuration and occupancy meet certain characteristics that together do not present an unacceptable fire risk to the buildings' occupants. Those structures that are relatively small in size have a shorter travel distance and fewer occupants; thus, having access to a single exit does not significantly compromise the safety of the occupants.

TABLE 1021.2(1)
STORIES WITH ONE EXIT OR ACCESS TO ONE EXIT FOR R-2 OCCUPANCIES

STORY	OCCUPANCY	MAXIMUM NUMBER OF DWELLING UNITS	MAXIMUM EXIT ACCESS TRAVEL DISTANCE
Basement, first, second or third story	R-2[a, b]	4 dwelling units	125 feet
Fourth story and above	NP	NA	NA

For SI: 1 foot = 304.8 mm.
NP – Not Permitted
NA – Not Applicable
a. Buildings classified as Group R-2 equipped throughout with an *automatic sprinkler system* in accordance with Section 903.3.1.1 or 903.3.1.2 and provided with *emergency escape and rescue openings* in accordance with Section 1029.
b. This table is used for R-2 occupancies consisting of *dwelling units*. For R-2 occupancies consisting of *sleeping units*, use Table 1021.2(2).

TABLE 1021.2(2)
STORIES WITH ONE EXIT OR ACCESS TO ONE EXIT FOR OTHER OCCUPANCIES

STORY	OCCUPANCY	MAXIMUM OCCUPANTS PER STORY	MAXIMUM EXIT ACCESS TRAVEL DISTANCE
First story or basement	A, B[b], E, F[b], M, U, S[b]	49 occupants	75 feet
	H-2, H-3	3 occupants	25 feet
	H-4, H-5, I, R-1, R-2[a,c], R-4	10 occupants	75 feet
	S	29 occupants	100 feet
Second story	B, F, M, S	29 occupants	75 feet
Third story and above	NP	NA	NA

For SI: 1 foot = 304.8 mm.
NP – Not Permitted
NA – Not Applicable
a. Buildings classified as Group R-2 equipped throughout with an *automatic sprinkler system* in accordance with Section 903.3.1.1 or 903.3.1.2 and provided with *emergency escape and rescue openings* in accordance with Section 1029.
b. Group B, F and S occupancies in buildings equipped throughout with an *automatic sprinkler system* in accordance with Section 903.3.1.1 shall have a maximum travel distance of 100 feet.
c. This table is used for R-2 occupancies consisting of *sleeping units*. For R-2 occupancies consisting of *dwelling units*, use Table 1021.2(1).

Table 1021.2(1) is only applicable to Group R-2 occupancies and allows a single exit from the basement, as well as the first, second and third stories under limited conditions. Table 1022.2(2) applies to all other occupancy groups and does not permit a single exit from the third story where serving such occupancies.

Topic: Three or More Exits
Reference: IBC 1021.4

Category: Means of Egress
Subject: Number of Exits

Code Text: *Three exits, or exit access stairways or ramps providing access to exits at other stories, shall be provided from any story or occupied roof with an occupant load of 501-1,000. Four exits, or exit access stairways or ramps providing access to exits at other stories, shall be provided from any story or occupied roof with an occupant load greater than 1,000.*

Discussion and Commentary: Although two exits per story is an acceptable minimum for most buildings, those stories with larger occupant loads (greater than 500) must be provided with at least three exits. Four exits are mandated where the occupant load exceeds 1,000. Under no circumstances does the IBC require more than four exits from any story based upon the number of persons present. It must be noted, however, that additional exits will sometimes be required to satisfy other egress requirements of Chapter 10, such as those addressing travel distance.

Although three or four exits may be required from a story due to occupant load, all occupants on the story may not necessarily need to be provided with access to all of the exits if the three conditions of Exception 7 to Section 1021.2 are met.

Topic: Construction
Reference: IBC 1022.2
Category: Means of Egress
Subject: Interior Exit Stairways and Ramps

Code Text: *Enclosures for interior exit stairways and ramps shall be constructed as fire barriers in accordance with Section 707 or horizontal assemblies constructed in accordance with Section 711, or both. Interior exit stairway and ramp enclosures shall have a fire-resistance rating of not less than 2 hours where connecting four stories or more and not less than 1 hour where connecting less than four stories. Interior exit stairways and ramps shall have a fire-resistance rating not less than the floor assembly penetrated, but need not exceed 2 hours.*

Discussion and Commentary: Vertical openings created for stairways must typically be enclosed with fire-resistance-rated construction. The presence of openings into and penetrations through interior exit stairways and ramps is strictly limited due to the high-level nature of the egress elements. Ventilation of an interior stairway or ramp is also highly regulated. The integrity of the means of egress system must be maintained until arrival at the exit discharge or public way.

Interior exit stairways and ramps must always be enclosed with fire-resistance-rated construction. Allowances for unenclosed stairways within the means of egress are established in Section 1009.3 under the provisions for exit access stairways.

Topic: Interior Stairway Exterior Walls
Reference: IBC 1022.7
Category: Means of Egress
Subject: Interior Exit Stairways and Ramps

Code Text: *Where nonrated walls or unprotected openings enclose the exterior of the stairway and the walls or openings are exposed to other parts of the building at an angle of less than 180 degrees (3.14 rad), the building exterior walls within 10 feet (3048 mm) horizontally of a nonrated wall or unprotected opening have a fire-resistance rating of not less than 1 hour. Openings within such exterior walls shall be protected by opening protectives having a fire protection rating of not less than $3/4$ hour. This construction shall extend vertically from the ground to a point 10 feet (3048 mm) above the topmost landing of the stairway or to the roof line, which ever is lower.*

Discussion and Commentary: Unless regulated according to fire separation distance or type of construction, the exterior wall of an interior exit stairway or ramp usually needs no fire-resistance rating. However, where exposure is possible from other exterior walls of the building in close proximity to the enclosure, a limited degree of fire separation is necessary.

1 foot = 304.8 mm, 1 degree = 0.01745 rad.

An alternative to the protection of exterior walls adjacent to an interior exit stairway or ramp is the protection of the exterior wall of the enclosure itself. Should a fire breach an adjacent exterior wall, its penetration of the stairway or ramp enclosure would be halted for an acceptable time period.

Topic: Smokeproof Enclosures
Reference: IBC 1022.10
Category: Means of Egress
Subject: Interior Exit Stairways and Ramps

Code Text: *Where required by Section 403.5.4 (high-rise buildings) or 405.7.2 (underground buildings), interior exit stairways and ramps shall be smokeproof enclosures or pressurized stairways or ramps in accordance with Section 909.20. Access to the stairway within a smokeproof enclosure shall be by way of a vestibule or an open exterior balcony.* See exception for stairways using the pressurization alternative complying with Section 909.20.5.

Discussion and Commentary: In those buildings where vertical egress travel is extensive, an additional level of protection is mandated, primarily to address the hazard of smoke and toxic gases that are produced in a fire. Through ventilation or pressurization, the potential for smoke and gases to enter the enclosure is dramatically reduced.

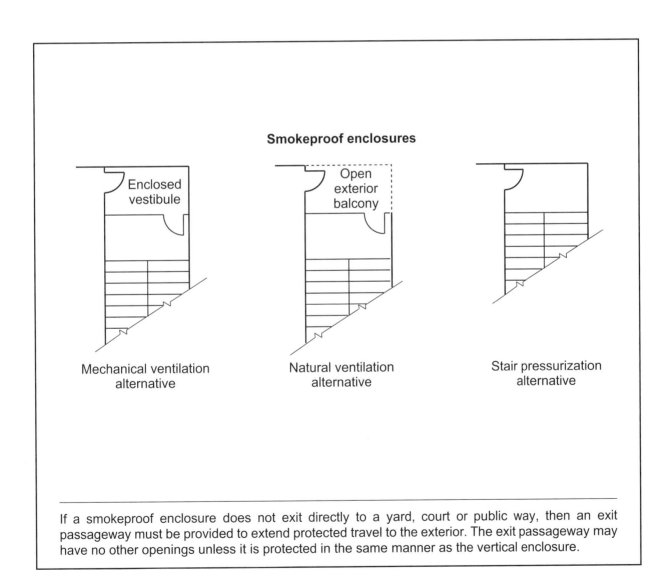

Topic: Construction
Reference: IBC 1023.1, 1023.3
Category: Means of Egress
Subject: Exit Passageways

Code Text: *An exit passageway shall not be used for any purpose other than a means of egress. Exit passageway enclosures shall have walls, floors and ceilings of not less than 1-hour fire-resistance rating, and not less than that required for any connecting interior exit stairway or ramp. Exit passageways shall be constructed as fire barriers in accordance with Section 707, or horizontal assemblies constructed in accordance with Section 711, or both.*

Discussion and Commentary: An exit passageway is defined as *an exit component that is separated from all other interior spaces of a building or structure by fire-resistance-rated construction and opening protectives, and provides for a protected path of egress travel in a horizontal direction to an exit or to the exit discharge.* It is an egress component of a higher level than a fire-resistance-rated corridor, based primarily on limited openings and penetrations, and increased fire ratings.

Once in an exit passageway, the building occupant is considered to be in a relatively safe location; thus, travel distances within the exit passageway are unregulated. Simply put, an exit passageway is a horizontal exit enclosure, with conditions and limitations similar to those required for an interior exit stairway.

Topic: Openings and Penetrations
Reference: IBC 1022.4, 1023.5
Category: Means of Egress
Subject: Exits

Code Text: *Except as permitted in Section 402.8.7 (exit passageways in covered mall buildings), openings in interior exit stairways and ramps (and exit passageways) other than unprotected exterior openings shall be limited to those necessary for exit access to the enclosure from normally occupied spaces and for egress from the enclosure (or exit passageway). Penetrations into and openings through interior exit stairways and ramps are prohibited except for required exit doors, equipment and duct work necessary for independent pressurization, sprinkler piping, standpipes, electrical raceway serving the fire department communication and electrical conduit serving the interior exit stairway or ramp (or exit passageway) and terminating at a steel box not exceeding 16 square inches (0.010 m^2).*

Discussion and Commentary: Given the importance of an exit enclosure in the means of egress system, no unnecessary openings or penetrations are permitted to breach the fire-resistant separation. The provisions are essentially the same for both interior exit stairways and exit passageways.

Elevation

Plan

Enclosure construction:
- Four or more stories—2-hour fire resistance
- Less than four stories—1-hour fire resistance

Openings and penetrations:
- Permitted exterior openings (705)
- Egress from normally occupied spaces
- Egress from enclosure
- Standpipe and sprinklers
- Ductwork for independent pressurization
- Limited electrical conduit

Doors: (716)
- Self-closing or automatic closing
- 1-hour rating in 1-hour construction
- 1$^1/_2$-hour rating in 2-hour construction
- Temperature rise limit of 450°F above ambient

Several methods are set forth in the code to provide for ventilation of an exit enclosure. In general, penetrations for ductwork must enter directly from the building's exterior or from an interior space separated from the remainder of the building by a shaft enclosure.

Topic: General Provisions
Reference: IBC 1024.1

Category: Means of Egress
Subject: Luminous Egress Path Markings

Code Text: *Approved luminous egress path markings delineating the exit path shall be provided in high-rise buildings of Groups A, B, E, I, M and R-1 occupancies in accordance with Sections 1024.1 through 1024.5. See exception for lobbies on the level of exit discharge.*

Discussion and Commentary: Photoluminescent or self-luminous materials are required in interior exit stairways and exit passageways of certain high-rise buildings in order to delineate the exit path. Improved safety for individuals negotiating stairs during the extended egress required in a high-rise building is provided by improving the visibility of stair treads and handrails under normal and emergency conditions.

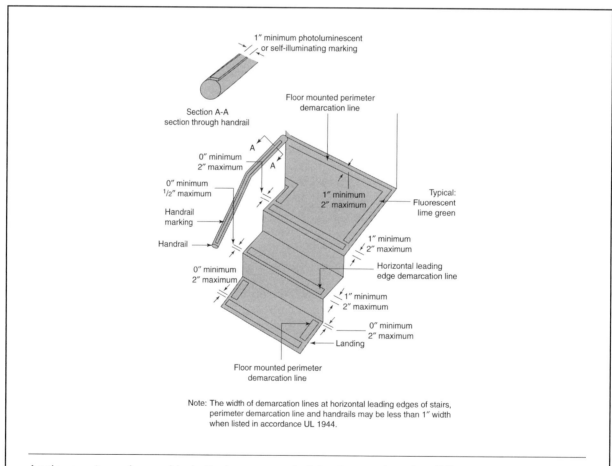

Analogous to rechargeable batteries, many photoluminescent and self-illuminating egress path markings require exposure to light to perform properly. Thus, such markings must be exposed to a minimum of 1 foot-candle of light energy at the walking surface for at least 60 minutes prior to the building being occupied.

Topic: General
Reference: IBC 1025
Category: Means of Egress
Subject: Horizontal Exits

Code Text: *A horizontal exit shall not serve as the only exit from a portion of a building, and where two or more exits are required, not more than one-half of the total number of exits or total width of exit width shall be horizontal exits.* See exceptions for Group I-2 and I-3 occupancies. *The refuge area of a horizontal exit shall be a space occupied by the same tenant or public areas and each such area of refuge shall be adequate to house the original occupant load of the refuge area plus the occupant load anticipated from the adjoining compartment.*

Discussion and Commentary: A horizontal exit is defined as *a path of egress travel from one building to an area in another building on approximately the same level, or a path of egress travel through or around a wall or partition to an area on approximately the same level in the same building, which affords safety from fire and smoke from the area of incidence and areas communicating therewith.* Constructed as a fire wall or a minimum 2-hour fire barrier, a horizontal exit is an exit component of the means of egress system.

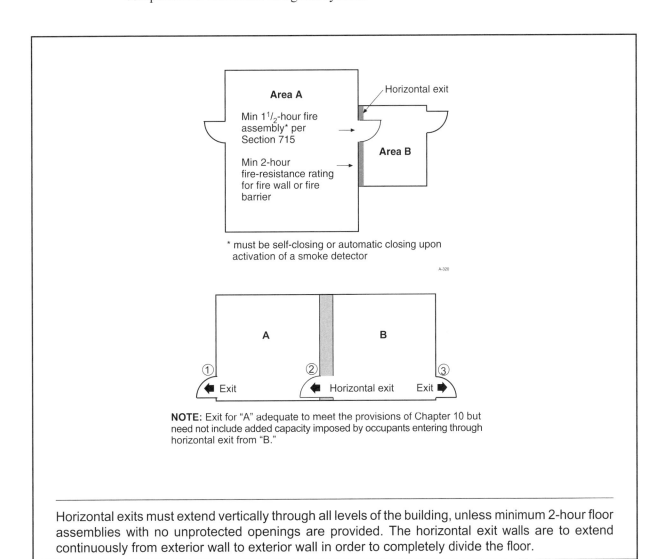

Horizontal exits must extend vertically through all levels of the building, unless minimum 2-hour floor assemblies with no unprotected openings are provided. The horizontal exit walls are to extend continuously from exterior wall to exterior wall in order to completely divide the floor.

Topic: Open Side
Reference: IBC 1026.3

Category: Means of Egress
Subject: Exterior Exit Ramps and Stairways

Code Text: *Exterior exit ramps and stairways serving as an element of a required means of egress shall be open on at least one side. An open side shall have a minimum of 35 square feet (3.3 m²) of aggregate open area adjacent to each floor level and the level of each intermediate landing. The required open area shall be located not less than 42 inches (1067 mm) above the adjacent floor or landing level.*

Discussion and Commentary: For a stairway or ramp to be considered exterior, it must be open enough to the outside so that smoke and toxic gases will not tend to corrupt the exit route. An exterior exit ramp or exterior exit stairway is considered an exit component and is permitted as an egress element in all occupancies except Group I-2. Where permitted as an element of a required means of egress, an exterior exit stairway is limited to buildings with a maximum of 6 stories and no more than 75 feet in height above the lowest level of fire department vehicle access.

For SI: 1 inch = 25.4 mm, 1 foot = 304.8 mm, 1 square foot = 0.093 m².

Consistent with the requirements for other exit components, an exterior exit ramp or stairway must be separated from the remainder of the building by fire-resistance-rated construction and protected openings. The IBC provides four exceptions where separation is not warranted.

Topic: Definition and Scope
Reference: IBC 1027.1, 1027.3, 202

Category: Means of Egress
Subject: Exit Discharge

Code Text: The exit discharge is *that portion of a means of egress system between the termination of an exit and a public way.* Exits shall discharge directly to the exterior of the building. See exceptions for discharge level spaces and vestibules. *The exit discharge shall be at grade or shall provide direct access to grade. The exit discharge shall not reenter a building. Exit discharge components shall be sufficiently open to the exterior so as to minimize the accumulation of smoke and toxic gases.*

Discussion and Commentary: Exit discharge travel typically takes place outside of the building, where hazards to the occupants are greatly reduced. Although concern over the accumulation of smoke and toxic gases is eliminated, there is still a risk to the occupants, which is eliminated only at a point of considerable distance from the structure, generally the public way.

Number of exits maintained until arrival at grade or public way

When specific conditions are met, up to 50 percent of the number and capacity of interior exit stairways may exit through a vestibule or an area on the discharge level, provided all the stated conditions have been met.

Topic: Egress Courts
Reference: IBC 1027.4

Category: Means of Egress
Subject: Exit Discharge

Code Text: *Where an egress court serving a building or portion thereof is less than 10 feet (3048 mm) in width, the egress court walls shall be not less than 1-hour fire-resistance-rated construction for a distance of 10 feet (3048 mm) above the floor of the court. Openings within such walls shall be protected by opening protectives having a fire protection rating of not less than $^3/_4$ hour. See exceptions for small occupant loads and Group R-3 occupancies.*

Discussion and Commentary: An egress court is defined as a *court or yard which provides access to a public way for one or more exits*. Because an egress court is an element of the exit discharge, occupants must be afforded sufficient protection from a fire within the structure to be reasonably sure that once outside they will reach the safety of a public way. Therefore, fire-resistance-rated exterior walls and openings must be provided where such occupants must travel adjacent to an exterior wall in order to reach the public way.

For SI: 1 foot = 304.8 mm

The minimum required width of an egress court is addressed in a manner similar to that of aisles, corridors and stairways. The width must accommodate the calculated capacity, based on occupant load served; however, in no case may it be less than a specified width of 44" (36" in Group R-3).

Topic: Access to a Public Way
Reference: IBC 1027.6
Category: Means of Egress
Subject: Exit Discharge

Code Text: *The exit discharge shall provide a direct and unobstructed access to a public way.* See exception for designation of a safe dispersal area where access to a public way cannot be provided.

Discussion and Commentary: The means of egress is not complete until the occupants of the building have reached a safe place. Such a place is typically a public way, in that it is relatively unobstructed, and more importantly, continuous. It is always possible to continue egress travel along a public way until the necessary level of safety is achieved. The path to reach the public way, as for all other components of the means of egress, must be free of obstructions and other concerns that would cause the occupants' egress travel to be delayed, restricted or unavailable. It is important that the travel path outside of the building be continuously maintained in order to keep the means of egress in a complying condition.

For SI: 1 foot = 304.8 mm, 1 square foot = 0.0929 m².

Safe dispersal areas

Occasionally it is impractical, and at times impossible, to provide a fully complying means of egress the entire distance to the public way. A safe dispersal area can be utilized where such conditions exist, such as on large industrial or educational campus sites.

Study Session 12

Topic: Assembly Main Exit
Reference: IBC 1028.2
Category: Means of Egress
Subject: Assembly Seating

Code Text: *In a building, room or space used for assembly purposes that has an occupant load of greater than 300 and is provided with a main exit, the main exit shall be of sufficient width to accommodate not less than one-half of the occupant load, but such width shall not be less than the total required width of all means of egress leading to the exit.*

Discussion and Commentary: In most assembly-type uses, the occupants tend to enter the room or space at a single location. It is expected that under emergency conditions, most of the occupants will attempt to exit at the same point. Therefore, the main entrance/exit must be wide enough to handle a sizeable percentage of the occupants. If there is no well-defined main exit or where multiple main entrance/exits are provided, the required exit width can be distributed among the exits around the perimeter of the building.

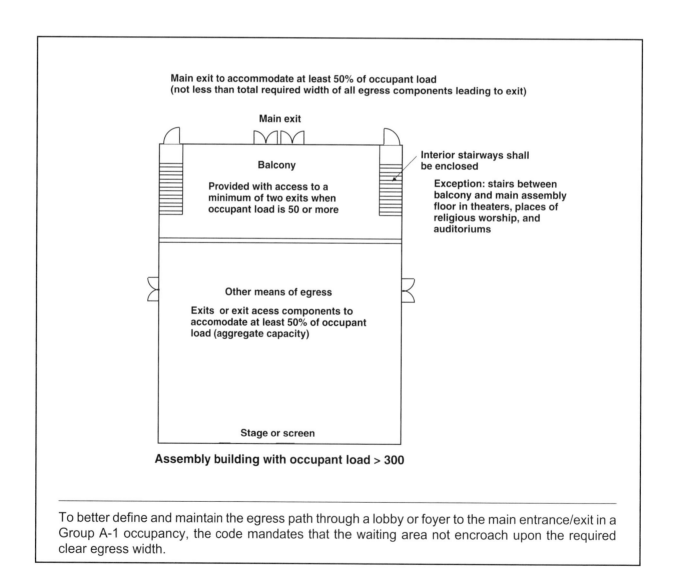

Assembly building with occupant load > 300

To better define and maintain the egress path through a lobby or foyer to the main entrance/exit in a Group A-1 occupancy, the code mandates that the waiting area not encroach upon the required clear egress width.

Topic: Seating at Tables
Reference: IBC 1028.10.1.1, 1028.10.1.2
Category: Means of Egress
Subject: Exit Access

Code Text: *Aisle accessways serving arrangements of seating at tables or counters shall have sufficient clear width to conform to the capacity requirements of Section 1005.1, but shall not have less than a minimum of 12 inches (305 mm) plus $^1/_2$ inch (12.7 mm) of width for each additional 1 foot (305 mm), or fraction thereof, beyond 12 feet (3658 mm) of aisle accessway length measured from the center of the seat farthest from an aisle.* See exception for aisle accessways of limited lengths and occupant loads. *The length of travel along the aisle accessway shall not exceed 30 feet (9144 mm) from any seat to the point where a person has a choice of two or more paths of egress travel to separate exits.*

Discussion and Commentary: To facilitate progress toward an established aisle, it is important that a minimum degree of egress width be established within areas furnished with tables and chairs. An aisle accessway, defined as *that portion of an exit access that leads to an aisle,* is thus regulated.

For SI: 1 inch = 25.4 mm.

The method of determining the clear width differs based on the type of seating that is provided. For fixed seats, the measurement is made from the back of the seats. Otherwise, the clear width is measured to a line 19 inches from the edge of the table or counter.

Study Session 12

Topic: Assembly Aisle Walking Surfaces
Reference: IBC 1028.11
Category: Means of Egress
Subject: Assembly Seating

Code Text: *Aisles with a slope not exceeding one unit vertical in eight units horizontal (12.5-percent slope) shall consist of a ramp having a slip-resistant walking surface. Aisles with a slope exceeding one unit vertical in eight units horizontal (12.5-percent slope) shall consist of a series of risers and treads that extends across the full width of aisles and complies with Sections 1028.11.1 through 1028.11.3. Tread depths shall be a minimum of 11 inches (279 mm) and shall have dimensional uniformity.* See exception allowing a 0.188 inch tolerance between adjacent treads. *Where the gradient of aisle stairs is to be the same as the gradient of adjoining seating areas, the riser height shall not be less than 4 inches (102 mm) nor more than 8 inches (203 mm) and shall be uniform within each flight.* See exceptions allowing riser height nonuniformity and maximum riser heights of 9 inches where needed to maintain sightlines.

Discussion and Commentary: Where aisles have a relatively flat slope, 1:8 or flatter, steps in the aisle are prohibited because occupants in low-slope aisles have a tendency to not notice steps as readily as they would in steeper aisles. Continuous surfaces provide a safer path of travel.

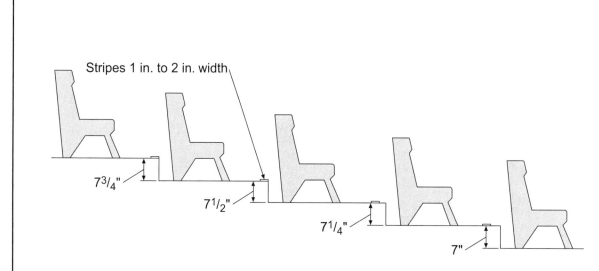

One provision that helps the user of an aisle notice a step requires that a contrasting strip or other approved marking be placed on the leading edge of each tread. Designed to identify the edge of the tread, this making strip may be omitted where it can be shown that the location of each tread is readily apparent when viewed in descent.

Topic: Handrails
Reference: IBC 1028.13
Category: Means of Egress
Subject: Assembly Seating

Code Text: *Ramped aisles having a slope exceeding one unit vertical in 15 units horizontal (6.7-percent slope) and aisle stairs shall be provided with handrails in compliance with Section 1012 located either at one or both sides of the aisle or within the aisle width. See exceptions for 1) ramped aisles with a slope no greater than 1:8 with seating on both sides, 2) guards that comply with the graspability requirements of handrails and 3) where crossovers are permitted within the aisles. Where there is seating on both sides of the aisle, the handrails shall be discontinuous with gaps or breaks at intervals not exceeding five rows to facilitate access to seating and to permit crossing from one side of the aisle to the other.*

Discussion and Commentary: Where seating is located on both sides of an aisle, the required handrails may be placed either on both sides of, or down the center of, the aisle served. Where seating is located on only one side of the aisle, a handrail is only required on one of the sides.

Where discontinuous handrails are provided, an intermediate handrail located 12 inches below the main handrail is required to prevent users from ducking under the handrail and hindering flow. It also provide a handrail for toddlers who may be using the aisle.

Topic: Required Openings
Reference: IBC 1029

Category: Means of Egress
Subject: Emergency Escape and Rescue

Code Text: *In addition to the means of egress required by Chapter 10, provisions shall be made for emergency escape and rescue in Group R-2 occupancies in accordance with Table 1021.2(1) and Table 1021.2(2) and Group R-3 occupancies. Basements and sleeping rooms below the fourth story above grade plane shall have at least one exterior emergency escape and rescue opening in accordance with Section 1029.* See three exceptions, including one for basements with a ceiling height of less than 80 inches and one for small basements without habitable spaces. *Such openings shall open directly into a public way or a yard or court that opens to a public way.*

Discussion and Commentary: In those occupancies where persons are sometimes sleeping, a fire will often spread quickly and block the normal egress routes. By requiring a sizeable opening directly from the sleeping room to the exterior, rescue can be more easily accomplished, or alternatively, the occupants may escape without having to travel through the building.

When operable windows are used for egress or rescue purposes, the intent is that they be double-hung, horizontal sliding or casement styles operated by a simple operation. Special types other than those listed must be evaluated for compliance with the operational constraint limitations.

Topic: Window Wells
Reference: IBC 1029.5
Category: Means of Egress
Subject: Emergency Escape and Rescue

Code Text: *An emergency escape and rescue opening with a finished sill height below the adjacent ground level shall be provided with a window well in accordance with Section 1029.5.1 and 1029.5.2. The minimum horizontal area of the window well shall be 9 square feet (0.84 m²), with a minimum dimension of 36 inches (914 mm). The area of the window well shall allow the emergency escape and rescue opening to be fully opened. Window wells with a vertical depth of more than 44 inches (1118 mm) shall be equipped with an approved permanently affixed ladder or steps.*

Discussion and Commentary: It is important that persons who travel through an emergency escape and rescue opening located below grade be provided with adequate space in the window well to allow for escape or rescue from the area. Therefore, a complying window well is mandated that provides an adequate cross-sectional area for escape and rescue operations.

For SI: 1 inch = 25.4 mm, 1 square foot = 0.093 m².

Window well ladders must have a minimum clear rung width of 12 inches with the rungs spaced at maximum 18-inch intervals vertically. The ladder or steps cannot encroach into the required dimensions of the window well more than 6 inches.

Quiz

Study Session 12
IBC Sections 1020 – 1029

1. A story having an occupant load of 720 persons shall be provided with a minimum of _____ exits or exit access stairways or ramps providing access to exits at other stories.

    a. two   b. three

    c. four   d. five

    Reference_____

2. A single exit is permitted from what type of parking garage?

    a. open   b. enclosed

    c. commercial trucks and buses   d. vehicles are mechanically parked

    Reference_____

3. A single exit from the second story of a nonsprinklered Group B office building is permitted if the maximum occupant load is _____ persons and the maximum travel distance is _____ feet.

    a. 10, 75   b. 29, 75

    c. 29, 100   d. 49, 200

    Reference_____

4. Enclosures for interior exit stairways connecting a minimum of _____ stories shall be 2-hour fire-resistance rated.

   a. three
   b. four
   c. five
   d. six

   Reference_____

5. A basement provided with one exit shall be located a maximum of _____ story (stories) below grade plane.

   a. 0, all basements require a minimum of two exits
   b. one
   c. two
   d. three

   Reference_____

6. Where a smokeproof enclosure is extended by an exit passageway, the exit passageway shall be separated from the remainder of the building by minimum _____ -hour fire-resistance-rated construction.

   a. ³/₄
   b. 1
   c. 2
   d. 3

   Reference_____

7. Stairway floor number signs shall be provided in interior exit stairways and ramps connecting a minimum of _____ stories.

   a. 2
   b. 3
   c. 4
   d. 6

   Reference_____

8. Smokeproof enclosures or pressurized stairways or ramps are only required in _____ buildings and _____ buildings.

   a. high-rise, underground
   b. high-rise, Group I-2
   c. Group I-2, Group I-3
   d. atrium, underground

   Reference_____

9. Exit passageways shall be constructed as minimum _____ .
   a. 1-hour fire partitions   b. 1-hour fire barriers
   c. 2-hour fire barriers   d. 2-hour fire walls

   Reference_____

10. In other than Group I-2 and I-3 occupancies, a refuge area serving a horizontal exit shall be sized for capacity based on _____ square feet for each occupant to be accommodated.
    a. 3   b. 6
    c. 15   d. 30

    Reference_____

11. In other than a Group I-2 occupancy, exterior exit stairways are permitted as means of egress elements for non-high-rise buildings a maximum of _____ stories above grade plane.
    a. three   b. four
    c. five   d. six

    Reference_____

12. An exterior exit stairway serving as an element of a required means of egress must be open on at least one side, with such open area a minimum of _____ square feet and located a minimum of _____ inches above the adjacent floor or landing level.
    a. 9, 36   b. 16, 42
    c. 20, 36   d. 35, 42

    Reference_____

13. Where a safe dispersal area is provided in lieu of access to a public way, the area shall be of a size to accommodate a minimum of _____ square feet for each person.
    a. 3   b. 5
    c. 7   d. 15

    Reference_____

14. What is the minimum required width of an egress court serving 40 occupants from an office building?

    a. 36 inches
    b. 44 inches
    c. 48 inches
    d. 60 inches

    Reference_____

15. Where seating is located at a table and is adjacent to an aisle, the measurement of required clear width of the aisle shall be made to a line _____ inches away from and parallel to the edge of the table.

    a. 15
    b. 19
    c. 21
    d. 24

    Reference_____

16. Where an assembly space with a main exit has a minimum occupant load of _____ persons, the main exit shall be of sufficient width to accommodate at least one-half of the occupant load.

    a. 51
    b. 301
    c. 501
    d. 1,001

    Reference_____

17. At least two means of egress shall be provided from assembly seating areas in balconies having a minimum seating capacity of _____ persons.

    a. 10
    b. 30
    c. 50
    d. 100

    Reference_____

18. In assembly areas not provided with smoke protection, the minimum width for stairs having 8-inch risers is based on a minimum of _____ inch of width for each occupant served.

    a. 0.2
    b. 0.22
    c. 0.3
    d. 0.35

    Reference_____

Study Session 12

19. In a 15,000-seat arena provided with smoke-protected assembly seating, the minimum width for a 1:12 ramped aisle is based on a minimum of _____ inch of width for each occupant served.

    a. 0.300
    b. 0.200
    c. 0.120
    d. 0.070

    Reference_____

20. The smoke level in a means of egress serving a smoke-protected assembly seating area is intended to be maintained a minimum of _____ feet above the floor of the means of egress.

    a. 6
    b. 8
    c. 12
    d. 15

    Reference_____

21. What is the maximum travel distance for seating in open-air assembly structures of Type I or II construction?

    a. 200 feet
    b. 300 feet
    c. 400 feet
    d. unlimited

    Reference_____

22. What is the minimum required clear width of aisle stairs having seating on only one side?

    a. 32 inches
    b. 36 inches
    c. 42 inches
    d. 44 inches

    Reference_____

23. In aisle stairs of assembly seating areas, the maximum tolerance permitted between adjacent treads is _____ inch.

    a. $1/8$
    b. $3/16$
    c. $1/4$
    d. $3/8$

    Reference_____

24. Where the foot of the aisle is more than 30 inches above the floor or grade below, a railing or fascia system at the foot of aisles in an assembly occupancy shall be a minimum of _____ in height.

   a. 26 inches
   b. 32 inches
   c. 36 inches
   d. 42 inches

   Reference_____

25. What is the minimum net clear opening required for emergency escape and rescue grade floor openings?

   a. 20 inches by 22 inches
   b. 20 inches by 24 inches
   c. 5.0 square feet
   d. 5.7 square feet

   Reference_____

26. Where horizontal exits are used in the means of egress in a Group I-2 hospital, the horizontal exits can be used for a maximum of _____ of the required exits.

   a. one
   b. two
   c. one-half
   d. two-thirds

   Reference_____

27. Along with other limitations, the exit discharge may occur through a vestibule, provided the vestibule has a maximum depth from the exterior of the building of _____ feet and a maximum length of _____ feet.

   a. 8, 15
   b. 10, 20
   c. 10, 30
   d. 15, 30

   Reference_____

28. Egress need not extend to a public way where a safe dispersal area is provided a minimum of _____ feet from the building.

   a. 30
   b. 50
   c. 75
   d. 100

   Reference_____

Study Session 12

29. An emergency escape and rescue opening shall be located so that the bottom of the clear opening is a maximum of _____ inches above the floor surface.

    a. 36   b. 42
    c. 44   d. 48

    Reference_____

30. Where a window well is provided to serve an emergency escape and rescue opening, it shall have a minimum horizontal area of _____ square feet.

    a. 5.0   b. 5.7
    c. 9.0   d. 10.0

    Reference_____

31. The separation between refuge areas connected by a horizontal exit will be provided by a fire wall or fire barrier having a minimum fire-resistance rating of _____ hour(s).

    a. 1   b. 2
    c. 3   d. 4

    Reference_____

32. Where seating rows in an assembly seating area have 14 or fewer seats, the clear aisle accessway width shall be a minimum of _____ inches measured horizontally from the back of the row ahead to the nearest projection of the row behind.

    a. 12   b. 16
    c. 19   d. 22

    Reference_____

33. Where there is assembly seating on both sides of an aisle, the handrails shall be discontinuous with gaps or breaks a minimum of _____ inches and a maximum of _____ inches in width, measured horizontally.

    a. 16, 32   b. 20, 34
    c. 22, 36   d. 27, 42

    Reference_____

34. An emergency escape and rescue opening is not required from a basement without habitable spaces, provided the floor area of the basement is a maximum of _____ square feet.

   a. 120   b. 200
   c. 240   d. 400

   Reference _____

35. An emergency escape and rescue opening shall be a minimum of _____ inches in net clear opening height and a minimum of _____ inches in net clear opening width.

   a. 24, 20   b. 24, 22
   c. 20, 26   d. 20, 28

   Reference _____

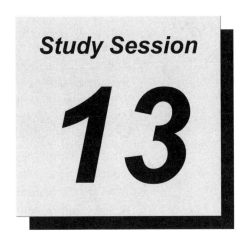

Study Session 13

2012 IBC Chapter 11
Accessibility

OBJECTIVE: To become familiar with the scoping provisions relating to the design and construction of accessible buildings, facilities and elements.

REFERENCE: Chapter 11, 2012 *International Building Code*

KEY POINTS:
- What is the scope of the provisions regulating accessibility and usability?
- How does ICC A117.1 relate to the IBC?
- Are temporary buildings regulated for accessibility?
- How are employee work areas viewed for accessibility?
- Which types of uses are not required to be accessible?
- Where within employee work areas is an accessible route not required?
- What is an accessible route? Which site elements must be connected by an accessible route to an accessible building entrance?
- Which areas within a building must be connected by an accessible route?
- Under which conditions is a vertical accessible route not required?
- How should the accessible route be provided in relationship to the general circulation path?
- How many entrances to a building must be accessible? To a tenant space within a building?
- What percentage of parking spaces must be designed as accessible?
- When is a van-accessible space required?
- Where must accessible parking spaces be located?
- When is an accessible loading zone required?
- What is a Type A dwelling unit? Type B unit? In which types of buildings are such units located?
- Where dwelling units and sleeping units are required to be accessible, what accessible features are required in each Accessible unit, Type A unit and Type B unit?
- In a theater, auditorium or similar assembly area, how is the minimum number of required wheelchair spaces determined?
- How shall wheelchair spaces be distributed throughout a multilevel assembly facility?

KEY POINTS:
(Cont'd)
- What is the function of an assistive listening system? Under which conditions are assistive listening systems required?
- Under what conditions must prerecorded or real-time captions of audible public announcements be provided?
- Which types of dining areas must be accessible?
- What specific areas of a judicial facility must be accessible?
- What percentage of toilet rooms are required to be accessible? Bathing facilities?
- What are family or assisted-use toilet rooms and bathing rooms? When are such rooms required?
- What are the features required in a family or assisted-use toilet room? Bathing room?
- Where drinking fountains are provided, how many must be accessible? Storage lockers? Fitting rooms? Check-out aisles? Saunas and steam rooms?
- Under what conditions is a platform (wheelchair) lift permitted to be a part of a required accessible route?
- Where are detectable warnings required?
- What type of operating mechanisms or controls are regulated for usability?
- When must a stairway be considered an accessible element?
- What is the International Symbol of Accessibility? Where are such signs required?
- Where is directional signage mandated?

Topic: Scope
Reference: IBC 1101
Category: Accessibility
Subject: General Provisions

Code Text: *The provisions of Chapter 11 shall control the design and construction of facilities for accessibility to physically disabled persons. Buildings and facilities shall be designed and constructed to be accessible in accordance with the IBC and ICC A117.1.*

Discussion and Commentary: Chapter 11 of the *International Building Code* sets forth the scoping provisions that identify where and to what degree access must be provided. Once it has been determined that accessible elements are required, the ICC design standard *Accessible and Usable Buildings and Facilities* sets forth the specific technical criteria. As with any other provision of the code, alternative designs, products or technologies that provide equivalent or superior compliance may be accepted by the building official.

Scope
The provisions of Chapter 11 shall control the design and construction of facilities for accessibility to physically disabled persons.

Design
Buildings and facilities shall be designed and constructed in accordance with the IBC and the ICC A117.1-2009 standard.

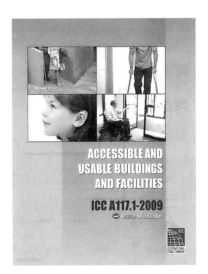

Although space requirements can vary greatly depending on the nature of the disability and the physical functions of the individual, it is generally accepted that spaces designed to accommodate persons using wheelchairs will be functional for most people.

Topic: Where Required
Reference: IBC 1103

Category: Accessibility
Subject: Scoping Requirements

Code Text: *Sites, buildings, structures, facilities, elements and spaces, temporary or permanent, shall be accessible to persons with physical disabilities.* See 15 general exceptions.

Discussion and Commentary: In general, all portions of all buildings are to be provided with elements that will make them fully accessible to individuals with disabilities. There are, however, a number of general exceptions that reduce or eliminate accessibility requirements. Specific areas that are not required to be accessible include: individual employee work areas; detached dwellings and their accessory structures; construction sites; raised security or safety areas, such as observation galleries or fire towers; nonoccupiable spaces and equipment spaces, including elevator pits and transformer vaults; and single-occupant structures accessed at other than grade, such as toll booths.

Specific requirements. Where not required per Sections 1104 through 1110.

Existing buildings. Existing buildings shall comply with Section 3411.

Employee work areas. Need only comply with fire alarm, accessible means of egress and common use circulation path provisions. Must be able to approach, enter and exit the work area (small raised work areas totally exempted).

Detached dwellings. Detached one- and two-family dwellings and accessory structures, and their associated sites and facilities.

Utility buildings. Group U are exempt except:
1. In agricultural buildings, access is required to paved work areas and areas open to the general public.
2. Private garages or carports that contain required accessible parking.

Construction sites. Structures, sites and equipment directly associated with the actual processes of construction.

Raised areas. Raised areas used primarily for purpose of security, life safety or fire safety.

Limited access spaces. Nonoccupiable spaces accessed only by ladders, catwalks, crawl spaces, freight elevators, very narrow passageways or tunnels.

Equipment spaces. Spaces frequented only by personnel for maintenance, repair or monitoring of equipment.

Single occupant structures. Single occupant structures accessed only by passageways below grade or elevated above grade.

Residential Group R-1. Buildings of Group R-1 containing not more than five sleeping units for rent or hire which are also occupied as the residence of the proprietor.

Day care facilities. Where part of a dwelling unit.

Live/work units. Nonresidential area to be accessible, while residential area only regulated by Sections 1107.6.2 and 1107.7.

Detention and correctional facilities. Common use areas not serving accessible cells.

Walk-in coolers and freezers. Coolers and freezers intended for employee use only.

Other than those residential occupancies exempt from the accessibility provisions, most buildings will require some level of accessibility. Only those specific areas identified by the code are exempt, whereas the remainder of the structure is regulated for complying accessibility and usability.

Topic: Connected Spaces
Reference: IBC 1104.3
Category: Accessibility
Subject: Accessible Route

Code Text: *When a building, or portion of a building, is required to be accessible, an accessible route shall be provided to each portion of the building, to accessible building entrances connecting accessible pedestrian walkways and the public way.* See exceptions for fixed-seating assembly areas and sleeping unit doors in Group I-2 occupancies.

Discussion and Commentary: An accessible route is defined as *a continuous, unobstructed path that complies with Chapter 11.* It potentially includes corridors, aisles, ramps, elevators, platform (wheelchair) lifts and clear floor space at fixtures. The general requirement for the connection of accessible spaces is modified for common use circulation paths within employee work areas and for access to press boxes. Chapter 4 of ICC A117.1 addresses the design and construction specifications for the elements of an accessible route.

Site arrival points and exterior elements on the site must also be provided with an accessible route to an accessible building entrance. The elements addressed may include public transportation stops, accessible passenger loading zones and accessible parking spaces.

Study Session 13

Topic: Multilevel Buildings
Reference: IBC 1104.4
Category: Accessibility
Subject: Accessible Route

Code Text: *At least one accessible route shall connect each accessible level, including mezzanines, in multistory buildings and facilities.* See exceptions for specific occupancies with small floor areas, areas without accessible elements, air traffic control towers, two-story buildings with a small occupant load on one story, and employee work stations within courtrooms.

Discussion and Commentary: Access must be provided both horizontally and vertically throughout a building. Under most conditions, multilevel facilities will contain an accessible elevator to extend an accessible route to the other levels. Ramps also can be used where the elevation change is not excessive. In some occupancies, an accessible route is not required for levels above or below an accessible level, provided that the aggregate size of the inaccessible levels does not exceed 3,000 square feet. It has been determined that it is not feasible to require elevator service for such small spaces.

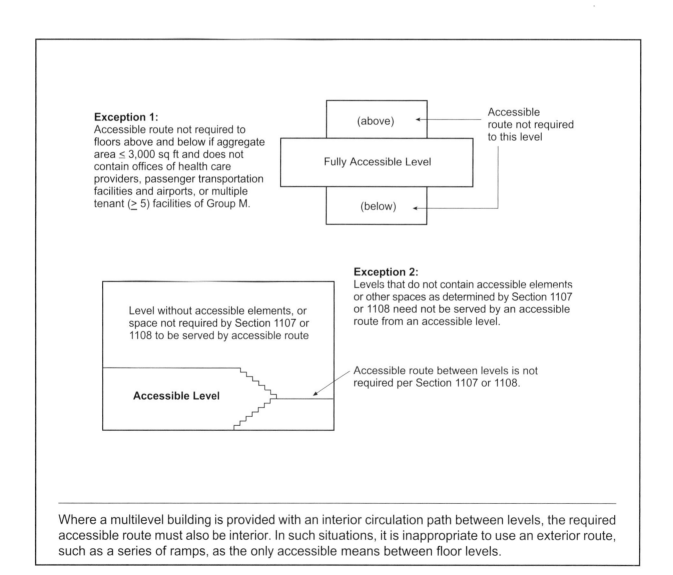

Topic: Where Required
Reference: IBC 1105.1

Category: Accessibility
Subject: Accessible Entrances

Code Text: *In addition to accessible entrances required by Sections 1105.1.1 through 1105.1.6* (parking garage entrances, entrances from tunnels or elevated walkways, restricted entrances, entrances for inmates or detainees, service entrances, and entrances to tenant spaces, dwelling units and sleeping units), *at least 60 percent of all public entrances shall be accessible.* See exceptions for entrances to areas not required to be accessible and loading/service entrances that are not the only tenant space entrance.

Discussion and Commentary: To provide accessibility to all buildings and tenant spaces, a minimum of one accessible entrance is required. Where additional entrances are provided, often for convenience purposes, at least 60 percent of the total number of entrances must be accessible. Elements to be considered at entrances include the slope of exterior surfaces, door hardware and clear floor space for maneuvering clearances.

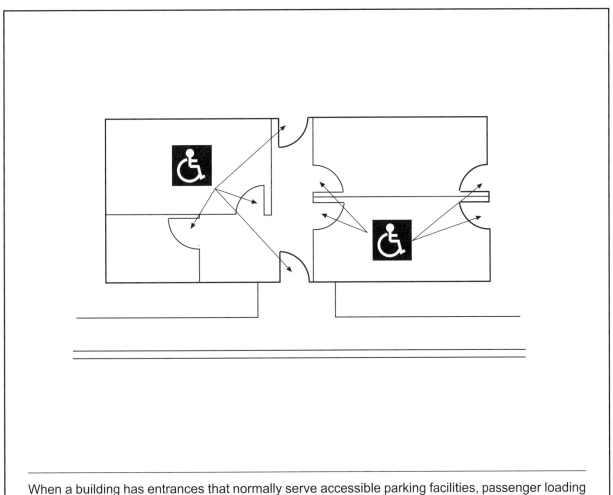

When a building has entrances that normally serve accessible parking facilities, passenger loading zones, public sidewalks and other site elements, then at least one of the entrances serving each of the functions shall comply with the accessible route provisions.

Topic: General Provisions
Reference: IBC 1106
Category: Accessibility
Subject: Parking Facilities

Code Text: *Where parking is provided, accessible parking spaces shall be provided in compliance with Table 1106.1 except as required by Sections 1106.2 (Groups R-2 and R-3), 1106.3, (hospital outpatient facilities) and 1106.4 (rehabilitation facilities). For every six or fraction of six accessible parking spaces, at least one shall be a van-accessible parking space. See exception for private garages serving Groups R-2 and R-3. Accessible parking spaces shall be located on the shortest accessible route of travel from adjacent parking to an accessible building entrance.*

Discussion and Commentary: The number of required accessible parking spaces is based on the total number of spaces in the lot or garage. In hospital outpatient facilities where it is anticipated that more accessible spaces will be needed, at least one in ten parking spaces must be accessible. All accessible parking spaces shall be located as close as possible to an accessible building entrance to reduce the distance of travel for those individuals with mobility limitations.

TABLE 1106.1
ACCESSIBLE PARKING SPACES

TOTAL PARKING SPACES PROVIDED	REQUIRED MINIMUM NUMBER OF ACCESSIBLE SPACES
1 to 25	1
26 to 50	2
51 to 75	3
76 to 100	4
101 to 150	5
151 to 200	6
201 to 300	7
301 to 400	8
401 to 500	9
501 to 1,000	2% of total
1,001 and over	20, plus one for each 100, or fraction thereof, over 1,000

Where parking is provided, accessible parking spaces shall be provided in compliance with Table 1106.1.

Exceptions:

1. Where Group R-2 and R-3 occupancies are required to have accessible dwelling units, 2% of the parking spaces shall be accessible. (Sec. 1106.2)

   Where parking is provided within or beneath a building, accessible parking spaces shall also be provided within or beneath the building.

2. Where care recipient and visitor parking spaces serve hospital outpatient facilities, 10% of the spaces shall be accessible. (Sec. 1106.3)

3. At rehabilitation facilities and outpatient physical therapy facilities, 20% of care recipient and visitor parking spaces shall be accessible.

Every parking facility with at least one accessible parking space must provide for accessible van parking. Based on a percentage of the total number of accessible spaces, the van space or spaces must have a vertical clearance of at least 98 inches and a minimum 8-foot access aisle.

Topic: Group I Occupancies
Reference: IBC 1107.5
Category: Accessibility
Subject: Dwelling Units and Sleeping Units

Code Text: *In Group I-1 occupancies, at least 4 percent, but not less than one, of the dwelling units and sleeping units shall be Accessible units. In nursing homes of Group I-2 occupancies, at least 50 percent but not less than one of each type of the dwelling units and sleeping units shall be Accessible units. In general-purpose hospitals, psychiatric facilities, detoxification facilities and residential care/assisted living facilities of Group I-2 occupancies, at least 10 percent, but not less than one, of the dwelling units and sleeping units shall be Accessible units. In hospitals and rehabilitation facilities of Group I-2 occupancies which specialize in treating conditions that affect mobility, or units within either which specialize in treating conditions that affect mobility, 100 percent of the dwelling and sleeping units shall be Accessible units. In Group I-3 occupancies, at least 2 percent, but not less than one, of the dwelling units and sleeping units shall be Accessible units.*

Discussion and Commentary: In Groups I-1, I-2 and I-3, a percentage of dwelling and sleeping units must be Accessible units as regulated by Section 1002 of ICC A117.1. The percentage of required Accessible units varies based on the anticipated need for such units because of the specifics of the institutional use.

Group I-2 nursing home Accessible units

Where Group I-1 occupancies and Group I-2 nursing homes and hospitals contain four or more dwelling or sleeping units intended to be occupied as a residence, all such units are required to be Type B units. The number of Type B units is permitted to be reduced by the general exceptions of Section 1107.7.

Topic: Group R Occupancies
Reference: IBC 1107.6
Category: Accessibility
Subject: Dwelling Units and Sleeping Units

Code Text: *In Group R-1 occupancies, Accessible dwelling units and sleeping units shall be provided in accordance with Table 1107.6.1.1. In Group R-2 occupancies* (limited to apartment houses, monasteries and convents) *containing more than 20 dwelling units or sleeping units, at least 2 percent but not less than one of the units shall be a Type A unit. Where there are four or more dwelling units or sleeping units intended to be occupied as a residence in a single structure, every dwelling and sleeping unit intended to be occupied as a residence shall be a Type B unit.* See general exceptions in Section 1107.7.

Discussion and Commentary: Accessible, Type A and Type B dwelling units and sleeping units are defined in IBC Section 202 and described in Chapter 10 of ICC A117.1. Accessible units are generally regarded as fully accessible. Type A units provide a considerable degree of accessibility, whereas Type B units are only required to have specific accessible elements. A dwelling unit designed and constructed as a Type B unit is intended to comply with the technical requirements for Fair Housing required by federal law.

In R-2 and R-3 occupancies where there are ≥ 4 dwelling units in a single structure, every unit shall be a Type B dwelling unit. (Type A units may be substituted for Type B.)

In R-2 occupancies containing > 20 dwelling units, at least 2 percent but not less than 1 shall be a Type A dwelling unit.

The general exceptions of Section 1107.7 selectively permit the required number of Type A units and Type B units to be reduced.

Where Group R-2 and R-3 occupancies contain public or common-use areas, such areas must be accessible if they serve accessible dwelling units. Any recreational facilities serving these occupancies must also must be accessible to a limited degree.

Topic: Assembly Area Seating
Reference: IBC 1108.2
Category: Accessibility
Subject: Special Occupancies

Code Text: *In theaters, bleachers, grandstands, stadiums, arenas and other fixed seating assembly areas, accessible wheelchair spaces complying with ICC A117.1 shall be provided in accordance with Sections 1108.2.2.1 through 1108.2.2.4. Wheelchair spaces shall be provided in accordance with Table 1108.2.2.1. In multilevel assembly seating areas, wheelchair spaces shall be provided on the main floor level and on one of each two additional floor or mezzanine levels. See two exceptions where all wheelchair spaces may be located on the main level.*

Discussion and Commentary: The unique features of assembly occupancies dictate special accessibility features. In addition to the requirements for wheelchair spaces and assistive listening devices, the code requires accessible seating throughout all dining areas. Specific provisions address fixed seating at booths and tables, as well as at counters.

In dining areas, the total floor area allotted for seating and tables shall be accessible.

Exception: An accessible route to a mezzanine seating area is not required, provided that the mezzanine contains less than 25 percent of the total area and the same services are provided in the accessible area.

100% of dining area required to be accessible

Mezzanine (see exception)

Under limited conditions, a dining area may have a mezzanine level that is not served by an accessible route. The mezzanine must be limited in size to 25 percent of the total floor area, and the services provided on the mezzanine must be available on the accessible level.

Topic: Assistive Listening Systems
Reference: IBC 1108.2.7
Category: Accessibility
Subject: Special Occupancies

Code Text: *Each building, room or space used for assembly purposes where audible communications are integral to the use of the space shall have an assistive listening system.* See exception for spaces where no audio amplification system is installed. *Receivers shall be provided for assistive listening systems in accordance with Table 1108.2.7.1.* See exceptions for 1) buildings with multiple assembly areas, and 2) assembly areas where all seats are served by an induction loop system.

Discussion and Commentary: These provisions are intended to accommodate people with a hearing impairment. In these assembly areas, audible communication is often integral to the use and full enjoyment of the space. This requirement offers the possibility for individuals with hearing impairments to attend functions in these facilities without having to give advance notice and without disrupting the event in order to have a portable assistive listening system set up and made ready for use.

International symbol of access for hearing loss

TABLE 1108.2.7.1
RECEIVERS FOR ASSISTIVE LISTENING SYSTEMS

CAPACITY OF SEATING IN ASSEMBLY AREAS	MINIMUM REQUIRED NUMBER OF RECEIVERS	MINIMUM NUMBER OF RECEIVERS TO BE HEARING-AID COMPATIBLE
50 or less	2	2
51 to 200	2, plus 1 per 25 seats over 50 seats*	2
201 to 500	2, plus 1 per 25 seats over 50 seats*	1 per 4 receivers*
501 to 1,000	20, plus 1 per 33 seats over 500 seats*	1 per 4 receivers*
1,001 to 2,000	35, plus 1 per 50 seats over 1,000 seats*	1 per 4 receivers*
Over 2,000	55, plus 1 per 100 seats over 2,000 seats*	1 per 4 receivers*

Note: * = or fraction thereof

There are three primary types of listening systems available: induction loop, AM/FM and infrared. Each type of system has certain advantages and disadvantages that should be taken into consideration when choosing the system that is most appropriate for the intended application.

Topic: General Provisions
Reference: IBC 1109
Category: Accessibility
Subject: Features and Facilities

Code Text: *Accessible building features and facilities shall be provided in accordance with Sections 1109.2 through 1109.15.* See exception for Type A and Type B dwelling and sleeping units.

Discussion and Commentary: Where elements such as sinks, drinking fountains, storage lockers, fitting rooms and check-out aisles are provided, a portion, but not less than one of each type of element, must be accessible. In general, all toilet rooms must be accessible. Within each toilet room, at least one water closet and lavatory must be accessible. When other elements are provided, such as mirrors and towel fixtures, at least one must be accessible. Operating mechanisms intended for occupant operation, such as light switches and convenience outlets, must also be usable by persons with physical disabilities.

In other than Type A and Type B dwelling units, accessible building features and facilities shall be provided as required in Section 1109. This includes:

- Toilet and bathing facilities
- Sinks
- Kitchens, kitchenettes and wet bars
- Drinking fountains
- Saunas and steam rooms
- Elevators
- Lifts
- Storage
- Detectable warnings
- Seating at tables, counters and work surfaces
- Service facilities
- Controls, operating mechanisms and hardware
- Fuel-dispensing systems
- Recreational and sports facilities

Certain elements addressed in ICC A117.1, including telephones and automatic teller machines, have not been included in the scoping provisions of Chapter 11. However, scoping requirements for such features are set forth in Appendix E of the IBC.

Topic: Toilet and Bathing Facilities
Reference: IBC 1109.2
Category: Accessibility
Subject: Features and Facilities

Code Text: *Each toilet room and bathing room shall be accessible. Where a floor level is not required to be connected by an accessible route, the only toilet rooms or bathing facilities provided within the facility shall not be located on the inaccessible floor. At least one of each type of fixture, element, control or dispenser in each accessible toilet room and bathing room shall be accessible.* See exceptions for toilet rooms and bathing rooms 1) accessed through a private office and intended for use by a single occupant; 2) that serve a dwelling unit or sleeping unit not required to be accessible; 3) clustered in a single location; 4) part of critical care or intensive care patient sleeping rooms and 5) primarily for children's use. An additional exception provides that where only one urinal is provided in a toilet room or bathing facility, it need not be accessible.

Discussion and Commentary: As a general rule, all toilet rooms and bathing facilities must provide for accessibility. There are limited conditions under which some facilities need not be made accessible.

Ambulatory accessible compartment

In those toilet rooms and bathing facilities where water closet compartments are provided, a minimum of one wheelchair-accessible compartment is required. If the total number of water closet compartments and urinals provided is six or more, an ambulatory-accessible water closet compartment is mandated in addition to the wheelchair-accessible compartment.

Topic: Family or Assisted-Use Toilet Rooms **Category:** Accessibility
Reference: IBC 1109.2.1 **Subject:** Features and Facilities

Code Text: *In assembly and mercantile occupancies, an accessible family or assisted-use toilet room shall be provided where an aggregate of six or more male and female water closets is required. In recreational facilities where separate-sex bathing rooms are provided, an accessible family or assisted-use bathing room shall be provided. See exception for single-fixture bathing rooms. Fixtures located within family or assisted-use toilet and bathing rooms shall be included in determining the number of fixtures provided in an occupancy.*

Discussion and Commentary: The primary issue relative to family or assisted-use toilet/bathing facilities is that some people with disabilities require assistance to utilize them. If the attendant is of the opposite sex, a facility that can accommodate both persons is required. The provisions are applicable only to those types of transient uses where it is expected such facilities are frequently required.

Family or assisted-use toilet room

Applicable only to assembly and mercantile occupancies

Family or assisted-use toilet rooms shall include only one water closet and one lavatory. A urinal is also permitted but not required. Doors to family or assisted-use toilet and bathing rooms must be securable from within the room.

Topic: Signs
Reference: IBC 1110
Category: Accessibility
Subject: Signage

Code Text: *Required accessible elements shall be identified by the International Symbol of Accessibility at the following locations: 1) accessible parking spaces required by Section 1106.1 except where the total number of parking spaces provided is four or less, 2) accessible passenger loading zones, 3) accessible rooms where multiple single-user toilet or bathing rooms are clustered at a single location, 4) accessible entrances where not all entrances are accessible, 5) accessible check-out aisles where not all aisles are accessible, 6) family or assisted-use toilet and bathing rooms, 7) accessible dressing, fitting, and locker rooms where not all such rooms are accessible, 8) accessible areas of refuge in accordance with Section 1007.9, and 9) exterior areas for assisted rescue in accordance with Section 1007.9.*

Discussion and Commentary: Those site or building elements that need to be identified as accessible for convenience, clarification or life safety purposes are specified in the code. Special signage is also required for assistive listening capabilities and at every exit stairway door.

International Symbol of Accessibility

Where building entrances are not accessible, directional signage must be installed indicating the travel route to the nearest accessible entrance. Signs must also be provided at inaccessible public toilets directing occupants to the nearest accessible toilet facilities.

Study Session 13
IBC Chapter 11

1. For the purpose of accessibility, which of the following areas in a bank is considered an employee work area?

    a. vault
    b. toilet room
    c. corridor
    d. break room

    Reference_____

2. A continuous, unobstructed path complying with the provisions of IBC Chapter 11 is considered a(n) _____ .

    a. accessible route
    b. accessible means of egress
    c. circulation path
    d. public entrance

    Reference_____

3. Which of the following Group U occupancies is exempt from the accessibility provisions?

    a. paved work areas at agricultural buildings
    b. private garages containing required accessible parking
    c. areas in agricultural buildings open to the public
    d. stables and/or livestock shelters

    Reference_____

4. When occupied as the residence for the proprietor, Group R-1 occupancies containing a maximum of _____ sleeping units for rent or hire are not required to be accessible.

    a. 5   b. 6

    c. 10   d. 15

Reference_____

5. In other than an Accessible, Type A or Type B dwelling unit, where only one accessible route is provided, the route is permitted to pass through a _____ .

    a. kitchen   b. laboratory

    c. storage room   d. restroom

Reference_____

6. Where four public entrances are provided to a building, a minimum of _____ entrance(s) shall be accessible.

    a. one   b. two

    c. three   d. four

Reference_____

7. Where a 1,625-space parking garage serves a covered mall building, a minimum of _____ accessible parking spaces must be provided.

    a. 20   b. 27

    c. 33   d. 42

Reference_____

8. Where a 282-space parking lot serves a cluster of Group R-2 apartment buildings, a minimum of _____ accessible parking spaces must be provided.

    a. 3   b. 6

    c. 7   d. 14

Reference_____

9. A minimum of _____ van-accessible parking space(s) shall be provided for a parking facility containing 16 accessible parking spaces.

   a. one
   b. two
   c. three
   d. six

   Reference_____

10. Where stadiums, arenas and grandstands have a minimum of _____ fixed seats and provide audible public announcements, pre-recorded or real-time captions of those audible public announcements shall be provided.

    a. 3,000
    b. 5,000
    c. 10,000
    d. 15,000

    Reference_____

11. Where cubicles are provided in the visiting area of a judicial facility, a minimum of _____ of the cubicles shall be accessible.

    a. 5 percent
    b. 10 percent
    c. 25 percent
    d. 50 percent

    Reference_____

12. An ambulatory-accessible water closet compartment is required in a toilet room where the total number of water closet compartments and urinals provided in the room is a minimum of _____ fixtures.

    a. 4
    b. 6
    c. 10
    d. 12

    Reference_____

13. In a Group R-2 fraternity house having a minimum of _____ sleeping units, every unit intended to be occupied as a residence shall be a Type B unit unless a reduction is permitted.

    a. 4
    b. 10
    c. 16
    d. 20

    Reference_____

Study Session 13

14. In a 250-seat motion picture theater, a minimum of _____ receivers shall be provided for the assistive listening system.

    a. 2   b. 6
    c. 10   d. 14

    Reference_____

15. In a Group I-2 nursing home containing 120 patient sleeping units, what is the minimum number of such units that must be Accessible units?

    a. 5   b. 12
    c. 60   d. 120

    Reference_____

16. In a 185-room Group R-1 hotel, a minimum of _____ sleeping units shall be Accessible units.

    a. two   b. four
    c. seven   d. eight

    Reference_____

17. A minimum of _____ Accessible units in a 410-room hotel shall be provided with bathing facilities other than roll-in showers.

    a. 4   b. 9
    c. 12   d. 13

    Reference_____

18. A Group R-2 apartment building having 16 dwelling units shall be provided with a minimum of _____ Type A dwelling units.

    a. zero (no Type A units are required)
    b. one
    c. two
    d. four

    Reference_____

19. A self-storage facility containing 160 storage spaces shall be provided with a minimum of _____ accessible individual self-storage spaces.

    a. one   b. two

    c. five   d. eight

Reference_____

20. Based on the number of required water closets, an accessible family or assisted-use toilet room may be required in which of the following occupancies?

    a. assembly and educational   b. assembly and mercantile

    c. business and mercantile   d. factory and storage

Reference_____

21. The maximum distance of the accessible route from any separate-sex toilet room to a family or assisted-use toilet room shall be _____.

    a. 200 feet   b. 300 feet

    c. 500 feet   d. unlimited

Reference_____

22. In a large manufacturing plant, a minimum of how many of the 24 drinking fountains shall be accessible?

    a. 3   b. 4

    c. 8   d. 24

Reference_____

23. A minimum of _____ percent of seating at fixed tables in an accessible space shall be accessible.

    a. 5   b. 10

    c. 20   d. 50

Reference_____

24. A 4,000-square-foot retail sales room with six check-out aisles requires a minimum of _____ accessible check-out aisle(s).

    a. one   b. two

    c. three   d. four

Reference_____

25. An International Symbol of Accessibility sign is not required at an accessible parking space in parking facilities having a maximum of _____ total parking spaces.

    a. one, the sign is always required

    b. two

    c. four

    d. five

Reference_____

26. Permanently defined common use circulation paths that are located within an employee work area are not required to be regulated as accessible routes, provided the work area is less than _____ square feet in floor area.

    a. 300   b. 500

    c. 1,000   d. 1,500

Reference_____

27. A press box serving bleachers is not required to be served by an accessible route where it is a maximum of _____ square feet in floor area and has its points of entry at only one level.

    a. 500   b. 1,000

    c. 1,500   d. 3,000

Reference_____

28. What is the minimum number of accessible parking spaces required to be provided in a 40-space parking lot serving an outpatient physical therapy facility?

    a. one   b. two

    c. four   d. eight

Reference_____

29. Where a dining surface is provided at a counter for the consumption of food or drink, a minimum of _____ seating and standing space(s) shall be accessible.

    a. zero
    b. one
    c. two
    d. four

    Reference_____

30. At which of the following locations is directional signage not required to indicate the route to the nearest like accessible element?

    a. inaccessible building entrances
    b. elevators not serving an accessible route
    c. inaccessible dressing rooms
    d. inaccessible public toilet facilities

    Reference_____

31. Where passenger loading zones are provided, at least one accessible passenger loading zone is required in every continuous _____ linear feet maximum of loading zone space.

    a. 100
    b. 150
    c. 200
    d. 300

    Reference_____

32. All required wheelchair spaces are permitted to be located on the main level in multilevel assembly spaces used for religious worship where the mezzanine level contains a maximum of _____ percent of the total seating capacity.

    a. 10
    b. 20
    c. 25
    d. $33\frac{1}{3}$

    Reference_____

33. Where multiple single-user toilet rooms are clustered at a single location, a minimum of _____ percent, but not less than one room, shall be accessible.

    a. 5
    b. 10
    c. 25
    d. 50

    Reference_____

Study Session 13

34. Where a total of eight accessible drinking fountains are provided, _____ drinking fountain(s) shall comply with the requirements for standing persons.

    a. one   b. two
    c. four   d. five

    Reference _____

35. In a 40-lane bowling facility, an accessible route shall be provided to a minimum of _____ bowling lane(s).

    a. one   b. two
    c. four   d. ten

    Reference _____

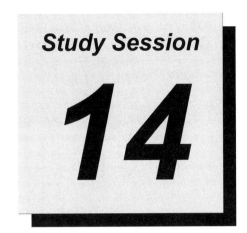

Study Session 14

2012 IBC Chapter 4
Detailed Occupancy Requirements

OBJECTIVE: To obtain an understanding of special building types, features and uses, including covered mall and open mall buildings, high-rise buildings, atriums, underground buildings, motor-vehicle-related occupancies, stages and platforms, concealed combustible storage areas, hazardous materials, dwelling unit and sleeping unit separations, live/work units, ambulatory care facilities, and Group I-2, I-3 and H occupancies.

REFERENCE: Chapter 4, 2012 *International Building Code*

KEY POINTS:
- How are covered mall and open mall buildings, anchor buildings and food courts defined?
- In a covered mall building, which special conditions relate to the automatic sprinkler system, standpipe system and smoke-control system?
- How is the occupant load determined for a covered mall building?
- Which specific provisions relate to covered mall and open mall buildings in determining the number and arrangement of means of egress?
- What qualifies a structure as a high-rise building?
- In a high-rise building, how are the type-of-construction provisions modified?
- What are the specific provisions relating to smoke detection, fire alarms and communication systems in high-rise buildings?
- What additional provisions are applicable to high-rise buildings more than 420 feet in height?
- What are the various required elements in a fire command center?
- In a high-rise structure, which types of standby power, light and emergency systems are required?
- What defines an atrium? What are the limits of an atrium's use? What fire protection features must be provided in a building containing an atrium?
- How must adjacent spaces be separated from an atrium?
- What is the maximum travel distance when a required exit path enters the atrium space?
- What are the conditions that create an underground building?
- In an underground building, when is compartmentalization required?
- What is the maximum floor area permitted for a private garage?

KEY POINTS:
(Cont'd)

- How must exterior openings be sized and distributed for a garage to qualify as open?
- How do the provisions differ for an open parking garage in either a single-use or multiple-occupancy building?
- What are the benefits of an open parking garage as opposed to an enclosed garage?
- Which special criteria must be applied to motor vehicle service stations? Repair garages?
- How does a platform differ from a stage?
- What are the minimum construction requirements for stages and platforms?
- What is a technical production area? What special provisions apply to such areas?
- When must a stage area be provided with a means for emergency ventilation?
- How shall the proscenium opening between a stage and an auditorium be protected?
- What is a special amusement building? What special provisions apply to such buildings? How are airport traffic control towers regulated?
- What specific requirements are applicable to aircraft hangars? What type of fire suppression is required? What unique requirements apply to residential aircraft hangars? Aircraft paint hangars?
- What type of fire separation is required for combustible storage in concealed spaces?
- When are smoke compartments required in Group I-2 occupancies?
- Which special considerations are given to detention facilities because of their unique characteristics?
- How are Group I-3 occupancies classified according to their occupancy condition?
- What is the function of a control area? How many control areas are permitted in a building?
- How must control areas be separated from other portions of the building?
- At what distance must Group H occupancies be set back from lot lines?
- Where the spraying of flammable finishes occurs, what special conditions must be met?
- What type of use is regulated as a live/work unit? What are its limitations? What special provisions are applicable?
- How must dwelling units or sleeping units be separated from other such units in Group I-1, R-1, R-2 and R-3 occupancies?
- Under what conditions are smoke barriers required in ambulatory care facilities? Automatic sprinkler system protection? Fire alarm system protection?
- What reference standard is applicable to storm shelters?
- In which occupancies are children's play structures regulated? What special requirements are applicable?

Topic: General Requirements
Reference: IBC 402

Category: Detailed Use Requirements
Subject: Covered Mall and Open Mall Buildings

Code Text: *A covered mall building (or open mall building) and attached anchor buildings and parking garages shall be surrounded on all sides by a permanent open space or not less than 60 feet (18 288 mm). The building area of any covered mall or open building, including anchor buildings, of Types I, II, III and IV construction shall not be limited provided the anchor buildings do not exceed three stories above grade plane. See exception for anchor buildings. Fire-resistance-rated separation is not required between tenant spaces and the mall.*

Discussion and Commentary: A covered mall building is defined as *a single building enclosing a number of tenants and occupants, such as retail stores, drinking and dining establishments, entertainment and amusement facilities, passenger transportation terminals, offices, and other similar uses wherein two or more tenants have a main entrance into one or more malls.* Because of its character, a covered mall building is uniquely regulated for fire protection and egress. An open mall building, where the pedestrian ways are unroofed, is regulated in a similar manner.

Mall width requirements

For SI: 1 foot = 304.8 mm

The provisions for covered mall and open mall buildings and their associated anchor buildings and parking structures allow for an alternative method of design for structures that have these specific features. Where compliant with the provisions of Section 402, similar requirements found elsewhere in the code can be superceded.

Topic: General Requirements
Reference: IBC 403, 202

Category: Detailed Use Requirements
Subject: High-Rise Buildings

Code Text: A high-rise building *is a building with an occupied floor located more than 75 feet (22 860 mm) above the lowest level of fire department vehicle access.* See five exceptions where high-rise provisions are not applicable. *Buildings and structures shall be equipped throughout with an automatic sprinkler system in accordance with Section 903.3.1.1 and a secondary water supply where required by Section 903.3.5.2.*

Discussion and Commentary: A high-rise building is characterized by several features: 1) it is impractical to completely evacuate the building in a timely manner,) prompt rescue and firefighting operations are difficult, 3) the occupant load is relatively high and 4) a potential exists for stack effect. The special provisions of Section 403 are designed to address these concerns. Additional provisions are applicable for those buildings more than 420 feet in height, sometimes referred to as *super high-rises*. Prohibited reductions in fire-resistance rating, additional criteria for structural integrity and bond strength of SFRM, and an additional required exit stairway are a few of the extra requirements for these super high-rise buildings.

For SI: 1 foot = 304.8 mm

Additional provisions for a high-rise building include the installation of a smoke detection system, emergency voice/alarm and fire department communications systems, a fire command center for use by fire department personnel, smokeproof exit stairway enclosures, luminous egress path markings, and standby power, light and emergency systems.

Topic: General Requirements
Reference: IBC 404
Category: Detailed Use Requirements
Subject: Atriums

Code Text: *In other than Group H occupancies, and where permitted by Section 712.1.6, the provisions of Section 404 shall apply to buildings or structures containing vertical openings defined herein as "Atriums." An approved automatic sprinkler system shall be installed throughout the entire building. A smoke-control system shall be installed in accordance with Section 909. Atrium spaces shall be separated from adjacent spaces by a 1-hour fire barrier.* See exceptions to the sprinkler system, smoke-control system and fire barrier separation requirements.

Discussion and Commentary: An atrium is defined as *an opening connecting two or more stories other than enclosed stairways, elevators, hoistways, escalators, plumbing, electrical, air-conditioning or other equipment, which is closed at the top and not defined as a mall.* The concept of developing atriums is to maintain equivalence in safety to that of an open court, as well as to provide protection of a shaft enclosure. The provisions for atriums are only applicable where Section 712.1.6 is utilized to address vertical openings. Where another method established in Section 712.1 is used, such as Section 712.1.8 for an opening connecting only two stories, the atrium provisions of Section 404 are not to be applied.

Atrium concept

Atriums are permitted based on alternative methods of protecting the building from vertical spread of fire, smoke and toxic gases. Additional protection is provided through 1) limited travel distance, 2) standby power, 3) smoke detection and 4) interior finish regulation.

Topic: General Requirements
Reference: IBC 405

Category: Detailed Use Requirements
Subject: Underground Buildings

Code Text: *The provisions of Section 405 apply to building spaces having a floor level used for human occupancy more than 30 feet (9144 mm) below the finished floor of the lowest level of exit discharge.* See six exceptions for uses not regulated as underground buildings. *The underground portion of the building shall be of Type I construction. The highest level of exit discharge serving the underground portions of the building and all levels below shall be equipped with an automatic sprinkler system installed in accordance with Section 903.3.1.1.*

Discussion and Commentary: An underground building is highly regulated for many of the same reasons as is a high-rise building. In the case of a structure substantially below ground level, fire department access and fire-fighting operations are often even more difficult. Therefore, the code mandates the installation of multiple fire protection systems, including a smoke-control system.

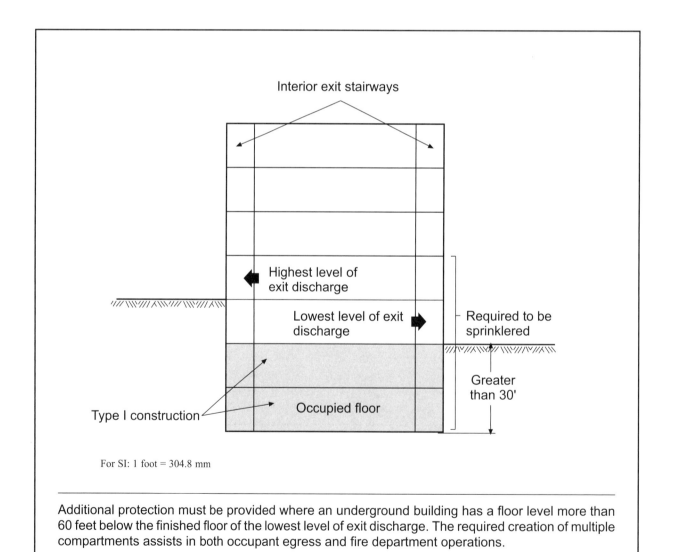

For SI: 1 foot = 304.8 mm

Additional protection must be provided where an underground building has a floor level more than 60 feet below the finished floor of the lowest level of exit discharge. The required creation of multiple compartments assists in both occupant egress and fire department operations.

Topic: General Requirements
Reference: IBC 406

Category: Detailed Use Requirements
Subject: Motor-Vehicle-Related Occupancies

Code Text: *Parking garages, other than private parking garages, shall be classified as public parking garages and shall comply with the provisions of Sections 406.4.2 through 406.4.8 and shall be classified as either an open parking garage or an enclosed parking garage. Motor fuel-dispensing facilities shall be constructed in accordance with the* International Fire Code *and Sections 406.7.1 and 406.7.2. Repair garages shall be constructed in accordance with the* International Fire Code *and Sections 406.8.1 through 406.8.6.*

Discussion and Commentary: Where motor vehicles are located within a structure, varying degrees of hazard are involved. In a small, private garage or carport, the hazard is relatively low. A moderate level of hazard exists in open parking garages, increasing where the parking garage is enclosed. In structures where vehicles are being fueled or repaired, a relatively high hazard exists. Specific provisions of the IBC address each type of motor-vehicle-related use.

Exterior walls must have uniformly distributed openings on two or more sides.

Interior wall and column lines shall be at least 20 percent open (area) with uniformly distributed openings.

General case

1. Area: A + B + C + D + E ≥ 20% total perimeter area of each tier
2. Length: A + B + C + D + E ≥ 40% total perimeter area of each tier

Open parking garages

Private garages classified as Group U occupancies are usually limited to 1,000 square feet in floor area. However, such structures are permitted to be 3,000 square feet maximum where no repair work is done and no fuel dispensed, and where the exterior wall and opening protection is specifically regulated.

Topic: Smoke Barriers
Reference: IBC 407.5
Category: Detailed Use Requirements
Subject: Group I-2 Occupancies

Code Text: *Smoke barriers shall be provided to subdivide every story used by persons receiving care, treatment or sleeping and to divide other stories with an occupant load of 50 or more persons, into no fewer than two smoke compartments. Such stories shall be divided into smoke compartments with an area of not more than 22,500 square feet (2092 m^2) and the travel distance from any point in a smoke compartment to a smoke barrier door shall not exceed 200 feet (60 960 mm).*

Discussion and Commentary: Hospitals, nursing homes and similar uses must have unique life-safety characteristics due to the immobility or limited mobility of most of the care recipients. By providing multiple refuge areas on each story of the building, occupants can be moved horizontally into an adjoining smoke compartment that provides protection from adjacent areas. In multistory buildings, smoke compartments are further enclosed through the use of horizontal assemblies that must be designed to resist the movement of smoke.

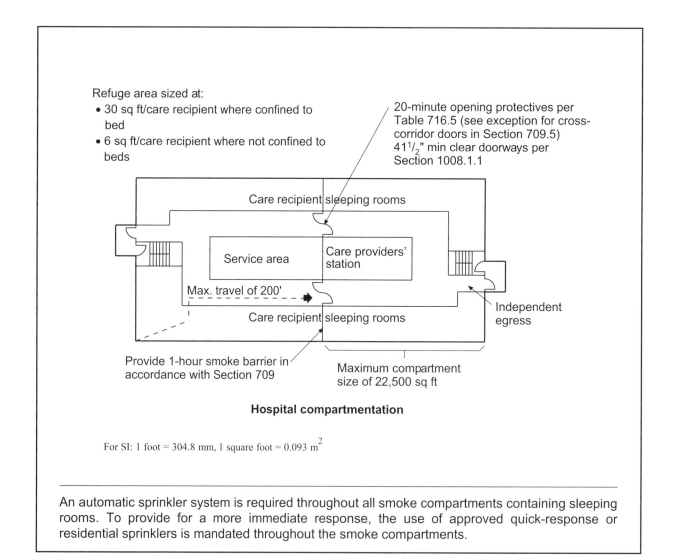

Hospital compartmentation

For SI: 1 foot = 304.8 mm, 1 square foot = 0.093 m^2

An automatic sprinkler system is required throughout all smoke compartments containing sleeping rooms. To provide for a more immediate response, the use of approved quick-response or residential sprinklers is mandated throughout the smoke compartments.

Topic: General Requirements
Reference: IBC 408.4

Category: Detailed Use Requirements
Subject: Group I-3 Occupancies

Code Text: In Group I-3 occupancies, *egress doors are permitted to be locked in accordance with the applicable use condition. Doors from a refuge area to the outside are permitted to be locked with a key in lieu of locking methods described in Section 408.4.1* (remote release). *The keys to unlock the exterior doors shall be available at all times and the locks shall be operable from both sides of the door. Remote release of locks on doors in a means of egress shall be provided with reliable means of operation, remote from the resident living areas, to release locks on all required doors.*

Discussion and Commentary: The need for restraint or security in specific types of uses such as jails, prisons and detention centers makes it necessary to install locking devices that are usually unacceptable for a means of egress. The code recognizes such a need and provides alternative design methods to balance the desire for both safety and security.

For SI: 1 foot = 304.8 mm

It is often important to ensure that several resident detention areas can be observed from a single location. The code permits the design of a vertical arrangement where several tiers of resident housing areas can be open to each other and to the supervisory area.

Topic: Stages
Reference: IBC 410.3, 202

Category: Detailed Use Requirements
Subject: Stages and Platforms

Code Text: *A stage is a space within a building utilized for entertainment or presentations, which includes overhead hanging curtains, drops, scenery or stage effects other than lighting or sound. Stages shall be constructed of materials as required for floors for the type of construction of the building in which such stages are located. See three exceptions. Where the stage height is greater than 50 feet (15 240 mm), all portions of the stage shall be completely separated from the seating area by a proscenium wall with not less than a 2-hour fire-resistance rating extending continuously from the foundation to the roof.*

Discussion and Commentary: Given the increased potential for fire hazards in an assembly occupancy with a stage, such a use is regulated for certain elements. The stage must be separated from accessory spaces by fire barriers; ventilation of the stage must be accomplished through smoke control or roof vents; and the proscenium opening must be protected with a curtain of approved materials or an approved water curtain.

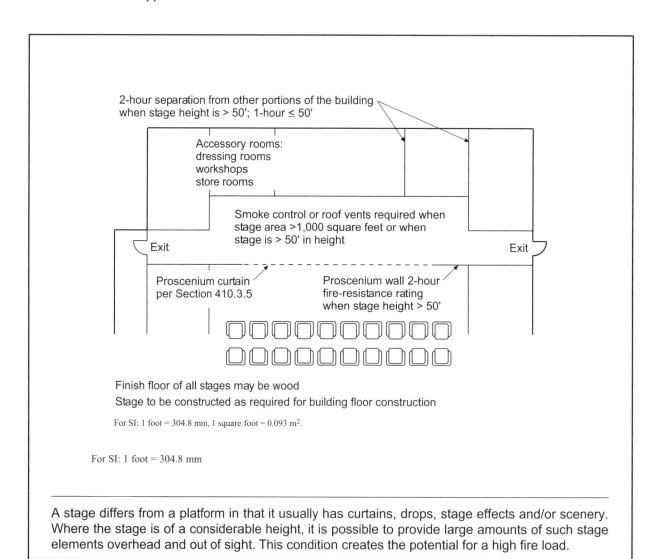

A stage differs from a platform in that it usually has curtains, drops, stage effects and/or scenery. Where the stage is of a considerable height, it is possible to provide large amounts of such stage elements overhead and out of sight. This condition creates the potential for a high fire load.

Topic: Platforms
Reference: IBC 410.4, 202
Category: Detailed Use Requirements
Subject: Stages and Platforms

Code Text: *A platform is a raised area within a building used for worship, the presentation of music, plays or other entertainment; the head table for special guests; the raised area for lecturers and speakers; boxing and wresting rings; theater-in-the-round stages; and similar purposes wherein there are no overhead hanging curtains, drops, scenery or stage effects other than lighting and sound. Permanent platforms shall be constructed of materials as required for the type of construction of the building in which the permanent platform is located.* See allowances for use of fire-retardant-treated wood in Types I, II or IV construction.

Discussion and Commentary: Few requirements are placed on platforms. However, a minimum 1-hour fire-resistant platform floor construction is required when the area below the platform is used for storage or a similar purpose, because of concern about combustibles being stored within a concealed space below a raised area.

For SI: 1 inch = 25.4 mm, 1 square foot = 0.093 m^2

Temporary platforms are those platforms used within an area for a period not to exceed 30 days. They may be constructed of any materials; however, the space between the floor and the platform cannot be used for any purpose other than wiring or plumbing for platform equipment.

Topic: General Requirements
Reference: IBC 411, 202

Category: Detailed Use Requirements
Subject: Special Amusement Buildings

Code Text: A special amusement building is *any temporary or permanent building or portion thereof that is occupied for amusement, entertainment or educational purposes and that contains a device or system that conveys passengers or provides a walkway along, around or over a course in any direction so arranged that the means of egress path is not readily apparent due to visual or audio distractions or is intentionally confounded or is not readily available because of the nature of the attraction or mode of conveyance through the building or structure.*

Discussion and Commentary: In most cases, an amusement building will be classified as a Group A occupancy. The hazards associated with such a unique use are addressed through provisions for the detection of fire, the illumination of the exit path, the presence of an alarm and emergency voice/alarm communications system and the sprinklering of the structure.

Rapid detection and notification of a fire condition, as well as the discernment of the exit path, are critical in an amusement building. Actuation of either the sprinkler system or the fire detection system shall automatically activate the approved egress directional markings.

Topic: Aircraft Hangars
Reference: IBC 412.4
Category: Detailed Use Requirements
Subject: Aircraft-Related Occupancies

Code Text: *Aircraft hangar exterior walls located less than 30 feet (9144 mm) from lot lines or a public way shall have a fire-resistance rating not less than 2 hours. Heating equipment shall be placed in another room separated by 2-hour fire barriers, or horizontal assemblies, or both. Aircraft hangars shall be provided with a fire suppression system designed in accordance with NFPA 409, based upon the classification for the hangar given in Table 412.4.6.*

Discussion and Commentary: Although most commercial aircraft hangars will not be limited in height (Section 504.1) or in area (Section 507) based on the presence of an automatic sprinkler system, they must be regulated in regards to exterior-wall fire-resistance ratings, basement limitations, floor surfaces, heating equipment separation and finishing restrictions. All of these provisions serve to abate the hazards associated with large aircraft and their integral fuel tanks to acceptable fire safety levels.

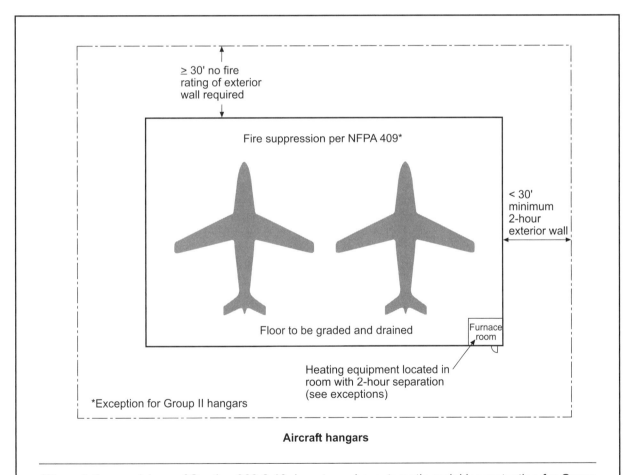

Aircraft hangars

Although the provisions of Section 903.2.10 do not require automatic sprinkler protection for Group S-2 occupancies (aircraft hangars), a fire suppression system per NFPA 409 is mandated. The foam suppression requirements are exempted for Group II hangars where no major maintenance is performed.

Topic: Attic and Under-Floor Spaces
Reference: IBC 413.2
Category: Detailed Use Requirements
Subject: Combustible Storage

Code Text: *Attic, under-floor and concealed spaces used for storage of combustible materials shall be protected on the storage side as required for 1-hour fire-resistant-rated construction. Openings shall be protected by assemblies that are self-closing and are of noncombustible construction or solid wood core not less than $1^3/_4$ inch (45 mm) in thickness. See exceptions for 1) areas protected by an automatic sprinkler system, and 2) Group R-3 and U occupancies.*

Discussion and Commentary: Those areas in a building that tend to be unoccupied present a potential fire hazard where combustible goods are being stored. The presence of a considerable fire load, coupled with the probable delay in recognition of the fire, makes it necessary to provide some degree of protection. A sprinkler system or a fire-resistant separation are considered acceptable methods to address any potential hazards.

The code does not mandate a full 1-hour fire-resistance-rated assembly to isolate the combustible storage area from the unoccupied space. Because the hazard presumably exists only on the inside of the storage space, that is the only side where the protection is required.

Topic: Control Areas
Reference: IBC 414.2.1, 202
Category: Detailed Use Requirements
Subject: Hazardous Materials

Code Text: *Control areas are spaces within a building where quantities of hazardous materials not exceeding the maximum allowable quantities per control area are stored, dispensed, used or handled. Control areas shall be separated from each other by fire barriers constructed in accordance with Section 707 or horizontal assemblies constructed in accordance with Section 711, or both.*

Discussion and Commentary: The use of control areas provides an alternative method for the use and storage of hazardous materials without classifying the building or structure as a high-hazard (Group H) occupancy. This concept is based on regulating the allowable quantities of hazardous materials per control area rather than per building area by giving credit for further compartmentation through the use of fire-resistance-rated fire barrier walls and horizontal assemblies.

*Exception allows for 1-hour in fully sprinklered Type IIA, IIIA and VA buildings no more than three stories in height.

Multistory control areas

The maximum quantities of hazardous materials within a given control area cannot exceed the quantities for a given material listed in either Table 307.1(1) for physical hazards and Table 307.1(2) for health hazards, as modified by Table 414.2.2 for location within the building.

Topic: Control Areas
Reference: IBC 414.2.2 – 414.2.4
Category: Detailed Use Requirements
Subject: Hazardous Materials

Code Text: *The percentage of maximum allowable quantities of hazardous materials per control area permitted at each floor level within a building shall be in accordance with Table 414.2.2. The maximum number of control areas within a building shall be in accordance with Table 414.2.2. The required fire-resistance rating for fire barriers shall be in accordance with Table 414.2.2. The floor assembly of the control area and the construction supporting the floor of the control area shall have a minimum 2-hour fire-resistance rating.* See exception permitting a 1-hour floor separation in fully-sprinklered Type IIA, IIIA and VA buildings not exceeding three stories in height above grade.

Discussion and Commentary: By distributing hazardous materials in multiple fire-resistant compartments throughout a structure, the amount of material exposed to an immediate fire event is limited. The code allows such limited quantities in buildings of other than Group H occupancy; hence, the use of control areas is an effective method for reducing the occupancy classification by reducing the hazard.

**[F] TABLE 414.2.2
DESIGN AND NUMBER OF CONTROL AREAS**

FLOOR LEVEL		PERCENTAGE OF THE MAXIMUM ALLOWABLE QUANTITY PER CONTROL AREA[a]	NUMBER OF CONTROL AREAS PER FLOOR	FIRE-RESISTANCE RATING FOR FIRE BARRIERS IN HOURS[b]
Above grade plane	Higher than 9	5	1	2
	7-9	5	2	2
	6	12.5	2	2
	5	12.5	2	2
	4	12.5	2	2
	3	50	2	1
	2	75	3	1
	1	100	4	1
Below grade plane	1	75	3	1
	2	50	2	1
	Lower than 2	Not Allowed	Not Allowed	Not Allowed

a. Percentages shall be of the maximum allowable quantity per control area shown in Tables 307.1(1) and 307.1(2), with all increases allowed in the notes to those tables.
b. Fire barriers shall include walls and floors as necessary to provide separation from other portions of the building.

The purpose of a control area is to allow the building to be classified according to its general occupancy instead of being classified as a Group H occupancy. A building may comprise a single control area where the amount of hazardous materials in the entire structure is compliant.

Topic: Distance to Lot Lines
Reference: IBC 415.5.1

Category: Detailed Use Requirements
Subject: Group H Occupancies

Code Text: *Regardless of any other provisions, buildings containing Group H occupancies shall be set back to the minimum fire separation distance as set forth in Sections 415.5.1.1 through 415.5.1.4. Distances shall be measured from the walls enclosing the occupancy to lot lines, including those on a public way.*

Discussion and Commentary: Because of the potentially volatile nature of hazardous materials, specific setback requirements are necessary for Group H occupancies. These provisions take precedence over Table 602 regarding the minimum fire separation distance based on building construction type and exposure. The listed conditions are dependent on the type of materials that are indicative of the specified Group H occupancies, the size of the hazardous material storage area and whether or not a detached building is required.

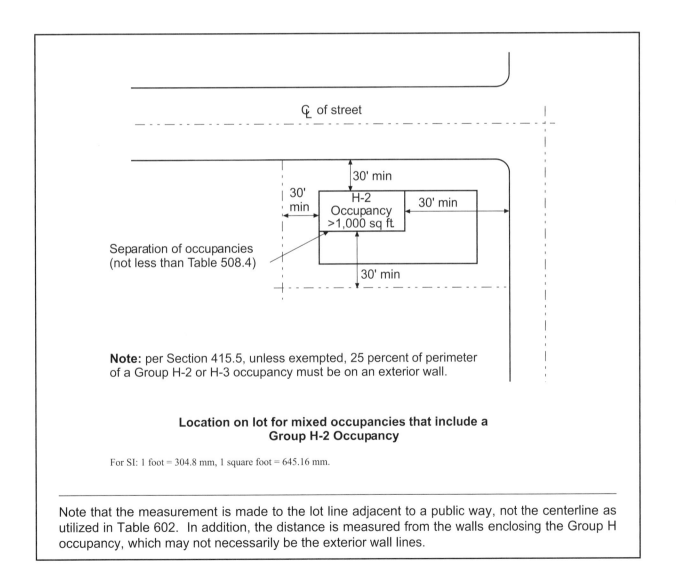

Location on lot for mixed occupancies that include a Group H-2 Occupancy

For SI: 1 foot = 304.8 mm, 1 square foot = 645.16 mm.

Note that the measurement is made to the lot line adjacent to a public way, not the centerline as utilized in Table 602. In addition, the distance is measured from the walls enclosing the Group H occupancy, which may not necessarily be the exterior wall lines.

Study Session 14

Topic: Detached Buildings
Reference: IBC 415.6, 415.7
Category: Detailed Use Requirements
Subject: Group H Occupancies

Code Text: *Group H-1 occupancies shall be in buildings used for no other purpose, shall not exceed one story in height and be without basements, crawl spaces or other under-floor spaces. Group H-2 and H-3 occupancies containing quantities of hazardous materials in excess of those set forth in Table 415.5.2 shall be in detached buildings used for manufacturing, processing, dispensing, use or storage of hazardous materials. Detached buildings shall not exceed one story in height and shall be without basements, crawl spaces or other under-floor spaces.*

Discussion and Commentary: Because of the explosion hazard potential associated with Group H-1 materials, Group H-1 occupancies are required to be in separate detached structures. The limitation of one story is based on the need for a building with a detonation hazard to be exited as soon as possible. Higher-level Group H-2 and Group H-3 occupancies where the quantities of hazardous materials pose an extremely high risk must also be located in buildings with no other uses.

[F] TABLE 415.5.2
DETACHED BUILDING REQUIRED

A DETACHED BUILDING IS REQUIRED WHEN THE QUANTITY OF MATERIAL EXCEEDS THAT LISTED HEREIN

Material	Class	Solids and Liquids (tons)[a, b]	Gases (cubic feet)[a, b]
Explosives	Division 1.1 Division 1.2 Division 1.3 Division 1.4 Division 1.4[c] Division 1.5 Division 1.6	Maximum Allowable Quantity Maximum Allowable Quantity Maximum Allowable Quantity Maximum Allowable Quantity 1 Maximum Allowable Quantity Maximum Allowable Quantity	Not Applicable
Oxidizers	Class 4	Maximum Allowable Quantity	Maximum Allowable Quantity
Unstable (reactives) detonable	Class 3 or 4	Maximum Allowable Quantity	Maximum Allowable Quantity
Oxidizer, liquids and solids	Class 3 Class 2	1,200 2,000	Not Applicable Not Applicable
Organic peroxides	Detonable Class I Class II Class III	Maximum Allowable Quantity Maximum Allowable Quantity 25 50	Not Applicable Not Applicable Not Applicable Not Applicable
Unstable (reactives) nondetonable	Class 3 Class 2	1 25	2,000 10,000
Water reactives	Class 3 Class 2	1 25	Not Applicable Not Applicable
Pyrophoric gases	Not Applicable	Not Applicable	2,000

For SI: 1 ton = 906 kg, 1 cubic foot = 0.02832 m³, 1 pound = 0.454 kg.

a. For materials that are detonable, the distance to other buildings or lot lines shall be in accordance with Chapter 33 of the *International Fire Code* based on trinitrotoluene (TNT) equivalence of the material. For materials classified as explosives, see Chapter 56 of the *International Fire Code*.
b. "Maximum Allowable Quantity" means the maximum allowable quantity per control area set forth in Table 307.1(1).
c. Limited to Division 1.4 materials and articles, including articles packaged for shipment, that are not regulated as an explosive under Bureau of Alcohol, Tobacco, Firearms and Explosives (BATF) regulations or unpackaged articles used in process operations that do not propagate a detonation or deflagration between articles, provided the net explosive weight of individual articles does not exceed 1 pound.

The need for detached storage is a function of the type, physical state and quantity of material. Because such a single-use structure must be located an adequate distance from surrounding lot lines and other buildings, exterior walls and exterior openings need not be protected for exposure.

Topic: General Requirements
Reference: IBC 416.1, 416.2
Category: Detailed Use Requirements
Subject: Application of Flammable Finishes

Code Text: *The provisions of Section 416 shall apply to the construction, installation and use of buildings and structures, or parts thereof, for the application of flammable finishes. Such construction and equipments shall comply with the* International Fire Code. *Spray rooms shall be enclosed with not less than 1-hour fire barriers constructed in accordance with Section 707 or horizontal assemblies constructed in accordance with Section 711, or both. Floors shall be waterproofed and drained in an approved manner. Mechanical ventilation and interlocks with the spraying operation shall be in accordance with the* International Mechanical Code.

Discussion and Commentary: The primary hazards associated with paint spraying and spray booths originate from the presence of flammable liquids or powders and their vapors or mists. The requirements address such issues as the ventilation, sprinkler protection and interior surfaces of spray rooms and spraying spaces.

Spray rooms	Spray booths	Limited spray space
• Designed and constructed per IBC Section 416 • Separated from remainder of building by 1-hour fire barriers • Automatic fire-extinguishing system required	• Designed and constructed per IFC Section 2404.3.2 • Constructed of approved noncombustible materials • Limited in size and location • Automatic fire-extinguishing system required	• Aggregate surface area to be sprayed limited to 9 sq ft • Spraying operations not to be continuous in nature • Mechanical ventilation and hazardous location wiring regulated

Chapter 24 of the *International Fire Code* provides comprehensive requirements for the application of flammable finishes, including detailed provisions for spray booths. Spray finishing operations in Group A, E, I or R occupancies shall be located in a spray room. In other occupancies, such operations may occur in a spray room, spray booth or spraying space approved for such use.

Topic: General Provisions
Reference: IBC 419, 202
Category: Detailed Use Requirements
Subject: Live/Work Units

Code Text: *A live/work unit is a dwelling unit or sleeping unit in which a significant portion of the space includes a nonresidential use that is operated by the tenant. A live/work unit shall comply with Sections 419.1 through 419.9. The following shall apply to all live/work areas: 1) the live/work unit is permitted to be not greater than 3,000 square feet (279 m^2) in area; 2) the nonresidential area is permitted to be not more than 50 percent of the area of each live/work unit; 3) the nonresidential area function shall be limited to the first or main floor only of the live/work unit; and 4) a maximum of five nonresidential workers or employees are allowed to occupy the nonresidential area at any one time.*

Discussion and Commentary: An occasional concept of building use combines a residential unit with a small business activity. Residential live/work units typically include a dwelling unit along with some retail or service business, such as an artist's studio, coffee shop or chiropractor's office. There may be a small number of employees working within the residence, and the public is able to enter the work area of the unit to acquire service.

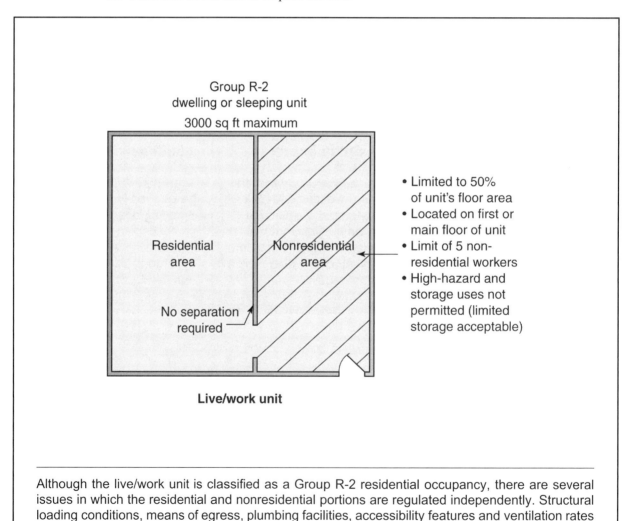

Although the live/work unit is classified as a Group R-2 residential occupancy, there are several issues in which the residential and nonresidential portions are regulated independently. Structural loading conditions, means of egress, plumbing facilities, accessibility features and ventilation rates are all to be based on the individual function of the space within the unit.

Topic: Unit Separations
Reference: IBC 420
Category: Detailed Use Requirements
Subject: Groups I-1, R-1, R-2 and R-3

Code Text: *Occupancies in Groups I-1, R-1, R-2 and R-3 shall comply with the provisions of Sections 420.1 through 420.5 and other applicable provisions of the IBC. Walls separating dwelling units in the same building, walls separating sleeping units in the same building and walls separating dwelling or sleeping units from other occupancies contiguous to them in the same building shall be constructed as fire partitions in accordance with Section 708. Floor assemblies separating dwelling units in the same buildings, floor assemblies separating sleeping units in the same building and floor assemblies separating dwelling or sleeping units from other occupancies contiguous to them in the same building shall be constructed as horizontal assemblies in accordance with Section 711.*

Discussion and Commentary: In residential-type occupancies, it is important that some degree of fire-resistive separation be provided to isolate each individual living unit from all others in the building. It is intended that should a fire initiate within one of the dwelling units or sleeping units, the occupants and contents of the other units would be adequately protected.

Section AA

All vertical (fire partitions) and horizontal (horizontal assemblies) separation elements are required to be of minimum one-hour fire-resistance-rated construction and provided with protected openings. Where the building is sprinklered throughout with an NFPA 13 system, the required separation may be reduced to $^{1}/_{2}$ hour.

Study Session 14

Topic: Smoke Barriers
Reference: IBC 422.2

Category: Detailed Use Requirements
Subject: Ambulatory Care Facilities

Code Text: *Ambulatory care facilities where the potential for four or more care recipients are to be incapable of self preservation at any time shall be separated from adjacent spaces, corridors or tenants with a fire partition installed in accordance with Section 708. Where the aggregate area of one or more ambulatory care facilities is greater than 10,000 square feet (929 m^2) on one story, the story shall be provided with a smoke barrier to subdivide the story into no fewer than two smoke compartments. The area of any one such smoke compartment shall be not greater than 22,500 square feet (2092 m^2).*

Discussion and Commentary: A building *used to provide medical, surgical, psychiatric, nursing or similar care on a less than 24-hour basis to individuals who are rendered incapable of self-preservation by the services provided* is considered an ambulatory care facility. Although classified as a Group B occupancy in the same manner as an outpatient clinic or other health care office, such facilities pose distinctly different hazards to life and fire safety because of the presence of individuals who are temporarily rendered incapable of self-preservation that is due to the application of nerve blocks, sedation or anesthesia.

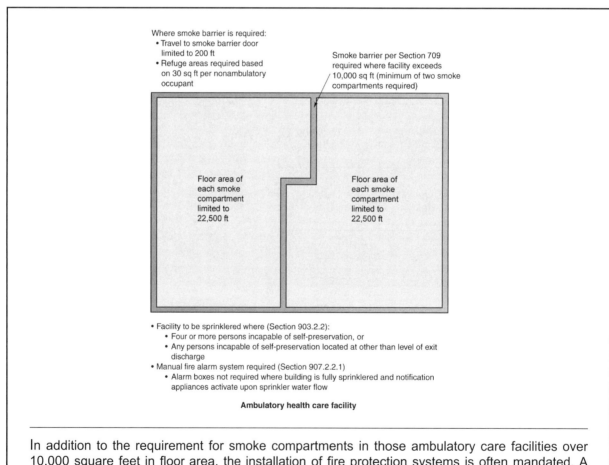

In addition to the requirement for smoke compartments in those ambulatory care facilities over 10,000 square feet in floor area, the installation of fire protection systems is often mandated. A manual fire alarm system is required in all ambulatory care facilities, and an automatic sprinkler system is required where there are four or more care recipients incapable of self-preservation, or if one or more such recipients is located on other than the level of exit discharge.

Study Session 14
IBC Chapter 4

1. The maximum distance from any point within a mall to an exit shall be _____ feet.

    a. 75   b. 200

    c. 250   d. 300

    Reference_____

2. In a covered mall building, groupings of kiosks shall be separated a minimum of _____ feet from other structures within the mall.

    a. 6   b. 10

    c. 20   d. 30

    Reference_____

3. A high-rise building is defined as a building having occupied floors located more than _____ feet above the lowest level of fire department vehicle access.

    a. 55   b. 75

    c. 120   d. 160

    Reference_____

4. In a high-rise building, if standby power is a generator within the building, it shall be located in a separate room enclosed by minimum _____ fire-resistance-rated assemblies or horizontal assemblies, or both.

   a. 1-hour fire barrier   b. 1 hour fire-partition
   c. 2-hour fire wall   d. 2- hour fire barrier

   Reference_____

5. What is the minimum required classification for the interior finish of atrium walls and ceilings?

   a. A   b. B
   c. C   d. DOC FF-1

   Reference_____

6. At other than the lowest level of the atrium, the maximum permitted travel distance within an atrium space is _____ feet.

   a. 75   b. 100
   c. 150   d. 200

   Reference_____

7. An underground building need not be divided into at least two compartments when a floor level is located a maximum of _____ feet below the lowest level of exit discharge.

   a. 30   b. 60
   c. 75   d. 120

   Reference_____

8. In parking facilities, vehicle barriers need not be provided where the difference in adjacent floor elevation is a maximum of _____ inches.

   a. 12   b. 30
   c. 48   d. 60

   Reference_____

9. Canopies under which fuels are dispensed shall have a minimum clearance of _____ above the surface of the drive-through area.

   a. 12 feet, 0 inches
   b. 13 feet, 6 inches
   c. 14 feet, 6 inches
   d. 16 feet, 0 inches

   Reference_____

10. Unless a permitted increase is applied, what is the maximum number of tiers permitted in a ramp-accessed single-use open parking garage of Type IIA construction?

    a. 8
    b. 10
    c. 12
    d. 15

    Reference_____

11. Smoke compartments in a Group I-2 occupancy shall have a maximum floor area of _____ square feet.

    a. 5,000
    b. 10,000
    c. 12,000
    d. 22,500

    Reference_____

12. On floors housing patients confined to beds, a minimum of _____ net square feet per care recipient shall be provided in refuge areas of Group I-2 occupancies.

    a. 3
    b. 6
    c. 15
    d. 30

    Reference_____

13. Doors to resident sleeping units in Group I-3 occupancies shall have a minimum clear width of _____ inches.

    a. 22
    b. 26
    c. 28
    d. 32

    Reference_____

14. In a Group I-3 occupancy, what is the maximum permitted number of residents in any single smoke compartment?

    a. 30   b. 50
    c. 100   d. 200

    Reference_____

15. The minimum permitted size of a regulated motion picture projection room containing 8 projecting machines shall be _____ square feet.

    a. 320   b. 360
    c. 600   d. 640

    Reference_____

16. An approved fire curtain or water curtain need not be provided at the proscenium opening of stages having a maximum height of _____ feet.

    a. 35   b. 50
    c. 55   d. 75

    Reference_____

17. Stages exceeding _____ square feet in floor area shall be provided with emergency ventilation.

    a. 200   b. 400
    c. 500   d. 1,000

    Reference_____

18. What is the minimum required interior finish classification for walls and ceilings in a special amusement building?

    a. A   b. B
    c. C   d. DOC FF-1

    Reference_____

19. An aircraft hangar of Type IIB construction may be classified as Group III for fire suppression purposes, provided the hangar has a maximum single fire area size of _____ square feet.

    a. 5,000   b. 12,000
    c. 20,000   d. 40,000

    Reference _____

20. What is the maximum number of control areas permitted on the sixth floor above grade plane of a research and development building?

    a. zero   b. one
    c. two   d. four

    Reference _____

21. A Group H-2 liquid use, dispensing and mixing room need not be located on the outer perimeter of a building when limited to a maximum floor area of _____ square feet.

    a. 100   b. 200
    c. 500   d. 1,000

    Reference _____

22. An ambulatory care facility shall be separated from adjacent spaces, corridors and tenants with a fire partition where a minimum of _____ care recipients are to be incapable of self-preservation at any time.

    a. 4   b. 10
    c. 20   d. 30

    Reference _____

23. Dead ends in service corridors of Group H-5 occupancies shall be a maximum of _____ feet in length.

    a. 4   b. 15
    c. 20   d. 30

    Reference _____

24. In a drying room, all overhead heating piping shall be located a minimum of _____ inches from combustible contents in the dryer.

    a. 1   b. 2
    c. 4   d. 6

    Reference_____

25. Nitrocellulose storage located in a building where organic coatings are manufactured shall be stored in a room enclosed with minimum _____ fire barriers and/or horizontal assemblies.

    a. 1-hour   b. 2-hour
    c. 3-hour   d. 4-hour

    Reference_____

26. Where the stage height exceeds 50 feet, workshops and storerooms serving the stage shall be separated from the stage by minimum _____ and/or horizontal assemblies.

    a. 1-hour fire partitions   b. 1-hour fire barriers
    c. 2-hour fire barriers   d. 2-hour fire walls

    Reference_____

27. In a nonsprinklered office building, an attic space used for the storage of combustible material shall be protected on the storage side with _____.

    a. $^1/_2$-inch gypsum board
    b. $^1/_2$-inch Type X gypsum board
    c. $^5/_8$-inch gypsum board
    d. any material approved for one-hour construction

    Reference_____

28. A minimum _____ fire-resistance rating is required for the floor construction separating multiple control areas within a Type IIIB building.

    a. $^1/_2$-hour   b. 1-hour
    c. 2-hour   d. 3-hour

    Reference_____

29. Where weather protection shelters an outdoor hazardous material storage area, the sheltered area can only be considered outdoor storage for purposes of the code where the overhead structure is of noncombustible construction and the maximum allowable area of the structure is _____ square feet, plus any permitted increases.

    a. 1,000   b. 1,500
    c. 2,000   d. 3,000

    Reference_____

30. What is the minimum level of wall construction required to enclose spray rooms used for the application of flammable paints?

    a. noncombustible, nonrated   b. 1-hour fire partitions
    c. 1-hour fire barriers   d. 2-hour fire barriers

    Reference_____

31. Where a building contains only a private garage or carport classified as a Group U occupancy, the exterior wall does not require a fire-resistance rating where the fire separation distance is a minimum of _____ feet.

    a. 3   b. 5
    c. 10   d. 20

    Reference_____

32. A temporary platform shall be installed for a maximum of _____ days.

    a. 10   b. 30
    c. 90   d. 180

    Reference_____

33. Unless located a minimum of _____ feet from lot lines or a public way, exterior walls of an aircraft hangar shall have a minimum 2-hour fire-resistance rating.

    a. 20   b. 30
    c. 40   d. 60

    Reference_____

34. Walls separating sleeping units in the same buildings shall comply with the provisions for _____.

    a. fire partitions   b. fire barriers
    c. smoke partitions   d. smoke barriers

    Reference _____

35. The gas detection system required in a hydrogen cutoff room shall be designed to activate when the level of flammable gas exceeds _____ percent of the lower flammability limit for the gas or mixtures present at their anticipated temperature and pressure.

    a. 5   b. 10
    c. 15   d. 25

    Reference _____

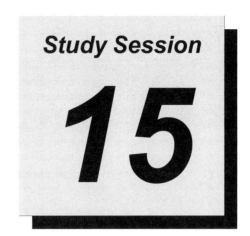

2012 IBC Chapters 14, 15 and 18
Exterior Wall Coverings, Roofs and Foundations

OBJECTIVE: To obtain an understanding of the requirements for exterior wall coverings, including weather-resistant coverings and veneer; roofing assemblies, roof coverings and rooftop structures; and footings and foundations.

REFERENCE: Chapters 14, 15 and 18, 2012 *International Building Code*

KEY POINTS:
- What are the components of a weather-resistant wall envelope?
- What material is considered acceptable as a water-resistive barrier?
- In what geographical areas are Class I or II vapor retarders required on the interior side of framed walls? Under what conditions are Class III vapor retarders permitted?
- Where is flashing to be installed?
- What is the minimum required thickness of wood veneers on exterior walls of Type I, II, III and IV buildings?
- What is anchored masonry veneer? Adhered masonry veneer?
- How must metal veneers be attached? Glass veneer? Stone veneer?
- What limitations are placed on the sill height of operable windows?
- Under what conditions is vinyl siding permitted?
- How is fiber cement siding to be applied?
- How are combustible exterior wall coverings regulated for ignition resistance? Fireblocking?
- What level of fire resistance is required of combustible balconies and similar projections?
- When is fire-retardant-treated wood permitted for the construction of balconies, porches, decks and exterior stairways in Types I and II construction?
- How shall bay windows and oriel windows be constructed?
- What is a roof assembly? Roof covering?
- What is the purpose of a roof assembly?
- How are roof assemblies classified?
- What is the effectiveness of a Class A roof assembly? Class B? Class C?
- What is a nonclassified roof? A special purpose roof? Where are such roofs permitted?
- How must roof covering materials be identified?

403

KEY POINTS:
(Cont'd)

- How shall asphalt shingles be installed? Clay and concrete tile? Wood shakes and shingles? Metal roof panels and roof shingles?
- What is considered a penthouse? When is a penthouse considered an additional story? How must a penthouse be constructed?
- What is the maximum permitted height of a penthouse? Tower or spire?
- How is reroofing addressed?
- What is the minimum depth of a footing below the undisturbed ground surface?
- How shall backfill in the excavated area adjacent to the foundation be placed?
- What special considerations are applicable to footings adjacent to ascending slopes? Descending slopes?
- If not specifically designed, how must footings supporting walls of light-frame construction be constructed?
- Where is dampproofing required? Waterproofing?
- What are deep foundations? Shallow foundations? How are such foundations regulated?

Topic: Weather Protection
Reference: IBC 1403.2
Category: Exterior Walls
Subject: Performance Requirements

Code Text: *Exterior walls shall provide the building with a weather-resistant exterior wall envelope. The exterior wall envelope shall include flashing, as described in Section 1405.4. The exterior wall envelope shall be designed and constructed in such a manner as to prevent the accumulation of water within the wall assembly by providing a water-resistive barrier behind the exterior veneer, as described in Section 1404.2, and a means for draining water that enters the assembly to the exterior. See three exceptions.*

Discussion and Commentary: The code considers it necessary to apply at least one layer of No. 15 asphalt felt, complying with ASTM D 226 for Type 1 felt, in order to provide the water-resistive barrier between the sheathing and exterior wall veneer. The use of flashing in conjunction with the felt provides a continuous barrier against water penetration.

TABLE 1405.2
MINIMUM THICKNESS OF WEATHER COVERINGS

COVERING TYPE	MINIMUM THICKNESS (inches)
Adhered masonry veneer	0.25
Aluminum siding	0.019
Anchored masonry veneer	2.625
Asbestos-cement boards	0.125
Asbestos shingles	0.156
Cold-rolled copper[d]	0.0216 nominal
Copper shingles[d]	0.0162 nominal
Exterior plywood (with sheathing)	0.313
Exterior plywood (without sheathing)	See Section 2304.6
Fiber cement lap siding	0.25[c]
Fiber cement panel siding	0.25[c]
Fiberboard siding	0.5
Glass-fiber reinforced concrete panels	0.375
Hardboard siding[c]	0.25
High-yield copper[d]	0.0162 nominal
Lead-coated copper[d]	0.0216 nominal
Lead-coated high-yield copper	0.0162 nominal
Marble slabs	1
Particleboard (with sheathing)	See Section 2304.6
Particleboard (without sheathing)	See Section 2304.6
Porcelain tile	.025
Precast stone facing[e]	0.625
Steel (approved corrosion resistant)	0.0149
Stone (cast artificial, anchored)	1.5

(Continued)

For all exterior walls other than those constructed of concrete or masonry, the IBC requires the installation of a weather-resistant exterior wall envelope. Any other approved method to resist condensation and moisture leakage is also acceptable.

Topic: Flashing
Reference: IBC 1405.4
Category: Exterior Walls
Subject: Installation of Wall Coverings

Code Text: *Flashing shall be installed in such a manner so as to prevent moisture from entering the wall or to redirect it to the exterior. Flashing shall be installed at the perimeters of exterior door and window assemblies, penetrations and terminations of exterior wall assemblies, exterior wall intersections with roofs, chimneys, porches, decks, balconies and similar projections and at built-in gutters and similar locations where moisture could enter the wall. Flashing with projecting flanges shall be installed on both sides and the ends of copings, under sills and continuously above projecting trim.*

Discussion and Commentary: In general, the code requires that all intersections of exterior surfaces and/or components be flashed to prevent water intrusion. Roof and wall intersections and parapets are especially troublesome, as are exterior wall openings exposed to weather and, in particular, wind-driven rain.

For SI: 1 inch = 25.4 mm. **Chimney flashing detail**

Where anchored masonry veneer is installed, the masonry must be provided with flashing and weepholes in the first course above finished ground level above the foundation wall or slab, as well as at other points of support.

Topic: Definitions
Reference: IBC 1405.5, 1405.11, 202
Category: Exterior Walls
Subject: Veneer

Code Text: *Veneer is a facing attached to a wall for the purpose of providing ornamentation, protection, or insulation, but not counted as adding strength to the wall. Adhered masonry veneer is veneer secured and supported through the adhesion of an approved bonding material applied to an approved backing. Anchored masonry veneer is veneer secured with approved mechanical fasteners to an approved backing. Wood veneers on exterior walls of buildings of Type I, II, III and IV construction shall be not less than 1 inch (25 mm) nominal thickness, 0.438-inch (11.1 mm) exterior hardboard siding or 0.375-inch (9.5 mm) exterior-type wood structural panels or particleboard. Metal veneers shall not be less than 0.0149-inch (0.378 mm) nominal thickness sheet steel mounted on wood or metal furring strips or approved sheathing on the wood construction.*

Discussion and Commentary: Years ago, veneer was considered an ornamental facing for a masonry wall. Today, the IBC regulates a variety of veneer materials: wood, anchored masonry, stone, slab-type, terra cotta, adhered masonry, metal, glass and, in Chapter 26, plastic. The code regulates material size, type and attachment, as well as other concerns that would cause the veneer to fail.

Generic application of anchored masonry veneer (5 in. max thickness)

For SI: 1 inch = 25.4 mm, 1 square foot = 0.093 m².

As they are constantly subjected to alternate cycles of wetting and drying, the anchors, ties or supports used in the attachment of veneer must be corrosion resistant. These materials, when used on the exterior of a building, must support the veneer properly for the life of the building.

Topic: Balconies
Reference: IBC 1406.3
Category: Exterior Walls
Subject: Combustible Materials on Exterior Walls

Code Text: *Balconies and similar projections of combustible construction other than fire-retardant-treated wood shall be fire-resistance rated where required by Table 601 for floor construction or shall be of Type IV construction in accordance with Section 602.4. The aggregate length of the projections shall not exceed 50 percent of the building's perimeter on each floor.* See four exceptions.

Discussion and Commentary: Although projections may extend beyond the floor area of a building, they still pose some degree of hazard due to their materials of construction. Combustible projections are generally limited to buildings of Types III, IV and V construction; however, fire-retardant-treated wood is acceptable for balconies, porches, decks and exterior stairways not used as required exits in Type I or II buildings that are three stories or less in height.

Balconies and similar projections, when combustible, to have fire-resistive rating as for floor, or be of Type IV construction, except:

- Fire-retardant-treated wood in Types I and II, limited to 3 stories above grade plane, and not used for exiting

- Permitted to be Type V in buildings of Type III, IV or V with no fire rating where sprinkler protection extended to projections

- Guard devices, such as pickets and rails, limited to 42 inches in height

Balconies and similar projections, when combustible, to have rating as for floor construction or of Type IV construction (See exceptions)

If not constructed of fire-retardant-treated wood, combustible projections must be of heavy timber or must provide the same degree of fire resistance as the building's floor construction. However, in sprinklered buildings of combustible construction, no fire-resistance rating is required.

Topic: Scope and Definitions
Reference: IBC 1501.1, 202

Category: Roof Assemblies and Rooftop Structures
Subject: Roof Assembly

Code Text: *The provisions of Chapter 15 shall govern the design, materials, construction and quality of roof assemblies and rooftop structures. A roof assembly is a system designed to provide weather protection and resistance to design loads. The system consists of a roof covering and roof deck or a single component serving as both the roof covering and the roof deck. A roof assembly includes the roof deck, vapor retarder, substrate or thermal barrier, insulation, vapor retarder and roof covering. Roof covering is the covering applied to the roof deck for weather resistance, fire classification or appearance.*

Discussion and Commentary: There are many components of a roof assembly. Viewed as a unit, a roof assembly is regulated for its resistance to wind, weathering, impact and fire. In addition, roof coverings must be designed, installed and maintained to protect the building from the weather.

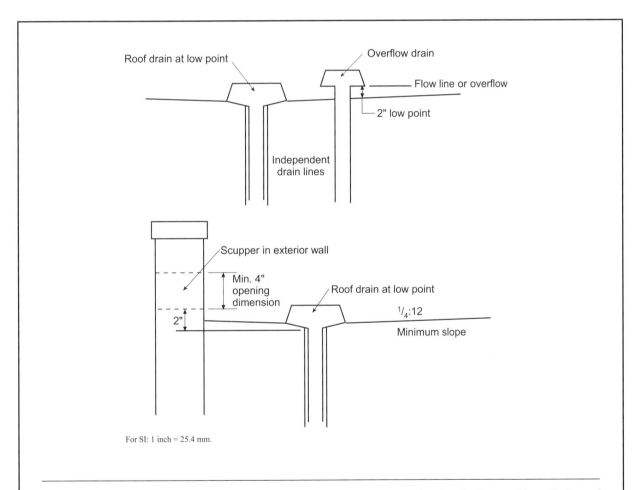

The *International Plumbing Code* is referenced in Section 1503.4 for the design and installation of roof drainage systems. Where water on the roof is not intended to flow over the roof edge, provisions for roof drainage will include primary roof drains supplemented by overflow drains.

Topic: General Requirements
Reference: IBC 1503, 1507

Category: Roof Assemblies and Rooftop Structures
Subject: Roof Coverings

Code Text: *Roof decks shall be covered with approved roof coverings secured to the building or structure in accordance with the provisions of Chapter 15. Roof coverings shall be designed and installed in accordance with the IBC and the approved manufacturer's installation instructions such that the roof covering shall serve to protect the building or structure. Roof coverings shall be applied in accordance with the applicable provisions of Section 1507 and the manufacturer's instructions.*

Discussion and Commentary: The IBC contains installation requirements for a number of types of roof covering materials and systems. Selectively included in the provisions are deck requirements, limitations on roof slope, underlayment, materials, fasteners and attachment, flashings and application methods.

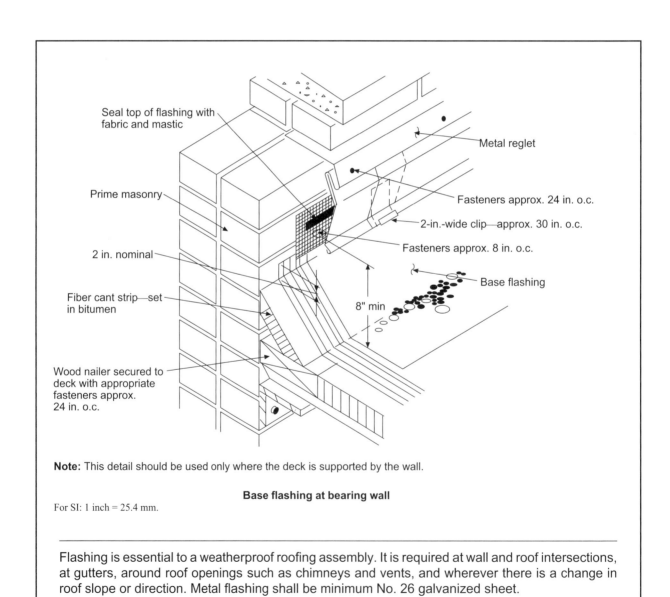

Note: This detail should be used only where the deck is supported by the wall.

Base flashing at bearing wall

For SI: 1 inch = 25.4 mm.

Flashing is essential to a weatherproof roofing assembly. It is required at wall and roof intersections, at gutters, around roof openings such as chimneys and vents, and wherever there is a change in roof slope or direction. Metal flashing shall be minimum No. 26 galvanized sheet.

Topic: Gravel and Stone Materials
Reference: IBC 1504.8
Category: Roof Assemblies and Rooftop Structures
Subject: Performance Requirements

Code Text: *Aggregate used as surfacing for roof coverings and aggregate, gravel or stone used as ballast shall not be used on the roof of a building located in a hurricane-prone region as defined in Section 202, or on any other building with a mean roof height exceeding that permitted by Table 1504.8 based on the exposure category and basic wind speed at the site.*

Discussion and Commentary: Field assessments of damage to buildings caused by high-wind events have shown that aggregate, gravel or stone blown from the roofs of buildings has increased the damage to other buildings due to the breakage of glass. Once the glass is broken, higher internal pressures are created within the building, often resulting in substantial structural damage. In addition, breakage of windows will generally result in considerable wind and water damage to the building's interior and contents.

TABLE 1504.8
MAXIMUM ALLOWABLE MEAN ROOF HEIGHT PERMITTED FOR BUILDINGS WITH AGGREGATE ON THE ROOF IN AREAS OUTSIDE A HURRICANE-PRONE REGION

NOMINAL DESIGN WIND SPEED, V_{asd} (mph)[b,d]	MAXIMUM MEAN ROOF HEIGHT (ft)[a,c]		
	Exposure category		
	B	C	D
85	170	60	30
90	110	35	15
95	75	20	NP
100	55	15	NP
105	40	NP	NP
110	30	NP	NP
115	20	NP	NP
120	15	NP	NP
Greater than 120	NP	NP	NP

For SI: 1 foot = 304.8 mm; 1 mile per hour = 0.447 m/s.
a. Mean roof height as defined in ASCE 7.
b. For intermediate values of V_{asd}, the height associated with the next higher value of V_{asd} shall be used, or direct interpolation is permitted.
c. NP = gravel and stone not permitted for any roof height.
d. V_{asd} shall be determined in accordance with Section 1609.3.1.

Aggregate, gravel and stone is prohibited on roofs of buildings in hurricane-prone regions, defined as areas along the U.S. Atlantic Ocean and Gulf of Mexico coasts where the ultimate design wind speed exceeds 115 miles per hour, as well as the islands of Hawaii, Puerto Rico, Guam, Virgin Islands and American Samoa.

Topic: General Requirements
Reference: IBC 1505.1

Category: Roof Assemblies and Rooftop Structures
Subject: Fire Classification

Code Text: *Roof assemblies shall be divided into the classes defined below. Class A, B and C roof assemblies and roof coverings required to be listed by Section 1505 shall be tested in accordance with ASTM E 108 or UL 790. The minimum roof coverings installed on buildings shall comply with Table 1505.1 based on the type of construction of the building. See exception for skylights and sloped glazing.*

Discussion and Commentary: The various required roof covering classifications are related directly to the type of construction of the building. Based on Table 1505.1, a minimum level of fire protection is assigned to address external fire exposures. The exposures are generally created by fires in adjoining structures, wild fires and fire from the subject building that may extend up the exterior wall and onto the top surface of the roof.

TABLE 1505.1[a, b]
MINIMUM ROOF COVERING CLASSIFICATION FOR TYPES OF CONSTRUCTION

IA	IB	IIA	IIB	IIIA	IIIB	IV	VA	VB
B	B	B	C[c]	B	C[c]	B	B	C[c]

For SI: 1 foot = 304.8 mm, 1 square foot = 0.0929 m^2.

a. Unless otherwise required in accordance with the *International Wildland-Urban Interface Code* or due to the location of the building within a fire district in accordance with Appendix D.

b. Nonclassified roof coverings shall be permitted on buildings of Group R-3 and Group U occupancies, where there is a minimum fire-separation distance of 6 feet measured from the leading edge of the roof.

c. Buildings that are not more than two stories above grade plane and having not more than 6,000 square feet of projected roof area and where there is a minimum 10-foot fire-separation distance from the leading edge of the roof to a lot line on all sides of the building, except for street fronts or public ways, shall be permitted to have roofs of No. 1 cedar or redwood shakes and No. 1 shingles.

In addition to the Class A, B and C listed roof assemblies, the IBC permits the use of nonclassified roofing and special purpose roofs under limited conditions. These types of roof coverings are limited, respectively, to Group R-3 occupancies and small nonfire-rated buildings.

Topic: Roof Assemblies
Reference: IBC 1505
Category: Roof Assemblies and Rooftop Structures
Subject: Fire Classification

Code Text: *Class A roof assemblies are those that are effective against severe fire-test exposure. Class B roof assemblies are those that are effective against moderate fire-test exposure. Class C roof assemblies are those that are effective against light fire-test exposure. Nonclassified roofing is approved material that is not listed as a Class A, B or C roof covering.*

Discussion and Commentary: Class A roof coverings include masonry, concrete, slate, tile and cement-asbestos, as well as ferrous or copper shingles or sheets, metal sheets and metal shingles. Additionally, any assembly that is tested as Class A in accordance with ASTM E 108 or UL 790 by an approved testing agency and is listed and identified by that agency is included. There are no specific materials that qualify as Class B or C; therefore all Class B and C roof covering are listed as such.

Brands for Classes A, B, and C Tests

Several test methods are included as parts of the fire-test-response standards ASTM E 108 and UL 790, including the intermittent flame exposure test, spread of flame test, burning brand test, flying brand test and rain test. It also is critical that the roof coverings do not slip from position.

Topic: Identification
Reference: IBC 1506.4
Category: Roof Assemblies and Rooftop Structures
Subject: Roof Covering Materials

Code Text: *Roof covering materials shall be delivered in packages bearing the manufacturer's identifying marks and approved testing agency labels required in accordance with Section 1505. Bulk shipments of materials shall be accompanied with the same information issued in the form of a certificate or on a bill of lading by the manufacturer.*

Discussion and Commentary: Roof covering materials must comply with the appropriate quality standards set forth in Section 1507 for each different type of material. The materials must be compatible with the building or structure to which they are applied. In addition, identification of the roof covering materials is mandatory to verify that they comply with the quality standard. Gardens and other landscaping installed on a roof are specifically regulated in regard to roof construction and structural integrity. For structural purposes, the provisions of Sections 1607.12.3 and 1607.12.3.1 addressing occupiable and landscaped roofs, respectively, are applicable.

Material	Section
Asphalt Shingles	Section 1507.2
Clay and Concrete Tile	Section 1507.3
Metal Roof Panels	Section 1507.4
Metal Roof Shingles	Section 1507.5
Mineral-surfaced Roll Roofing	Section 1507.6
Slate Shingles	Section 1507.7
Wood Shingles	Section 1507.8
Wood Shakes	Section 1507.9
Built-up Roofs	Section 1507.10
Modified Bitumen Roofing	Section 1507.11
Thermoset Single-ply Roofing	Section 1507.12
Thermoplastic Single-ply Roofing	Section 1507.13
Sprayed Polyurethane Foam Roofing	Section 1507.14
Liquid-applied Roofing	Section 1507.15

Where there are not applicable standards for a specific roof covering material, or where the materials are of questionable suitability, the building official must ask for testing by an approved agency to determine the material's character, quality and limitations of application.

Topic: Asphalt Shingles
Reference: IBC 1507.2
Category: Roof Assemblies and Rooftop Structures
Subject: Roof Coverings

Code Text: *Asphalt shingles shall be fastened to solidly sheathed decks. Asphalt shingles shall only be used on roof slopes of two units vertical in 12 units horizontal (17-percent slope) or greater. Asphalt shingles shall have the minimum number of fasteners required by the manufacturer, but not less than four fasteners per strip shingle or two fasteners per individual shingle. For roof slopes from two units vertical in 12 units horizontal (17-percent slope), up to four units vertical in 12 units horizontal (33-percent slope), underlayment shall be two layers . . . . Distortions in the underlayment of the shingles shall not interfere with the ability of the shingles to seal. Provide drip edge at eaves and gables of shingle roofs.*

Discussion and Commentary: There are two fundamental types of asphalt shingles: strip shingles (the most common type) such as three-tab shingles, and individual interlocking shingles such as t-lock shingles. In addition to three-tab, other strip shingles include random or multi-tab, no-cut-out and laminated architectural.

Note: In areas where there has been a history of ice forming along the eaves causing a backup of water, felt plies of underlayment should be cemented up from eaves far enough to overlie a point 24 in. inside the wall line of the building.

source NRCA

For SI: 1 inch = 25.4 mm, °C = [(°F)-32/1.8].

Application of asphalt shingle on slopes between 2:12 and 4:12

Asphalt shingles are typically classified in two types, either cellulose felt reinforced (i.e., organic shingles) and fiberglass mat reinforced (i.e., fiberglass shingles). Consistent with the requirements of Section 1203.2, the roofing industry recommends the attic space below asphalt shingle roofs be properly ventilated.

Topic: Wood Shakes
Reference: IBC 1507.9
Category: Roof Assemblies and Rooftop Structures
Subject: Roof Coverings

Code Text: *Wood shakes shall only be used on solid or spaced sheathing. Wood shakes shall only be used on slopes of four units vertical in 12 units horizontal (33-percent slope) or greater. Interlayment shall comply with ASTM D 226, Type I. Fasteners for wood shakes shall be corrosion resistant with a minimum penetration of $^3/_4$ inch (19.1 mm) into the sheathing. Wood shakes shall be laid with a side lap not less than $1^1/_2$ inches (38 mm) between joints in adjacent courses. Spacing between shakes in the same course shall be $^3/_8$ to $^5/_8$ (9.5 to 15.9 mm) inches for shakes and taper sawn shakes of naturally durable wood and shall be $^1/_4$ to $^3/_8$ inch (6.4 to 9.5 mm) for preservative taper sawn shakes. Weather exposure for wood shakes shall not exceed those set in Table 1507.9.8.*

Discussion and Commentary: Wood shakes, which are defined as roofing products split from logs and then shaped as required by the individual manufacturers, differ from wood shingles in that shingles are defined as sawed wood products featuring a uniform butt thickness per individual length.

Wood shake application

For SI: 1 inch = 25.4 mm.

Both wood shakes and wood shingles are required to be labeled by an approved third-party inspection agency. The applicable set of grading rules, required of the quality control program, is typically prescribed by the Cedar Shake and Shingle Bureau.

Topic: Penthouses
Reference: IBC 1509.2, 202

Category: Roof Assemblies and Rooftop Structures
Subject: Rooftop Structures

Code Text: *A penthouse is an enclosed, unoccupied rooftop structure used for sheltering mechanical and electrical equipment, tanks, elevators and related machinery, and vertical shaft openings. Penthouses in compliance with Sections 1509.2.1 through 1509.2.5 shall be considered as a portion of the story directly below the roof deck on which such penthouses are located. All other penthouses shall be considered as an additional story of the building. Penthouses shall not be used for purposes other than the shelter of mechanical or electrical equipment, tanks, or vertical shaft openings in the roof assembly.*

Discussion and Commentary: The general premise is that a penthouse be treated no differently than any other portion of the building. However, the reductions in the general requirements for a story recognize the lack of occupant load or fire loading, as well as the reduced exposure of penthouses when the exterior wall is recessed from the exterior wall of the building. Penthouses do not contribute to the building area, number of stories or fire area of the buildings on which they are located.

Penthouse limitations

For SI: 1 foot = 304.8 mm

Special allowances are provided for towers, spires, domes and cupolas. It is important, however, to limit the height of such structures where constructed of combustible materials. Limited provisions also regulate the installation of tanks and cooling towers on buildings.

Topic: Foundation Elevation
Reference: IBC 1808.7.4

Category: Soils and Foundations
Subject: Foundations

Code Text: *On graded sites, the top of any exterior foundation shall extend above the elevation of the street gutter at point of discharge or the inlet of an approved drainage device a minimum of 12 inches (305 mm) plus 2 percent. Alternate elevations are permitted subject to the approval of the building official, provided it can be demonstrated that required drainage to the point of discharge and away from the structure is provided at all locations on the site.*

Discussion and Commentary: Where natural drainage away from a building is not available, the site must be graded so that water will not drain toward, or accumulate at, the exterior foundation wall. A prescriptive elevation is set forth that will ensure positive drainage to a street gutter or other drainage point; however, any other method that moves water away from the building can be accepted by the building official.

Footing elevation on graded sites

Foundation clearances from slopes

For SI: 1 foot = 304.8 mm, 1 inch = 25.4 mm.

Where footings are located adjacent to a slope steeper than 1:3 (1 vertical to 3 horizontal), either at the top or the bottom, special clearances between the building and the sloping surfaces are required to protect against slope drainage, erosion and shallow failures.

Topic: Frost Protection
Reference: IBC 1809.2, 1809.3
Category: Soils and Foundations
Subject: Footings and Foundations

Code Text: *Shallow foundations shall be built on undisturbed soil, compacted fill material, or controlled low-strength material (CLSM). The top surface of footings shall be level. The bottom surface of footings are permitted to have a slope not exceeding 1 unit vertical in 10 units horizontal (10-percent slope). Footings shall be stepped where it is necessary to change the elevation of the top surface of the footing or where the surface of the ground slopes more than 1 unit vertical in 10 units horizontal (10-percent slope).*

Discussion and Commentary: If compacted fill material is used to support a footing, the material must be in compliance with the provisions of an approved report. The code identifies a number of issues that must be addressed in the report, including specifications for both the site preparation and the material to be used as fill.

For SI: 1 foot = 304.8 mm.

Footings must be placed at least 12 inches below the undisturbed ground surface. Where the site is recognized to contain shifting or moving soils, the footings must be extended to a sufficient depth to ensure stability.

Topic: Frost Protection
Reference: IBC 1809.4, 1809.5

Category: Soils and Foundations
Subject: Shallow Foundations

Code Text: *The minimum depth of footings below the undisturbed ground surface shall be 12 inches (305 mm). Except where otherwise protected from frost, foundations and other permanent supports of buildings and structures shall be protected from frost by one or more of the following methods: 1) extending below the frost line of the locality, 2) construction in accordance with ASCE-32* (Design and Construction of Frost Protected Shallow Foundations), *or 3) erecting on solid rock.* See exception for small free-standing structures. *Shallow footings shall not bear on frozen soil unless such frozen condition is of a permanent character.*

Discussion and Commentary: In winter, frost action can raise the ground level (frost heave), whereas in springtime, the same area will soften and settle back. If foundations are constructed on soils that can freeze, then the heave or vertical movement of the ground, which is rarely uniform, can cause serious damage to buildings and other structures.

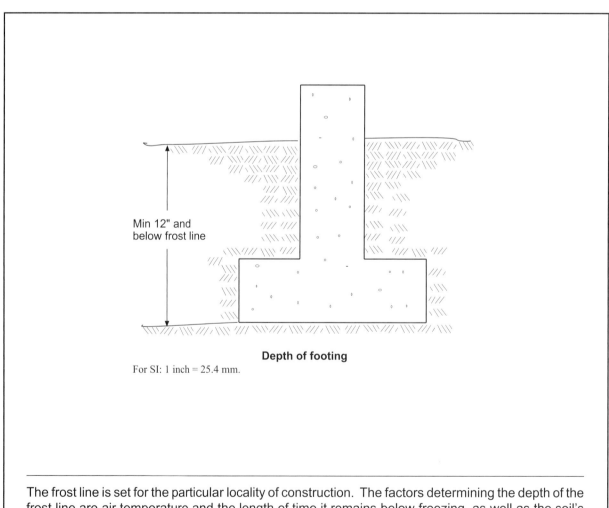

Depth of footing

For SI: 1 inch = 25.4 mm.

The frost line is set for the particular locality of construction. The factors determining the depth of the frost line are air temperature and the length of time it remains below freezing, as well as the soil's level of thermal conductivity and its ability to conduct heat.

Topic: Light-Frame Construction
Reference: IBC 1809.7
Category: Soils and Foundations
Subject: Footings

Code Text: *Where a specific design is not provided, concrete or masonry-unit footings supporting walls of light frame construction are permitted to be designed in accordance with Table 1809.7.*

Discussion and Commentary: In lieu of an engineered design, Table 1809.7 provides a prescriptive method for determining footing size criteria that can be used in conjunction with conventional light-framed construction. The minimum thickness of the foundation wall, as well as the minimum width and thickness of the footing, are specified based on the number of floors supported. The minimum depth below undisturbed ground surface is also addressed. Unless protected from frost or erected on solid rock, the footings must also extend below the frost line. The table is based on anticipated loads on the footings and foundations due to wall, floor and roof systems.

TABLE 1809.7
PRESCRIPTIVE FOOTINGS SUPPORTING WALLS OF LIGHT-FRAME CONSTRUCTION[a, b, c, d, e]

NUMBER OF FLOORS SUPPORTED BY THE FOOTING[f]	WIDTH OF FOOTING (inches)	THICKNESS OF FOOTING (inches)
1	12	6
2	15	6
3	18	8[g]

For SI: 1 inch = 25.4 mm, 1 foot = 304.8 mm.

a. Depth of footings shall be in accordance with Section 1809.4.
b. The ground under the floor shall be permitted to be excavated to the elevation of the top of the footing.
c. Interior stud-bearing walls shall be permitted to be supported by isolated footings. The footing width and length shall be twice the width shown in this table, and footings shall be spaced not more than 6 feet on center.
d. See Section 1908 for additional requirements for concrete footings of structures assigned to Seismic Design Category C, D, E or F.
e. For thickness of foundation walls, see Section 1807.1.6.
f. Footings shall be permitted to support a roof in addition to the stipulated number of floors. Footings supporting roof only shall be as required for supporting one floor.
g. Plain concrete footings for Group R-3 occupancies shall be permitted to be 6 inches thick.

Although Table 1809.7 is normally used for continuous footings, it can also be used for isolated footings that support interior-stud bearing walls. The footings shall be spaced a maximum of 6 feet on center, with their widths and lengths being twice that shown in the table.

Study Session 15

Study Session 15
IBC Chapters 14, 15 and 18

1. An exterior wall is defined as a building enclosing wall that has a minimum slope of _____ degrees with the horizontal plane.

   a. 45   b. 60
   c. 75   d. 90

   Reference_____

2. Veneer secured and supported through the adhesion of an approved bonding material applied to an approved backing is considered _____ masonry veneer.

   a. adhered   b. anchored
   c. attached   d. spandrel

   Reference_____

3. A minimum of _____ shall be attached to exterior sheathing in order to provide a continuous water-resistive barrier behind the exterior wall veneer.

   a. one layer of No. 15 asphalt felt
   b. one layer of No. 30 asphalt felt
   c. two layers of No. 15 asphalt felt applied shingle fashion
   d. two layers of No. 30 asphalt felt applied shingle fashion

   Reference_____

4. Precast stone facing shall be a minimum of _____ inch in thickness in order to be acceptable as an approved weather covering.

   a. 0.50   b. 0.625
   c. 0.75   d. 1.00

   Reference_____

5. Where thin exterior structural glass veneer is located no more than 15 feet above grade, the area of a single veneer section is limited to a maximum of _____ square feet.

   a. 6   b. 8
   c. 10   d. 12

   Reference_____

6. What is the minimum permitted thickness for metal veneer mounted on approved sheathing on wood construction?

   a. 0.0149 inch   b. 0.0224 inch
   c. 0.0478 inch   d. 0.0630 inch

   Reference_____

7. Combustible exterior wall coverings are permitted in Type II buildings if located a maximum height of _____ feet above grade plane.

   a. 25   b. 30
   c. 35   d. 40

   Reference_____

8. Roofing interlayment shall have a minimum width of _____ inches.

   a. 12   b. 18
   c. 34   d. 36

   Reference_____

9. What is the minimum required roof covering classification for a roof assembly on a building of Type IA construction?

    a. Class A   b. Class B

    c. Class C   d. nonclassified

Reference_____

10. Roof assemblies consisting of metal sheets or shingles installed on noncombustible decks are considered _____ roof assemblies.

    a. Class A   b. Class B

    c. Class C   d. special purpose

Reference_____

11. Double underlayment application is required beneath asphalt shingles on roofs having a slope of 2:12 to _____.

    a. $2^{1}/_{2}$:12   b. 3:12

    c. 4:12   d. 5:12

Reference_____

12. In areas subject to high winds, underlayment beneath asphalt shingles shall be fastened along the overlap at a maximum spacing of _____ inches on center.

    a. 2   b. 18

    c. 24   d. 36

Reference_____

13. Fasteners for concrete or clay roof tiles shall penetrate the deck a minimum of _____ inch or through the thickness of the deck, whichever is less.

    a. $^{1}/_{2}$ b. $^{5}/_{8}$

    c. $^{3}/_{4}$ d. 1

Reference_____

14. What is the minimum permitted roof slope for the installation of metal roof shingles?

    a. 2:12   b. 3:12

    c. 4:12   d. 5:12

Reference_____

15. Wood shakes shall be applied to a roof with a minimum side lap of _____ inch(es) between joints in adjacent courses.

    a. $^3/_8$
    b. $^1/_2$
    c. $^3/_4$
    d. $1^1/_2$

    Reference_____

16. Where 18-inch-long No. 1 wood shingles are installed on a roof of 12:12 pitch, the maximum weather exposure shall be _____ inches.

    a. $3^3/_4$
    b. $4^1/_4$
    c. $4^1/_2$
    d. $5^1/_2$

    Reference_____

17. Sprayed polyurethane foam roofs shall have a minimum design slope of _____ for drainage purposes.

    a. $^1/_8$:12
    b. $^1/_4$:12
    c. 1:12
    d. 2:12

    Reference_____

18. In other than Type I construction, a penthouse shall extend a maximum of _____ feet above the roof deck when used for the protection of rooftop mechanical equipment.

    a. 8
    b. 12
    c. 18
    d. 28

    Reference_____

19. The bottom surface of footings shall have a maximum slope of _____.

    a. $^1/_2$:12
    b. 1:12
    c. 1:10
    d. 1:20

    Reference_____

Study Session 15

20. The bottom of a footing shall be located a minimum of _____ inches below the undisturbed ground surface.

    a. 6   b. 12
    c. 15   d. 18

    Reference_____

21. Unless data is submitted that substantiates the use of a higher value, the presumptive vertical foundation pressure for silty gravel supporting soil shall be _____ pounds per square foot.

    a. 12,000   b. 4,000
    c. 3,000   d. 2,000

    Reference_____

22. Where a swimming pool is located adjacent to a slope, any pool wall located a maximum distance of _____ feet from the top of a slope shall be capable of supporting the water in the pool without soil support.

    a. 5   b. 7
    c. 10   d. 15

    Reference_____

23. Concrete footings shall be protected from freezing during depositing and for a minimum time period of _____ thereafter.

    a. 12 hours   b. 24 hours
    c. 3 days   d. 5 days

    Reference_____

24. Unless a specific design is provided, concrete footings supporting two floors of light-frame construction shall be a minimum of _____ inches in width and _____ inches in thickness.

    a. 12, 6   b. 15, 6
    c. 15, 7   d. 18, 8

    Reference_____

25. At the girder supports at the top of hollow masonry foundation walls, a minimum of _____ inches of solid masonry shall be provided.

    a. 3   b. 4

    c. 6   d. 8

Reference_____

26. Where glass veneer extends to the surface of a sidewalk, each section shall be set a minimum of _____ inch above the highest point of the sidewalk.

    a. $1/4$ b. $1/2$

    c. $5/8$ d. $3/4$

Reference_____

27. The construction of a special purpose wood shake roof mandates a minimum underlayment of _____ placed under the roof sheathing.

    a. $5/8$-inch wood structural panel

    b. $1/2$-inch gypsum wallboard

    c. $5/8$-inch Type X water resistant gypsum backer board or sheathing

    d. $1/2$-inch water-resistant gypsum backing board

Reference_____

28. Slate shingles shall be installed only on roof decks having a slope of _____ or greater.

    a. 2:12   b. $2^1/_2$:12

    c. 3:12   d. 4:12

Reference_____

29. A cupola used as an architectural embellishment, where built of combustible construction, is limited to a maximum height of _____ feet above grade plane.

    a. 45   b. 50

    c. 60   d. 85

Reference_____

30. Unless warranted by climatic or soil conditions, a minimum ground slope of _____ is required away from a building's foundation wall for a minimum distance of _____ feet.

    a. 1:48, 10
    b. 1:20, 10
    c. 1:15, 5
    d. 1:12, 5

    Reference _____

31. Sheet polyethylene and nonperforated aluminum foil are considered to be Class _____ vapor retarders.

    a. 0
    b. I
    c. II
    d. III

    Reference _____

32. Fiber cement horizontal lap siding shall be installed with a minimum lap of _____ inch(es).

    a. $3/4$
    b. 1
    c. $1 1/4$
    d. $1 1/2$

    Reference _____

33. Fire-retardant-treated wood is permitted for the construction of balconies, porches, decks and exterior stairways of Type I and II buildings, provided the building is a maximum of _____ in height above grade plane.

    a. 3 stories
    b. 4 stories
    c. 55 feet
    d. 75 feet

    Reference _____

34. Where a building is located on a site with a nominal design wind speed of 90 mph and an exposure category of B, aggregate is permitted on the roof, provided the building has a maximum height of _____ feet.

    a. 0 (aggregate covering materials prohibited in Exposure Category B)
    b. 15
    c. 35
    d. 110

    Reference _____

35. The steel pipe or tube used for a micropile shall have a minimum wall thickness of _____ inch.

    a. $1/8$ b. $3/16$

    c. $1/4$ d. $5/16$

Reference _____

Study Session 16

2012 IBC Chapters 16, 17, 19, 21, 22 and 23
Special Inspections, Concrete, Masonry and Wood

OBJECTIVE: To identify the provisions relating to general structural forces and engineered design; applicable structural tests; special inspections and structural observation; and specific materials of construction, including concrete, masonry, steel and wood.

REFERENCE: Chapters 16, 17, 19, 21, 22 and 23, 2012 *International Building Code*

KEY POINTS:
- In structural design, what is considered a live load? A dead load?
- How is the minimum design live load for a floor system determined? Concentrated loads? Partition loads?
- How shall design roof loads be determined? When should snow loads be considered?
- What is the design criteria for wind loads? Seismic loads? Flood loads?
- What is special inspection? What are the duties and responsibilities of a special inspector?
- Which types of work shall be inspected by a special inspector?
- What is structural observation? When is structural observation required?
- What information is required in the statement of special inspections?
- How does Chapter 19 (Concrete) relate to the provisions of ACI 318?
- How shall concrete be evaluated and accepted?
- Where are the general requirements for concrete mixing, conveying, depositing and curing located?
- In masonry construction, how shall mortar and grout be regulated?
- How shall masonry be prepared, constructed and protected in cold weather? In hot weather?
- How must a masonry chimney be designed, anchored, supported and reinforced?
- What are the minimum required thicknesses of masonry fireplace walls and firebox walls?
- What is the minimum clearance required between combustible materials and fireplace or chimney walls? Between combustible materials and the fireplace opening?
- How must the hearth be constructed for a masonry fireplace?
- Hearth extensions must be of what minimum size?

KEY POINTS:
(Cont'd)
- At what minimum height must a masonry chimney terminate?
- How can the minimum capacity of structural wood-framing members be established?
- For decay and termite protection, what manner of under-floor clearance is required between exposed ground and wood girders, joists or structural floors?
- How must structural floor and roof sheathing be designed?
- What is considered "conventional light-frame construction"?
- How shall girders be supported?
- What are the limitations on the notching and boring of holes in floor joists, ceiling joists and roof rafters?
- What are braced wall panels? Where are such panels required?
- How shall rafters be framed at the ridge?
- What is a purlin? How shall a purlin system be constructed?

Topic: Uniform and Concentrated Loads **Category:** Structural Design
Reference: IBC 1607.3, 1607.4 **Subject:** Live Loads

Code Text: *The live loads used in the design of buildings and other structures shall be the maximum loads expected by the intended use or occupancy but shall in no case be less than the minimum uniformly distributed live loads given in Table 1607.1. Floors and other similar surfaces shall be designed to support the uniformly distributed live loads prescribed in Section 1607.3 or the concentrated live loads, in pounds (kilonewtons), given in Table 1607.1, whichever produces the greater load effects.*

Discussion and Commentary: The anticipated live loads are based on the daily use of the building, as well as any temporary loading conditions such as remodeling activities, large group gatherings and short-term storage. They also reflect that, within the general use category, changes will likely occur in furniture layout, traffic patterns, etc. Concentrated loads take into account more specific types of loading consistent with the use of the building.

TABLE 1607.1
MINIMUM UNIFORMLY DISTRIBUTED LIVE LOADS, L_o, AND MINIMUM CONCENTRATED LIVE LOADS[g]

OCCUPANCY OR USE	UNIFORM (psf)	CONCENTRATED (lbs.)
1. Apartments (see residential)	—	—

2. Access floor systems
   Office use
   Computer use
	
	50
	100
		
		2,000
		2,000

3. Armories and drill rooms	150[m]	—

4. Assembly areas
   Fixed seats (fastened to floor)
   Follow spot, projections and control rooms
   Lobbies
   Movable seats
   Stage floors
   Platforms (assembly)
   Other assembly areas
	
	60[m]
	
	50
	100[m]
	100[m]
	150[m]
	100[m]
	100[m]	

5. Balconies and decks[h]	Same as occupancy served	—
6. Catwalks	40	300
7. Cornices	60	—

8. Corridors
   First floor
   Other floors
	
	100
	Same as occupancy served except as indicated
		—

9. Dining rooms and restaurants	100[m]	—

OCCUPANCY OR USE	UNIFORM (psf)	CONCENTRATED (lbs.)

23. Penal institutions
   Cell blocks
   Corridors
	
	40
	100
		
		—

24. Recreational uses:
   Bowling alleys, poolrooms and similar uses
   Dance halls and ballrooms
   Gymnasiums
   Reviewing stands, grandstands and bleachers
   Stadiums and arenas with fixed seats (fastened to floor)
	
	75[m]
	100[m]
	100[m]
	
	100[c, m]
	
	60[c, m]
		—

25. Residential
   One- and two-family dwellings
     Uninhabitable attics without storage[i]
     Uninhabitable attics with storage [i, j, k]
     Habitable attics and sleeping areas[k]
     All other areas
   Hotels and multifamily dwellings
     Private rooms and corridors serving them
     Public rooms[m] and corridors serving them
	
	10
	20
	30
	40
	
	40
	
	100	

26. Roofs
   All roof surfaces subject to maintenance workers
   Awnings and canopies:
		
		300

(continued)

The code does not require the concurrent application of uniform live load and concentrated live load. The load to be utilized in the structural design of the building would be of the type that produces the greater stress in the structural elements.

Topic: General Requirements
Reference: IBC 1610.1
Category: Structural Design
Subject: Soil Lateral Loads

Code Text: *Foundation walls and retaining walls shall be designed to resist lateral soil loads. Soil loads specified in Table 1610.1 shall be used as the minimum design lateral soil loads unless determined otherwise by a geotechnical investigation in accordance with Section 1803. Design lateral pressure from surcharge loads shall be added to the lateral earth pressure load. Design lateral pressure shall be increased if soils at the site are expansive.*

Discussion and Commentary: Design lateral loads are established in Table 1610.1 for various soil types. The indicated loads address both at-rest pressure and active pressure conditions. It is noted that expansive soils are not to be used as backfill, as such materials can exert very high pressures against walls. Special soil testing is required to determine the magnitude of these pressures.

TABLE 1610.1
LATERAL SOIL LOAD

DESCRIPTION OF BACKFILL MATERIAL[c]	UNIFIED SOIL CLASSIFICATION	DESIGN LATERAL SOIL LOAD[a] (pound per square foot per foot of depth)	
		Active pressure	At-rest pressure
Well-graded, clean gravels; gravel-sand mixes	GW	30	60
Poorly graded clean gravels; gravel-sand mixes	GP	30	60
Silty gravels, poorly graded gravel-sand mixes	GM	40	60
Clayey gravels, poorly graded gravel-and-clay mixes	GC	45	60
Well-graded, clean sands; gravelly sand mixes	SW	30	60
Poorly graded clean sands; sand-gravel mixes	SP	30	60
Silty sands, poorly graded sand-silt mixes	SM	45	60
Sand-silt clay mix with plastic fines	SM-SC	45	100
Clayey sands, poorly graded sand-clay mixes	SC	60	100
Inorganic silts and clayey silts	ML	45	100
Mixture of inorganic silt and clay	ML-CL	60	100
Inorganic clays of low to medium plasticity	CL	60	100
Organic silts and silt clays, low plasticity	OL	Note b	Note b
Inorganic clayey silts, elastic silts	MH	Note b	Note b
Inorganic clays of high plasticity	CH	Note b	Note b
Organic clays and silty clays	OH	Note b	Note b

For SI: 1 pound per square foot per foot of depth = 0.157 kPa/m, 1 foot = 304.8 mm.

a. Design lateral soil loads are given for moist conditions for the specified soils at their optimum densities. Actual field conditions shall govern. Submerged or saturated soil pressures shall include the weight of the buoyant soil plus the hydrostatic loads.
b. Unsuitable as backfill material.
c. The definition and classification of soil materials shall be in accordance with ASTM D 2487.

Basement and foundation walls that have restricted horizontal movement at the top shall be designed for at-rest pressure. Retaining walls free to move and rotate at the top are permitted to be designed for active pressure. Where basement walls extend a maximum of 8 feet below grade and support a flexible floor system, they also may be designed for active pressure.

Topic: Scope and Definitions
Reference: IBC 1701.1, 202
Category: Structural Tests and Special Inspections
Subject: Inspections and Observations

Code Text: *The provisions of* Chapter 17 *shall govern the quality, workmanship and requirements for materials covered. Special inspection* (continuous and periodic) *is the inspection of construction requiring the expertise of an approved special inspector in order to ensure compliance with this code and the approved construction documents. Structural observation is the visual observation of the structural system by a registered design professional for general conformance to the approved construction documents. Structural observation does not include or waive the responsibility for the inspection required by Section 110, Section 1705 or other sections of* the IBC.

Discussion and Commentary: In addition to the general inspections called for in Section 110 (footings, frame, final, etc.), it is often necessary to call for a more exacting review of the construction process. Through special inspections and structural observation, the work can be evaluated more closely for compliance with the approved construction documents.

Special Cases	Section 1705.1.1
Steel Construction	Section 1705.2
Concrete Construction	Section 1705.3
Masonry Construction	Section 1705.4
Wood Construction	Section 1705.5
Soils	Section 1705.6
Driven Deep Foundations	Section 1705.7
Cast-in-Place Deep Foundations	Section 1705.8
Helical Pile Foundations	Section 1705.9
Wind Resistance	Section 1705.10
Seismic Resistance	Section 1705.11
Testing and Qualification for Seismic Resistance	Section 1704.12
Sprayed Fire-resistant Materials	Section 1705.13
Mastic and Intumescent Fire-Resistant Coatings	Section 1705.14
Exterior Insulation and Finish Systems (EIFS)	Section 1705.15
Fire-Resistant Penetrations and Joints	Section 1705.16
Smoke Control	Section 1705.17

A statement of special inspections must be prepared by the registered design professional in responsible charge where special inspection or testing is required. The code specifies the content required to be provide in the statement. The code also sets forth unique provisions for special inspections and structural testing for seismic-resistance and wind-resistance considerations.

Topic: General Requirements
Reference: IBC 1704.2
Category: Structural Tests and Special Inspections
Subject: Special Inspections

Code Text: *Where application is made for construction as described in Section 1704, the owner or the registered design professional in responsible charge acting as the owner's agent shall employ one or more approved agencies to perform inspections during construction on the types of work listed under Section 1705.* See three exceptions for work, components or occupancies where special inspection is not required. *These inspections are in addition to the inspections identified in Section 110.*

Discussion and Commentary: Most building departments do not have the staff of inspectors to provide detailed inspections on large and complex projects. There are also projects where the nature of construction is such that extra care in quality control must be exercised to assure compliance. For these reasons, the code mandates continuous or periodic inspection by special inspectors for certain types of work.

TABLE 1705.3
REQUIRED VERIFICATION AND INSPECTION OF CONCRETE CONSTRUCTION

VERIFICATION AND INSPECTION	CONTINUOUS	PERIODIC	REFERENCED STANDARD[a]	IBC REFERENCE
1. Inspection of reinforcing steel, including prestressing tendons, and placement.	—	X	ACI 318: 3.5, 7.1-7.7	1910.4
2. Inspection of reinforcing steel welding in accordance with Table 1705.2.2, Item 2b.	—	—	AWS D1.4 ACI 318: 3.5.2	—
3. Inspection of anchors cast in concrete where allowable loads have been increased or where strength design is used.	—	X	ACI 318: 8.1.3, 21.2.8	1908.5, 1909.1
4. Inspection of anchors post-installed in hardened concrete members[b].	—	X	ACI 318: 3.8.6, 8.1.3, 21.2.8	1909.1
5. Verifying use of required design mix.	—	X	ACI 318: Ch. 4, 5.2-5.4	1904.2, 1910.2, 1910.3
6. At the time fresh concrete is sampled to fabricate specimens for strength tests, perform slump and air content tests, and determine the temperature of the concrete.	X	—	ASTM C 172 ASTM C 31 ACI 318: 5.6, 5.8	1910.10
7. Inspection of concrete and shotcrete placement for proper application techniques.	X	—	ACI 318: 5.9, 5.10	1910.6, 1910.7, 1910.8
8. Inspection for maintenance of specified curing temperature and techniques.	—	X	ACI 318: 5.11-5.13	1910.9
9. Inspection of prestressed concrete: a. Application of prestressing forces. b. Grouting of bonded prestressing tendons in the seismic force-resisting system.	X X		ACI 318: 18.20 ACI 318: 18.18.4	—
10. Erection of precast concrete members.	—	X	ACI 318: Ch. 16	—
11. Verification of in-situ concrete strength, prior to stressing of tendons in post-tensioned concrete and prior to removal of shores and forms from beams and structural slabs.	—	X	ACI 318: 6.2	—
12. Inspect formwork for shape, location and dimensions of the concrete member being formed.	—	X	ACI 318: 6.1.1	—

For SI: 1 inch = 25.4 mm.

a. Where applicable, see also Section 1705.11, Special inspections for seismic resistance.
b. Specific requirements for special inspection shall be included in the research report for the anchor issued by an approved source in accordance with ACI 355.2 or other qualification procedures. Where specific requirements are not provided, special inspection requirements shall be specified by the registered design professional and shall be approved by the building official prior to the commencement of the work.

A statement of special inspections containing the work requiring special inspection, the specific inspections to be performed, and the individuals or firms to be retained for conducting special inspections must be submitted by the permit applicant prior to issuance of the building permit.

Topic: Inspector Qualifications
Reference: IBC 1704.2.1, 1704.2.4
Category: Structural Tests and Special Inspections
Subject: Special Inspections

Code Text: *The special inspector shall provide written documentation to the building official demonstrating his or her competence and relevant experience or training. Experience or training shall be considered relevant when the documented experience or training is related in complexity to the same type of special inspection activities for projects of similar complexity and material qualities. Special inspectors shall keep records of inspections. The special inspector shall furnish inspection reports to the building official, and to the registered design professional in responsible charge.*

Discussion and Commentary: It is the duty of the special inspector not only to observe the work, but also to furnish inspection reports indicating that the work inspected was done in accordance with the approved construction documents. If discrepancies are found in the work, the inspector should bring them to the immediate attention of the contractor for correction. If the discrepancies are not corrected, the building official and registered design professional in responsible charge should be notified.

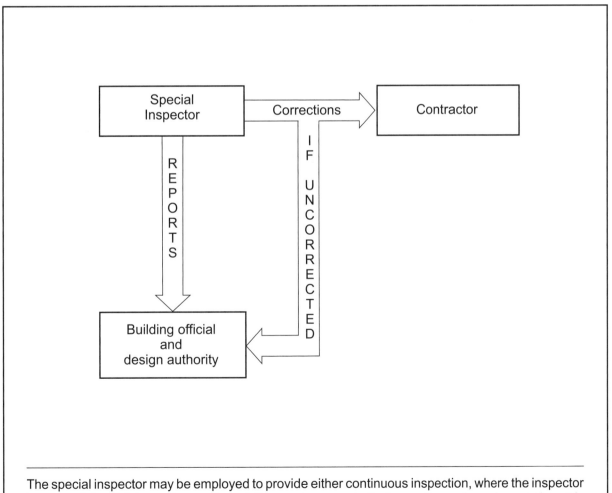

The special inspector may be employed to provide either continuous inspection, where the inspector is present for full-time observation of the work, or periodic inspection, where he or she only intermittently observes the work.

Study Session 16

Topic: Seismic Resistance
Reference: IBC 1704.5
Category: Structural Tests and Special Inspections
Subject: Structural Observations

Code Text: *Where required by the provisions of Section 1704.5.1 (seismic resistance) or 1704.5.2 (wind requirements), the owner shall employ a registered design professional to perform structural observations as defined in Section 202. At the conclusion of the work included in the permit, the structural observer shall submit to the building official a written statement that the site visits have been made and identify any reported deficiencies which, to the best of the structural observer's knowledge, have not been resolved. Structural observations shall be provided for those structures assigned to Seismic Design Category D, E or F, as determined in Section 1613, where one or more of five listed conditions exist.*

Discussion and Commentary: Structural observations not only are required for the higher seismic design categories but also are required in high-wind areas (basic wind speed > 110 mph) where the structure has a height greater than 75 feet or is classified as Risk Category III or IV. The building official or registered design professional in responsible charge can also mandate such observation.

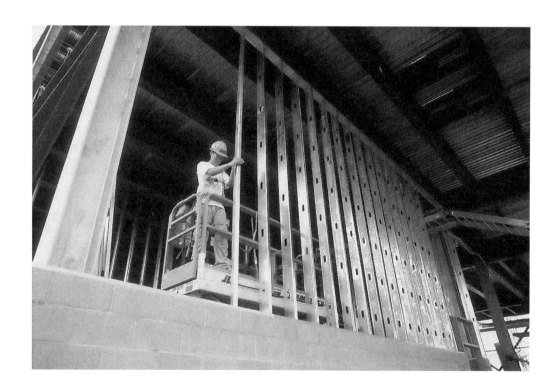

Structural observation is not to be confused with the mandated inspections specified in Section 110, or with the special inspections listed in Section 1705. This activity is intended to provide an additional level of expertise in the review of structures posing a very high level of complexity.

Topic: Scope
Reference: IBC 1901

Category: Concrete
Subject: Plain and Reinforced Concrete

Code Text: *The provisions of Chapter 19 shall govern the materials, quality control, design and construction of concrete use in structures. Structural concrete shall be designed and constructed in accordance with the requirements of Chapter 19 and ACI 318 as amended in Section 1905 of the IBC. The special inspection of concrete elements of buildings and structures and concreting operations shall be as required by Chapter 17.*

Discussion and Commentary: Requirements for the design, testing, mixing, placing and protection of concrete construction are located in ACI 318, *Building Code Requirements for Structural Concrete*, a publication published and copyrighted by the American Concrete Institute. Section 1905 contains the primary modifications to the provisions of ACI 318.

Construction Requirements for Concrete Work

Specifications for Tests and Materials (Chapter 3)	Section 1903
Durability Requirements (Chapter 4)	Section 1904

Where modifications have been made to the text of ACI 318, the section designations of IBC Section 1905 are followed by those found in ACI 318. Italics are used to indicate where the IBC differs substantially from the ACI standard.

Topic: Exposure Categories and Classes   **Category:** Concrete
Reference: IBC 1904.1   **Subject:** Durability Requirements

Code Text: *Concrete shall be assigned to exposure classes in accordance with ACI 318 based on: 1) exposure to freezing and thawing in a moist condition or deicer chemicals; 2) exposure to sulfates in water or soil; 3) exposure to water where the concrete is intended to have low permeability; and 4) exposure to chlorides from deicing chemicals, salt, saltwater, brackish water, seawater or spray from these sources, where the concrete has steel reinforcement.*

Discussion and Commentary: Concrete exposed to the conditions listed in this section must comply with the limitations on water-cementitious materials, or specified compressive strength, or both. The intent of the upper limit on the water-cementitious materials ratio and lower limit on the specified compressive strength is to achieve dense, impermeable or watertight concrete.

TABLE 1904.2
MINIMUM SPECIFIED COMPRESSIVE STRENGTH (f'_c)

TYPE OR LOCATION OF CONCRETE CONSTRUCTION	MINIMUM SPECIFIED COMPRESSIVE STRENGTH (f'_c at 28 days, psi)		
	Negligible exposure	Moderate exposure	Severe exposure
Basement walls[c] and foundations not exposed to the weather	2,500	2,500	2,500[a]
Basement slabs and interior slabs on grade, except garage floor slabs	2,500	2,500	2,500[a]
Basement walls[c], foundation walls, exterior walls and other vertical concrete surfaces exposed to the weather	2,500	3,000[b]	3,000[b]
Driveways, curbs, walks, patios, porches, carport slabs, steps and other flatwork exposed to the weather, and garage floor slabs	2,500	3,000[b, d]	3,500[b, d]

For SI: 1 pound per square inch = 0.00689 MPa.

a. Concrete in these locations that can be subjected to freezing and thawing during construction shall be of air-entrained concrete in accordance with Section 1904.2.
b. Concrete shall be air entrained in accordance with ACI 318.
c. Structural plain concrete basement walls are exempt from the requirements for exposure conditions of Section 1904.2.
d. For garage floor slabs where a steel trowel finish is used, the total air content required by ACI 318 is permitted to be reduced to not less than 3 percent, provided the minimum specified compressive strength of the concrete is increased to 4,000 psi.

In climatic areas where concrete construction is exposed to freeze-thaw cycles or deicer chemicals, concrete must be of the quality necessary to resist the harmful effects of such weather. In low-rise Group R occupancies, the compressive strength need only comply with Table 1904.2, which mandates a minimum compressive strength as a function of the specific concrete element under consideration and the exposure of such element.

Topic: Cold Weather and Hot Weather
Reference: IBC 2104.3, 2104.4
Category: Masonry
Subject: Construction Methods

Code Text: *The cold weather construction provisions of TMS 602/ACI 530.1/ASCE 6, Article 1.8 C, shall be implemented when the ambient temperature falls below 40°F (4°C). The hot weather construction provisions of TMS 602/ACI 530.1/ASCE 6, Article 1.8 D, shall be implemented when the ambient temperature exceeds 100°F (37.8°C), or 90°F (32.2°C) with a wind velocity greater than 8 mph (12.9 km/hr).*

Discussion and Commentary: During cold-weather construction, particularly in temperatures below freezing, the curing and subsequent performance of masonry are influenced by the properties and temperature of both the mortar and the masonry units. Temperature, solar radiation, humidity and wind influence the rate of absorption of masonry units and the rate of mortar set. During hot-weather conditions, special precautions must be taken for adequate strength gain of the masonry.

Masonry units to be used during cold-weather conditions should be dry. In hot-weather masonry construction, however, proper wetting is acceptable. Clay brick laid on fresh mortar generally absorbs water from the mortar along with fine particles of cementitious materials, which help to bond the mortar with the brick. Care must be taken because if the brick absorbs too much water, the mortar may have insufficient water for hydration.

Topic: Firebox Walls
Reference: IBC 2111.5

Category: Masonry
Subject: Fireplaces

Code Text: *Masonry fireboxes shall be constructed of solid masonry units, hollow masonry units grouted solid, stone, or concrete. When a lining of firebrick at least 2 inches (51 mm) in thickness or other approved lining is provided, the minimum thickness of back and sidewalls shall each be 8 inches (203 mm) of solid masonry, including the lining. The width of joints between firebricks shall not be greater than $^3/_4$ inch (6.4 mm). When no lining is provided, the total minimum thickness of back and sidewalls shall be 10 inches (254 mm) of solid masonry.*

Discussion and Commentary: Firebox thickness is regulated in order to insulate surrounding construction, both exposed and concealed, from excessive temperature levels. It is important that the walls be constructed in such a fashion that they are solid throughout.

For SI: 1 inch = 25.4 mm.

**Lined and unlined masonry chimneys - residential
Greater thicknesses are required for unburned clay units and stone**

The use of steel fireplace units is permitted under specific conditions. The unit must have a minimum $^1/_4$-inch steel firebox liner and an air chamber. The total thickness at the back and sides must be at least 8 inches, with no less than 4 inches of solid masonry.

Topic: Hearth Extensions
Reference: IBC 2111.10, 2111.9.2
Category: Masonry
Subject: Fireplaces

Code Text: *Hearth extensions shall extend at least 16 inches (406 mm) in front of, and at least 8 inches (203 mm) beyond, each side of the fireplace opening. Where the fireplace opening is 6 square feet (0.557 m^2) or larger, the hearth extension shall extend at least 20 inches (508 mm) in front of, and at least 12 inches (305 mm) beyond, each side of the fireplace opening. The minimum thickness of hearth extensions shall be 2 inches (51 mm). See exception for raised firebox openings.*

Discussion and Commentary: Hearth extensions are necessary to keep sparks and embers that fly from the firebox from igniting combustible material, such as carpet, on the floor. Radiated heat from the fireplace can also ignite combustible flooring materials located adjacent to the fireplace opening and adjacent fireplace walls.

For SI: 1 inch = 25.4 mm, 1 square foot = 0.093 m².

For obvious reasons, the hearth and the hearth extension must be constructed of, and supported by, noncombustible materials. Hearths must be specifically constructed of either concrete or masonry. The minimum required thickness of fireplace hearths is 4 inches.

Topic: Clearance to Combustible Material **Category:** Masonry
Reference: IBC 2111.11 **Subject:** Fireplaces

Code Text: *Any portion of a masonry fireplace located in the interior of a building or within the exterior wall of a building shall have a clearance to combustibles of not less than 2 inches (51 mm) from the front faces and sides of masonry fireplaces and not less than 4 inches (102 mm) from the back faces of masonry fireplaces. The airspace shall not be filled, except to provide fireblocking in accordance with Section 2111.12. See four exceptions for alternate methods to the required clearances.*

Discussion and Commentary: The radiant heat transfer through the materials used to construct a masonry fireplace and/or chimney necessitates a minimum separation between the masonry and combustible materials, such as wood floor, wall or ceiling framing. The depth of the noncombustible fireblocking, placed on metal strips or lath, is to be 1 inch.

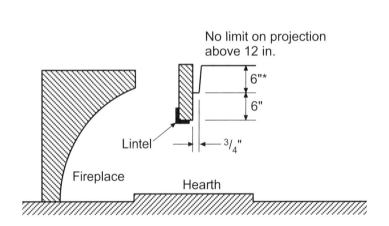

* Combustible materials may project $1/8$ in. for each 1 in. clearance. No combustible materials permitted within 6 in. of opening.

For SI: 1 inch = 25.4 mm.

Combustible materials projection from fireplace

No combustible materials, such as trim and ornamentation, are permitted within 6 inches directly above the opening at the face of the fireplace. Combustibles placed less than 12 inches from the opening, while permitted, are very limited in their projection from the fireplace opening.

Topic: Identification
Reference: IBC 2303.1.1, 2303.1.4
Category: Wood
Subject: Minimum Standards and Quality

Code Text: *Sawn lumber used for load-supporting purposes, including end-jointed or edge-glued lumber, machine stress-rated or machine-evaluated lumber, shall be identified by the grade mark of a lumber grading or inspection agency that has been approved by an accreditation body that complies with DOC PS 20 or equivalent procedures. Wood structural panels, when used structurally (including those used for siding, roof and wall sheathing, subflooring, diaphragms and built-up members), shall conform to the requirements for its type in DOC PS 1, DOC PS 2 or ANSI/APA PRP 210.*

Discussion and Commentary: Obviously, the proper use of a wood structural member cannot be determined unless it has been identified. Grade marks, identification marks, certificates of inspection and quality marks are various methods of indicating the type and quality of wood members.

Visually Graded Lumber

Machine Stress-rated lumber

Lumber and, particularly, wood structural panels are highly variable in strengths and other mechanical properties; hence, such materials must conform to the applicable standards or grading rules specified in the code.

Topic: Protection Against Decay
Reference: IBC 2304.11.2
Category: Wood
Subject: General Construction Requirements

Code Text: *Wood used above ground in the locations specified in Sections 2304.11.2.1 through 2304.11.2.7, 2304.11.3 and 2304.11.5 shall be naturally durable wood or preservative-treated wood using water-borne preservatives, in accordance with AWPA U1 (Commodity Specifications A or F) for above-ground use.*

Discussion and Commentary: To protect against decay and termite infestation, the code addresses those members for which care must be taken, including: joists, girders and subfloor adjacent to exposed ground in crawl spaces; framing members and wall sheathing that rest on exterior foundation walls; sleepers and sills on a concrete slab in direct contact with earth; girder ends in masonry or concrete walls; wood siding adjacent to the ground; and posts and columns supported by a concrete slab.

For SI: 1 inch = 25.4 mm.

Under-floor clearance

The term "naturally durable wood" describes the heartwood of a select group of wood species that provide a natural resistance to decay (redwood, cedar, black locust and black walnut) and/or termites (redwood, Alaska yellow cedar, Eastern red cedar and Western red cedar).

Topic: Scope
Reference: IBC 2308.1

Category: Wood
Subject: Conventional Light-Frame Construction

Code Text: *The requirements of Section 2308 are intended for conventional light-frame construction. Other methods are permitted to be used provided a satisfactory design is submitted showing compliance with other provisions of the IBC. Interior nonload-bearing partitions, ceilings and curtain walls of conventional light-frame construction are not subject to the limitations of Section 2308.*

Discussion and Commentary: Conventional light-frame wood construction is considered *a type of construction whose primary structural elements are formed by a system of repetitive wood-framing members.* The provisions of Section 2308 are based on experience gained over the last several decades. This experience has resulted in the prescriptive requirements contained in this section. Compliance with AF&PA's *Wood Frame Construction Manual* is considered equivalent to Section 2308.

For SI: 1 foot = 304.8 mm.

Framing around openings - header span > 6'

Floor framing at openings

The limitations placed on buildings of conventional light-frame construction are based on limits to: number of stories; floor-to-floor height of bearing walls; live, dead and snow loads; wind speeds; rafter span; and seismic design categories.

Topic: Foundation Plates or Sills
Reference: IBC 2308.6
Category: Wood
Subject: Conventional Light-Frame Construction

Code Text: *Foundation plates or sills shall be bolted or anchored to the foundation with not less than $^1/_2$-inch (12.7 mm) diameter steel bolts or approved anchors spaced to provide equivalent anchorage as the steel bolts. Bolts shall be embedded at least 7 inches (178 mm) into concrete or masonry, and spaced not more than 6 feet (1829 mm) apart. There shall be a minimum of two bolts or anchor straps per piece with one bolt or anchor strap located not more than 12 inches (305 mm) or less than 4 inches (102 mm) from each end of each piece. A properly sized nut and washer shall be tightened on each bolt to the plate.*

Discussion and Commentary: The prescriptive requirements for the installation of anchor bolts are applicable to all buildings of conventional light-frame wood construction. The provisions set forth the necessary criteria to adequately tie the framing system to the foundation.

For SI: 1 inch = 25.4 mm, 1 foot = 304.8 mm.

In Seismic Design Categories D and E, it has been shown that additional anchorage methods are necessary to resist the lateral forces being applied at the foundation connection. Thus, minimum 0.229-inch by 3-inch by 3-inch steel plate washers are mandated between the sill plate and the nut, in accordance with Section 2308.12.8.

Study Session 16
IBC Chapters 16, 17, 19, 21, 22 and 23

1. Construction documents shall indicate the flat-roof snow load, snow exposure factor, snow load importance factor and thermal factor unless the ground snow load in the area is a maximum of _____ pounds per square foot.

    a. 5   b. 10
    c. 20   d. 30

    Reference_____

2. The minimum uniformly distributed live load used in the design of a fixed-seat arena is determined to be _____ pounds per square foot.

    a. 50   b. 60
    c. 100   d. 125

    Reference_____

3. For structural design purposes, buildings used for fire, rescue, ambulance and police stations and emergency vehicle garages shall be assigned a Risk Category _____.

    a. I   b. II
    c. III   d. IV

    Reference_____

4. Unoccupied landscaped areas on roofs shall be designed for a uniform live load of _____ pounds per square foot.

   a. 20	b. 40
   c. 60	d. 100

   Reference_____

5. For wind design purposes, Surface Roughness Category _____ is considered to be flat open country and grasslands.

   a. A	b. B
   c. C	d. D

   Reference_____

6. Foundation walls in which horizontal movement is restricted at the top shall be designed for _____ pressure.

   a. at-rest	b. controlled
   c. flexible	d. active

   Reference_____

7. Special inspection thickness testing for sprayed fire-resistant material applied to structural framing members shall be performed on a minimum of _____ percent of the structural members on each floor.

   a. 10	b. 25
   c. 50	d. 100

   Reference_____

8. Structures a minimum of _____ stories in height above grade plane shall be provided with structural observation where assigned to Seismic Design Category E and classified as Risk Category I or II.

   a. two	b. three
   c. four	d. five

   Reference_____

9. A structural plain concrete footing is permitted to be 6 inches in thickness in a building of other than Group R-3 occupancy, provided the building is a maximum of _____ above grade plane.

   a. 0 stories, structural plain concrete footings are only permitted in Group R-3 occupancies

   b. 1 story

   c. 2 stories

   d. 3 stories

   Reference _____

10. As a modification to the text of ACI 318, a wall pier is defined as a wall segment with a horizontal length-to-thickness ratio between 2.5 and 6, with a clear height at least _____ times its horizontal length.

    a. 2   b. 3

    c. 4   d. 6

    Reference _____

11. Coarse aggregate used in shotcrete shall be a maximum of _____ inch.

    a. $^3/_8$ b. $^1/_2$

    c. $^3/_4$ d. 1

    Reference _____

12. The minimum thickness of a concrete floor slab supported directly on the ground shall be _____ inches.

    a. $3^1/_2$ b. 4

    c. 5   d. 6

    Reference _____

13. What is the term used to describe masonry in which the tensile resistance of the masonry is taken into consideration and the effects of stresses in reinforcement are neglected?

    a. ashlar masonry   b. plain masonry

    c. reinforced masonry   d. solid masonry

    Reference _____

14. The cold weather masonry construction provisions of TMS 602 need not be implemented when the ambient air temperature is a minimum of _____ °F.

    a. 32   b. 35
    c. 40   d. 45

    Reference _____

15. Where the ambient temperature is 95°, the hot weather masonry construction provisions of TMS 602 are not required to be implemented when the wind velocity is a maximum of _____ mph.

    a. 8   b. 10
    b. 15   d. 20

    Reference _____

16. A hearth extension for a masonry fireplace with a 24-inch by 42-inch opening shall extend a minimum of _____ inches in front of, and a minimum of _____ inches beyond, each side of the fireplace opening.

    a. 16, 8   b. 16, 10
    c. 20, 12   d. 24, 12

    Reference _____

17. Combustible materials located 10 inches directly above the opening of a masonry fireplace shall have a maximum projection of _____ inch(es).

    a. 0, no projection is permitted   b. $^3/_4$
    c. 1   d. $1^1/_4$

    Reference _____

18. Masonry chimneys shall extend at least 2 feet higher than any portion of the building within 10 feet, with a minimum height of _____ feet above the point where the chimney passes through the roof.

    a. 2   b. $2^1/_2$
    c. 3   d. 4

    Reference _____

19. Unless naturally durable or preservative-treated wood is used, what is the minimum air space required on the top, sides and end of wood girders entering exterior concrete walls?

    a. ½ inch
    b. 1 inch
    c. 1½ inch
    d. 2 inches

    Reference_____

20. Which of the following types of naturally durable wood cannot qualify as decay resistant?

    a. redwood
    b. cedar
    c. black walnut
    d. red oak

    Reference_____

21. What is the minimum number and size of common nails required to connect a wood jack rafter to a hip?

    a. two 10d, face nail
    b. three 10d, toenail
    c. two 16d, toenail
    d. three 16d, face nail

    Reference_____

22. In conventional light-frame construction, buildings shall be provided with exterior and interior braced wall lines at maximum intervals of _____ feet in both the longitudinal and transverse directions in each story.

    a. 20
    b. 25
    c. 35
    d. 50

    Reference_____

23. In conventional light-frame construction, trimmer and header joists in framing around openings shall be doubled where the span of the header exceeds a minimum of _____ feet.

    a. 4
    b. 6
    c. 10
    d. 12

    Reference_____

24. In conventional light-frame wall construction, the edge of a bored hole shall be a minimum of _____ inch(es) from the edge of the stud.

    a. $^5/_8$ b. $^3/_4$
    c. 1   d. $1^1/_4$

    Reference_____

25. In conventional light-frame roof construction, the maximum span of a 2-inch by 6-inch purlin shall be _____ feet.

    a. 4   b. 6
    c. 8   d. 12

    Reference_____

26. In the design of stairway treads, a minimum concentrated live load of _____ pounds must be used, determined on an area of 2 inches by 2 inches.

    a. 40   b. 100
    c. 200   d. 300

    Reference_____

27. For Group R occupancies less than four stories above grade plane in climates determined to be of moderate exposure, concrete foundation walls exposed to weather shall have a minimum specified compressive strength of _____ psi at 28 days.

    a. 2,000   b. 2,500
    c. 3,000   d. 3,500

    Reference_____

28. Gypsum sheathing, where used to enclose wood-framed buildings, shall have a minimum thickness of _____ inch and the wall studs shall be spaced at a maximum of _____ inches on center.

    a. $^1/_2$, 16   b. $^1/_2$, 24
    c. $^5/_8$, 16   d. $^5/_8$, 24

    Reference_____

29. Masonry nonstructural floor surfacing a maximum of _____ inches in thickness is permitted to be supported by wood members without checking for the effects of long-term loading.

    a. 2
    b. 4
    c. 6
    d. 8

    Reference_____

30. In conventional light-frame construction, foundation cripple walls exceeding _____ in height shall be framed of wood studs having the size required for an additional story.

    a. 14 inches
    b. 20 inches
    c. 30 inches
    d. 48 inches

    Reference_____

31. Unless the specified live load exceeds 80 psf, the partition load in an office building shall be a minimum uniformly distributed live load of _____ pounds.

    a. 5
    b. 10
    c. 15
    d. 20

    Reference_____

32. Grab bars shall be designed to resist a single minimum concentrated load of _____ pounds applied in any direction at any point.

    a. 160
    b. 225
    c. 250
    d. 300

    Reference_____

33. Areas along the U.S. Gulf of Mexico coast are not considered hurricane-prone regions where the maximum ultimate design wind speed is _____ miles per hour.

    a. 100
    b. 110
    c. 115
    d. 130

    Reference_____

34. Where a cold-formed steel truss has a minimum clear span of _____ feet, the special inspector shall verify that the temporary installation restraint bracing is installed in accordance with the approved truss submittal package.

    a. 40   b. 50

    c. 60   d. 80

    Reference _____

35. Hollow glass masonry units shall have a minimum average glass face thickness of _____ inch.

    a. $1/8$ b. $3/16$

    c. $1/4$ d. $1/2$

    Reference _____

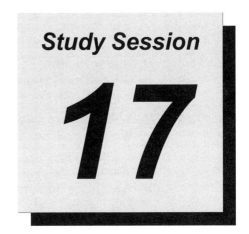

2012 IBC Chapters 8, 12, 25 and 30

Interior Finishes, Interior Environment, Gypsum Board and Elevators

OBJECTIVE: To gain an understanding of the limitations on interior wall and ceiling finishes; the installation requirements for gypsum board, lath and plaster; the important issues concerning the interior environment, including light, ventilation and sound transmission; and the provisions for elevators and their hoistways.

REFERENCE: Chapters 8, 12, 25 and 30, 2012 *International Building Code*

KEY POINTS:
- Which building elements are considered to be interior wall and ceiling finishes?
- Which types of materials are not regulated as wall or ceiling finishes?
- What is the standard of quality that addresses interior wall and ceiling finishes?
- Based on flame-spread index, what are the various classes of finish materials?
- When tested in a manner consistent with their use, what is the maximum smoke density index permitted for finish materials?
- In general, what is the maximum thickness of an interior wall or ceiling finish that must be applied directly against a noncombustible backing?
- How is Table 803.9 used to regulate the finish materials on ceilings and walls?
- Under which conditions are textile wall coverings permitted as wall and ceiling finishes?
- In what portions of a building are the interior floor finishes regulated? In which specific occupancies?
- What are the limitations on combustible trim and decorative materials?
- How must the occupiable portions of buildings be illuminated? Ventilated?
- When yards and courts are adjacent to exterior openings providing natural light and ventilation, how shall the openings be located?
- In which type of occupancy is the transmission of sound regulated?
- How is air-borne sound to be controlled? Structure-borne sound?
- What are the minimum room widths of habitable spaces? Minimum ceiling heights of occupiable spaces?
- How are efficiency dwelling units regulated for interior environment?
- What minimum size access opening is required for a crawl space? An attic?
- How must walls and floors in toilet rooms and bathing rooms be surfaced?

KEY POINTS:
(Cont'd)

- What are the limitations for gypsum wallboard in regard to exterior installation and weather protection?
- How shall fasteners for gypsum wallboard be applied?
- What type of gypsum board assembly requires treated joints and fasteners?
- Where is water-resistant gypsum backing board required?
- Water-resistant gypsum backing board is not permitted for use in which three locations?
- Where used to provide a horizontal diaphragm, how shall gypsum board be installed in a ceiling application?
- What level of fire resistance is required for elevator shaft enclosures?
- What is the maximum number of elevator cars that may be located in a single elevator hoistway? At what point are two hoistways required?
- What types of doors are prohibited at the point of access to an elevator car?
- When must elevator hoistways be vented? Where are the vents to be located?
- What is a fire service access elevator? In what buildings is such an elevator required? What special features are provided?
- Where are occupant evacuation elevators required? Can such elevators be used as a portion of the means of egress? What special features are provided?

Topic: Definition and Classifications
Reference: IBC 801.2, 202
Category: Interior Finishes
Subject: Wall and Ceiling Finishes

Code Text: *The provisions of Section 803 shall limit the allowable fire performance and smoke development of interior wall and ceiling finish materials based upon occupancy classification. Interior wall and ceiling finish includes the exposed interior surfaces of buildings including, but not limited to: fixed or movable walls and partitions; toilet room privacy partitions; columns; ceilings; and interior wainscotting, paneling, or other finish applied structurally or for decoration, acoustical correction, surface insulation, structural fire resistance or similar purposes, but not including trim.*

Discussion and Commentary: It is the intent of the IBC to govern those materials applied to walls or ceilings that could contribute to the spread of flame or the development of smoke. Floor finishes are regulated in a different manner as set forth in Section 804.

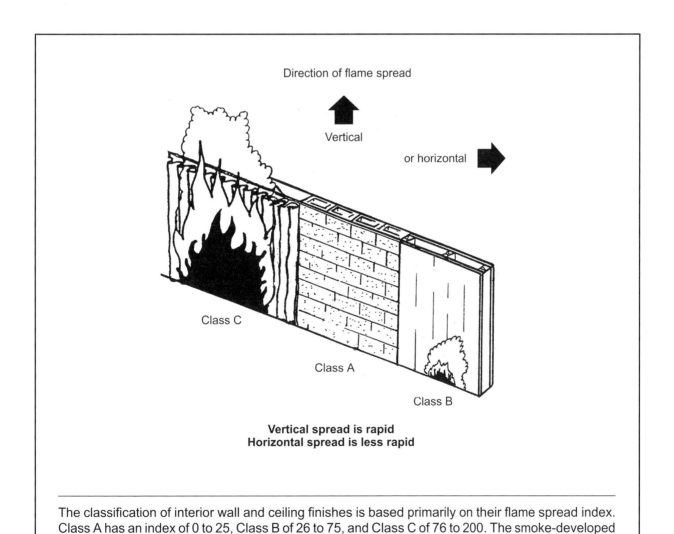

The classification of interior wall and ceiling finishes is based primarily on their flame spread index. Class A has an index of 0 to 25, Class B of 26 to 75, and Class C of 76 to 200. The smoke-developed index for all three classifications is limited to 450.

Topic: Flame-Spread Classification
Reference: IBC 803.9
Category: Interior Finishes
Subject: Finish Requirements by Occupancy

Code Text: *Interior wall and ceiling finish shall have a flame spread index not greater than that specified in Table 803.9 for the group and location designated.*

Discussion and Commentary: Based on fire statistics, the rapid spread of fire across an interior finish material has been second only to vertical fire spread through openings between floors as a cause of life loss during building fires. Therefore, limitations are placed on the materials that are used to cover the walls and ceilings of rooms and other enclosed spaces, corridors, and vertical exits and exit passageways. The rapid spread of fire and the increased contribution of fuel to the fire are the major reasons why finish materials must meet stringent criteria to gain acceptance.

TABLE 803.9
INTERIOR WALL AND CEILING FINISH REQUIREMENTS BY OCCUPANCY[k]

GROUP	SPRINKLERED[l]			NONSPRINKLERED		
	Exit enclosures and exit passageways[a, b]	Corridors	Rooms and enclosed spaces[c]	Exit enclosures and exit passageways[a, b]	Corridors	Rooms and enclosed spaces[c]
A-1 & A-2	B	B	C	A	A[d]	B[e]
A-3[f], A-4, A-5	B	B	C	A	A[d]	C
B, E, M, R-1	B	C	C	A	B	C
R-4	B	C	C	A	B	B
F	C	C	C	B	C	C
H	B	B	C[g]	A	A	B
I-1	B	C	C	A	B	B
I-2	B	B	B[h, i]	A	A	B
I-3	A	A[j]	C	A	A	B
I-4	B	B	B[h, i]	A	A	B
R-2	C	C	C	B	B	C
R-3	C	C	C	C	C	C
S	C	C	C	B	B	C
U	No restrictions			No restrictions		

For SI: 1 inch = 25.4 mm, 1 square foot = 0.0929m².
a. Class C interior finish materials shall be permitted for wainscoting or paneling of not more than 1,000 square feet of applied surface area in the grade lobby where applied directly to a noncombustible base or over furring strips applied to a noncombustible base and fireblocked as required by Section 803.11.1.
b. In exit enclosures of buildings less than three stories above grade plane of other than Group I-3, Class B interior finish for nonsprinklered buildings and Class C interior finish for sprinklered buildings shall be permitted.
c. Requirements for rooms and enclosed spaces shall be based upon spaces enclosed by partitions. Where a fire-resistance rating is required for structural elements, the enclosing partitions shall extend from the floor to the ceiling. Partitions that do not comply with this shall be considered enclosing spaces and the rooms or spaces on both sides shall be considered one. In determining the applicable requirements for rooms and enclosed spaces, the specific occupancy thereof shall be the governing factor regardless of the group classification of the building or structure.
d. Lobby areas in Group A-1, A-2 and A-3 occupancies shall not be less than Class B materials.
e. Class C interior finish materials shall be permitted in places of assembly with an occupant load of 300 persons or less.
f. For places of religious worship, wood used for ornamental purposes, trusses, paneling or chancel furnishing shall be permitted.
g. Class B material is required where the building exceeds two stories.
h. Class C interior finish materials shall be permitted in administrative spaces.
i. Class C interior finish materials shall be permitted in rooms with a capacity of four persons or less.
j. Class B materials shall be permitted as wainscoting extending not more than 48 inches above the finished floor in corridors.
k. Finish materials as provided for in other sections of this code.
l. Applies when the exit enclosures, exit passageways, corridors or rooms and enclosed spaces are protected by an automatic sprinkler system installed in accordance with Section 903.3.1.1 or 903.3.1.2.

Textile materials, where applied to walls or ceilings, must meet additional criteria prior to approval. Finishes that have napped, tufted, looped, nonwoven, woven or similar surface characteristics present a unique hazard on account of their contribution to extremely rapid fire spread.

Topic: Floor Finish Requirements
Reference: IBC 804.4.2

Category: Interior Finishes
Subject: Interior Floor Finish

Code Text: *In all occupancies, interior floor finish and floor covering materials in enclosures for stairways and ramps, exit passageways, corridors and rooms or spaces not separated from corridors by partitions extending from the floor to the underside of the ceiling shall withstand a minimum critical radiant flux not less than Class I in Groups I-1, I-2 and I-3 and not less than Class II in Groups A, B, E, H, I- 4, M, R-1, R-2 and S.* See exception for the permitted reduction of classification in fully sprinklered buildings.

Discussion and Commentary: Although there are many different types of floor finishes and floor coverings, only those flooring materials composed of fibers are regulated by Section 804. Where required to be classified as Class I or Class II materials, the floor covering materials must be tested by an approved agency in accordance with NFPA 253. In order to verify compliance, the materials must be identified by a hang tag or other suitable method that identifies the manufacturer or supplier, style and finish classification.

Types of classifications: (in terms of heat flux, Sec. 804.2)

- Class I: Minimum 0.45 watts/cm^2 per NFPA 253
- Class II: Minimum 0.22 watts/cm^2 per NFPA 253
- DOC FF-1: Minimum 0.04 watts/cm^2

Required classifications: (Sec. 804.4)

	Nonsprinklered[a]		Sprinklered (NFPA 13 only)	
	Exits/Corr.[b]	Other Areas	Exits/Corr.[b]	Other Areas
Groups I-1, I-2 and I-3	Class I	DOC FF-1[c]	Class II	DOC FF-1[c]
Groups F, R-3, R-4 and U	DOC FF-1[c]	DOC FF-1[c]	DOC FF-1[c]	DOC FF-1[c]
Other Groups	Class II	DOC FF-1[c]	DOC FF-1[c]	DOC FF-1[c]

Note: [a]Section 903.2 requires sprinklers in various occupancies
[b]Includes enclosures for stairways and ramps, exit passageways, corridors and rooms or spaces not separated from corridors by full-height partitions.
[c]Compliance with ASTM D 2859 also permitted

DOC FF-1, often referred to as the Methenamine Pill Test, essentially evaluates the floor covering when subjected to a cigarette-type ignition by using a small methenamine tablet. All carpeting sold in the United States is required by federal law to pass this test procedure.

Topic: Decorative Materials
Reference: IBC 806.1
Category: Interior Finishes
Subject: Decorations and Trim

Code Text: *In occupancies of Groups A, E, I and R-1 and dormitories in Group R-2, curtains, draperies, hangings and other decorative materials suspended from walls or ceilings shall meet the flame propagation performance criteria of NFPA 701 in accordance with Section 806.2 or be noncombustible. See exceptions for Group R-2 dormitories. In Groups I-1 and I-2, combustible decorations shall meet the flame propagation criteria of NFPA 701 unless the decorative materials, such as photographs and paintings, are of such limited quantities that a hazard of fire spread or development is not present. In Group I-3, combustible decorative materials are prohibited.*

Discussion and Commentary: In occupancies where large, concentrated occupant loads are expected, or where occupants have limited mobility due to physical limitations or restraint, the contribution of decorative materials to a fire condition is of concern. Therefore, the amount and characteristics of such materials are regulated to limit their impact on the fire severity.

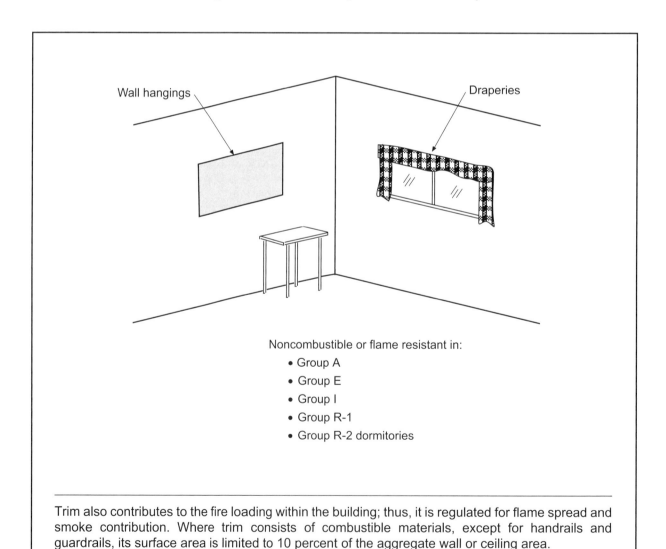

Noncombustible or flame resistant in:
- Group A
- Group E
- Group I
- Group R-1
- Group R-2 dormitories

Trim also contributes to the fire loading within the building; thus, it is regulated for flame spread and smoke contribution. Where trim consists of combustible materials, except for handrails and guardrails, its surface area is limited to 10 percent of the aggregate wall or ceiling area.

Topic: General Requirements
Reference: IBC 1203.1, 1203.4
Category: Interior Environment
Subject: Ventilation

Code Text: *Buildings shall be provided with natural ventilation in accordance with Section 1203.4 or mechanical ventilation in accordance with the* International Mechanical Code. *Natural ventilation of an occupied space shall be through windows, doors, louvers or other openings to the outdoors. The minimum openable area to the outdoors shall be 4 percent of the floor area being ventilated.*

Discussion and Commentary: To obtain a minimum level of environmental comfort, as well as to maintain sanitary conditions, some form of ventilation must be provided to portions of a building that are normally occupied. The *International Mechanical Code* will usually be used to determine the minimum acceptable ventilation methods and quantities.

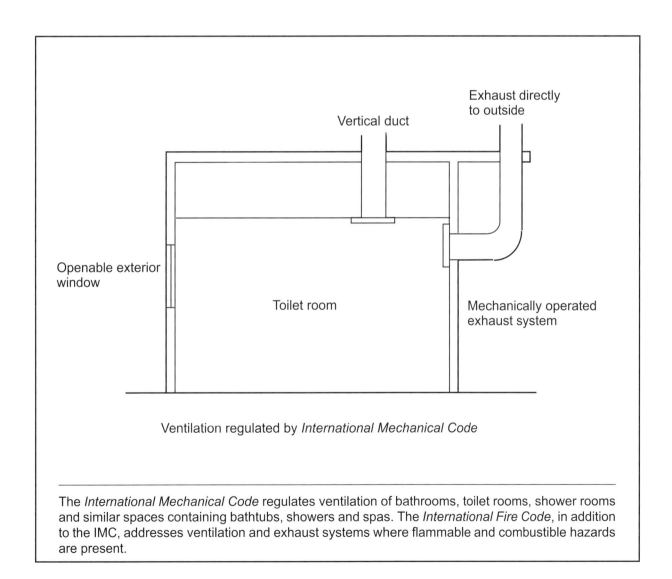

Ventilation regulated by *International Mechanical Code*

The *International Mechanical Code* regulates ventilation of bathrooms, toilet rooms, shower rooms and similar spaces containing bathtubs, showers and spas. The *International Fire Code*, in addition to the IMC, addresses ventilation and exhaust systems where flammable and combustible hazards are present.

Topic: Attic and Under-Floor Ventilation **Category:** Interior Environment
Reference: IBC 1203.2, 1203.3 **Subject:** Ventilation

Code Text: *Enclosed attics and enclosed rafter spaces formed where ceilings are applied directly to the underside of roof framing members shall have cross ventilation for each separate space by ventilating openings protected against the entrance of rain and snow. The space between the bottom of the floor joists and the earth under any building except spaces occupied by a basement or cellar shall be provided with ventilation openings through foundation walls or exterior walls. Such openings shall be placed so as to provide cross-ventilation of the under-floor space. See exceptions for reductions and eliminations of required ventilation.*

Discussion and Commentary: Ventilation of the attic and under-floor spaces prevents moisture condensation, which can have adverse effects on the materials of construction located in those spaces. Various exceptions are available that provide equivalent results.

Attic ventilation - calculations

For SI: 1 foot = 304.8 mm.

The general requirement for under-floor ventilation mandates a minimum net area of ventilation openings of $1/150$ of the area of the space ventilated. In addition, all openings to the exterior must be screened to prevent the entry of birds, rodents and similar creatures.

Topic: General Requirements
Reference: IBC 1205
Category: Interior Environment
Subject: Lighting

Code Text: *Every space intended for human occupancy shall be provided with natural light by means of exterior glazed openings in accordance with Section 1205.2 or shall be provided with artificial light in accordance with Section 1205.3. Exterior glazed openings shall open directly onto a public way or onto a yard or court in accordance with Section 1206. The minimum net glazed area shall not be less than 8 percent of the floor area of the room served. Artificial light shall be provided that is adequate to provide an average illumination of 10 foot-candles (107 lux) over the area of the room at a height of 30 inches (762 mm) above the floor level.*

Discussion and Commentary: It is fundamental that all occupiable areas of a building be provided with adequate illumination. The use of artificial light to satisfy the code is acceptable because it can produce the light necessary for occupancy at any time of the day or night.

For SI: 1 square foot = 0.093 m².

Where natural light is used to satisfy the provisions of the IBC, it can be shared by two rooms. The common wall between the rooms must be adequately open, and the total floor area of both rooms shall be used to calculate the minimum glazed area.

Topic: General Requirements
Reference: IBC 1207.2, 1207.3

Category: Interior Environment
Subject: Sound Transmission

Code Text: *Walls, partitions and floor/ceiling assemblies separating dwelling units from each other or from public or service areas shall have a sound transmission class (STC) of not less than 50 (45 if field tested) for air-borne noise when tested in accordance with ASTM E 90. Floor/ceiling assemblies between dwelling units or between a dwelling unit and a public or service area within the structure shall have an impact insulation class (IIC) rating of not less than 50 (45 if field tested) when tested in accordance with ASTM E 492.*

Discussion and Commentary: To control sound transmission between areas of a residential building, insulated walls and floor/ceiling assemblies are necessary. The regulations address air-borne sound that may be carried throughout the structure, as well as impact noise created on the floor of a floor/ceiling assembly.

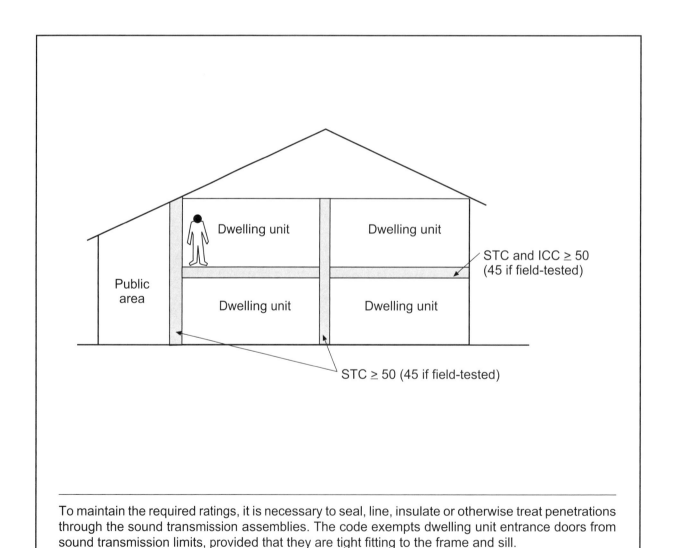

To maintain the required ratings, it is necessary to seal, line, insulate or otherwise treat penetrations through the sound transmission assemblies. The code exempts dwelling unit entrance doors from sound transmission limits, provided that they are tight fitting to the frame and sill.

Topic: Room Size and Ceiling Height
Reference: IBC 1208
Category: Interior Environment
Subject: Interior Space Dimensions

Code Text: *Habitable spaces, other than a kitchen, shall not be less than 7 feet (2134 mm) in any plan dimension. Kitchens shall have a clear passageway of not less than 3 feet (914 mm) between counter fronts and appliances or counter fronts and walls. Occupiable spaces, habitable spaces and corridors shall have a ceiling height of not less than 7 feet 6 inches (2286 mm). Bathrooms, toilet rooms, kitchens, storage rooms and laundry rooms shall be permitted to have a ceiling height of not less than 7 feet (2134 mm).* See exceptions for ceilings with exposed beams, sloped ceilings and mezzanines. *Every dwelling unit shall have at least one room that shall have not less than 120 square feet (13.9 m^2) of net floor area. Other habitable rooms shall have a net floor area of not less than 70 square feet (6.5 m^2).* See exception for kitchens in one-and two-family dwellings.

Discussion and Commentary: For fundamental usability and environmental purposes, it is necessary to mandate minimum requirements for the size and height of occupiable spaces.

For SI: 1 inch = 25.4 mm, 1 foot = 304.8 mm.

Efficiency dwelling units, often referred to as studio apartments, typically consist of a single room used as a combination living/sleeping/dining/cooking area, and a bathroom. The code regulates room size based on occupant load, as well as the closet, bathroom and kitchen area.

Topic: Attics and Crawl Spaces
Reference: IBC 1209.1, 1209.2

Category: Interior Environment
Subject: Access to Unoccupied Spaces

Code Text: *Crawl spaces shall be provided with a minimum of one access opening not less than 18 inches by 24 inches (457 mm by 610 mm). An opening not less than 20 inches by 30 inches (559 mm by 762 mm) shall be provided to any attic area having a clear height of over 30 inches (762 mm). A 30-inch (762 mm) minimum clear headroom in the attic space shall be provided at or above the access opening.*

Discussion and Commentary: Items such as plumbing and wiring installations pass through crawl space at times. Required initial and periodic inspections and maintenance and repairs cannot be carried out without access to such crawl spaces. Attic access is also required for similar reasons. Although uncommon, access to the attic for fire department purposes can also be accomplished through such openings. The required openings are a convenient and nondestructive means for any user to access such concealed spaces.

The *International Mechanical Code* regulates access to both underfloor and attic spaces for the inspection, service, repair or replacement of any mechanical equipment. In addition to the access opening, the passageway and service area sizes are also addressed.

Topic: Wall and Floor Finishes
Reference: IBC 1210
Category: Interior Environment
Subject: Surrounding Materials

Code Text: *In other than dwelling units, toilet, bathing and shower room floor finish materials shall have a smooth, hard, nonabsorbent surface. The intersections of such floors with walls shall have a smooth, hard, nonabsorbent vertical base that extends upward onto the walls at least 4 inches (102 mm). Walls and partitions within 2 feet (610 mm) of urinals and water closets shall have a smooth, hard, nonabsorbent surface, to a height of 4 feet (1219 mm) above the floor, and except for structural elements, the materials used in such walls shall be of a type that is not adversely affected by moisture. See exceptions for dwelling units and private toilet rooms. Accessories such as grab bars, towel bars, paper dispensers and soap dishes, provided on or within walls, shall be installed and sealed to protect structural elements from moisture.*

Discussion and Commentary: For obvious sanitary reasons, it is necessary to provide surfaces in bath and toilet areas that are easily cleaned and maintained.

Surrounding materials

For SI: 1 inch = 25.4 mm, 1 foot = 304.8 mm.

Shower stalls and compartments must be enclosed with smooth, hard, nonabsorbent surfaces to a minimum height of 70 inches above the drain inlet. This requirement is also applicable to those bathtubs that are provided with shower heads.

Topic: Installation
Reference: IBC 2508.2, 2508.3
Category: Gypsum Board and Plaster
Subject: Gypsum Construction

Code Text: *Gypsum wallboard or gypsum plaster shall not be used in any exterior surface where such gypsum construction will be exposed directly to the weather. Gypsum wallboard, gypsum lath or gypsum plaster shall not be installed until weather protection for the installation is provided. Edges and ends of gypsum board shall occur on the framing members, except those edges and ends that are perpendicular to the framing members.*

Discussion and Commentary: Gypsum wallboard, like gypsum plaster, is subject to deterioration from moisture. Accordingly, the code does not permit such gypsum materials to be installed on weather-exposed surfaces, as defined in Section 202. Gypsum materials shall not be installed on interior surfaces until adequate protection from the weather has been provided.

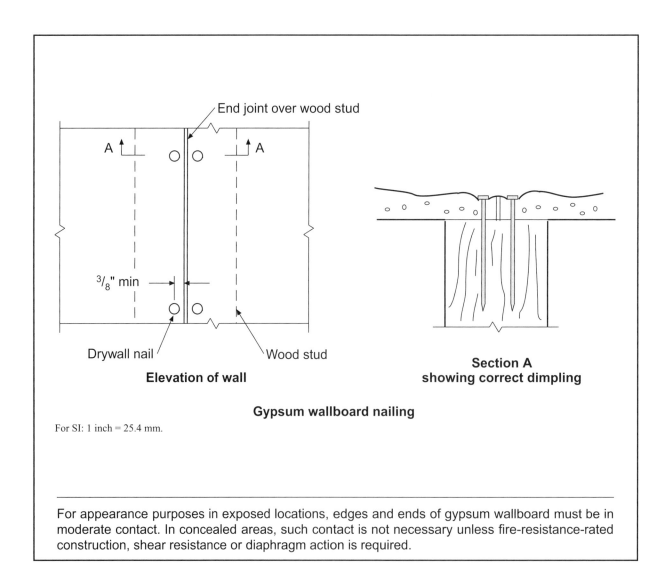

For appearance purposes in exposed locations, edges and ends of gypsum wallboard must be in moderate contact. In concealed areas, such contact is not necessary unless fire-resistance-rated construction, shear resistance or diaphragm action is required.

Topic: Base for Tile
Reference: IBC 2509.2

Category: Gypsum Board and Plaster
Subject: Gypsum Board in Showers

Code Text: *Glass mat water-resistant gypsum backing panels, discrete nonasbestos fiber-cement interior substrate sheets or nonasbestos fiber-mat reinforced cementitious backer units in compliance with ASTM C 1178, C 1288 or C 1325 and installed in accordance with manufacturer recommendations shall be used as a base for wall tile in tub and shower areas and wall and ceiling panels in shower areas. Water-resistant gypsum backing board shall be used as a base for tile in water closet compartment walls when installed in accordance with GA-216 or ASTM C 840 and manufacturer recommendations. Regular gypsum wallboard is permitted under tile or wall panels in other wall and ceiling areas when installed in accordance with GA-216 or ASTM C 840.*

Discussion and Commentary: Because of their moisture-resistant qualities, special types of panels or sheets are required when used as a backing material for tile in high-moisture areas. Although water-resistant gypsum backing board is required as a base for tile on public water closet compartment walls, such gypsum board is prohibited for use in three locations: 1) over a vapor retarder in tub or shower compartments, 2) in areas subject to continuous high humidity or where there will be direct exposure to water, and 3) on ceilings with excessive spacing between framing members.

Water-resistant gypsum backing board tends to sag when installed as ceiling material; therefore, its use is limited only to those applications where the ceiling framing is spaced at no more than 12 inches on center for $1/2$-inch board, or a maximum of 16 inches on center for $5/8$-inch board.

Topic: Weep Screeds
Reference: IBC 2512.1.2
Category: Gypsum Board and Plaster
Subject: Exterior Plaster

Code Text: *A minimum 0.019-inch (0.48 mm) (No. 26 galvanized sheet gage), corrosion-resistant weep screed with a minimum vertical attachment flange of $3^1/_2$ inches (89 mm) shall be provided at or below the foundation plate line on exterior stud walls in accordance with ASTM C 926. the weep screed shall be placed a minimum of 4 inches (102 mm) above the earth or 2 inches (51 mm) above paved areas and be of a type that will allow trapped water to drain to the exterior of the building.*

Discussion and Commentary: Water can penetrate exterior plaster walls for a variety of reasons. Once it penetrates the plaster, the water will run down the exterior face of the weather-resistive barrier until it reaches the sill plate or mudsill. At this point, the water will seek exit from the wall and, if the exterior plaster is not applied to allow the water to escape, it will exit through the inside of the wall and into the interior of the building. Therefore, a weep screed, when properly installed, will permit the water's escape to the exterior of the building.

For SI: 1 inch = 25.4 mm.

To allow the water to escape away from the building, the required water-resistive barrier in the wall assembly must lap the attachment flange of the weep screed. In addition, the exterior lath shall cover and terminate on the attachment flange.

Topic: Protection
Reference: IBC 3002.1, 3002.6
Category: Elevators and Conveying Systems
Subject: Hoistways Enclosures

Code Text: *Elevator, dumbwaiter and other hoistway enclosures shall be shaft enclosures complying with Section 713. Openings in hoistway enclosures shall be protected as required in Chapter 7. Doors, other than hoistway doors and the elevator car door, shall be prohibited at the point of access to an elevator car unless such doors are readily openable from the car side without a key, tool, special knowledge or effort.*

Discussion and Commentary: An elevator shaft is regulated under the shaft enclosure provisions of Section 713. Generally, an elevator enclosure must be of 2-hour fire-resistance-rated construction in Type I buildings or where four or more stories are connected. A 1-hour rating is permitted where the shaft enclosure connects three stories or less.

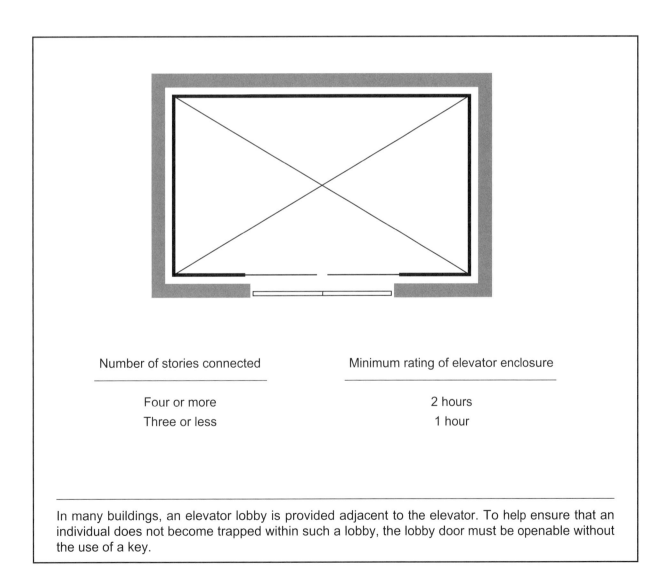

Number of stories connected	Minimum rating of elevator enclosure
Four or more	2 hours
Three or less	1 hour

In many buildings, an elevator lobby is provided adjacent to the elevator. To help ensure that an individual does not become trapped within such a lobby, the lobby door must be openable without the use of a key.

Topic: Number of Cars in Hoistway
Reference: IBC 3002.2, 3002.7

Category: Elevators and Conveying Systems
Subject: Hoistway Enclosures

Code Text: *Where four or more elevator cars serve all or the same portion of a building, the elevators shall be located in at least two separate hoistways. Not more than four elevator cars shall be located in any single hoistway enclosure. Elevators shall not be in a common shaft enclosure with a stairway.* See exception for open parking garages.

Discussion and Commentary: The basis for limiting the number of elevator cars in a single hoistway is to provide a reasonable level of assurance that a multilevel building served by several elevators would not have all of its elevator cars disabled by a single fire incident. The provisions increase the chance that some of the elevators would remain operational during an emergency situation.

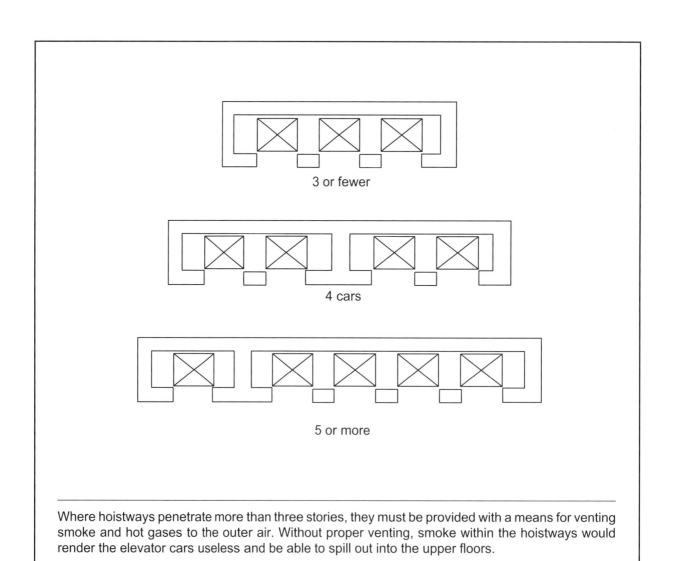

Where hoistways penetrate more than three stories, they must be provided with a means for venting smoke and hot gases to the outer air. Without proper venting, smoke within the hoistways would render the elevator cars useless and be able to spill out into the upper floors.

Topic: Fire Department Emergency Access
Reference: IBC 3002.4
Category: Elevators and Conveying Systems
Subject: Hoistway Enclosures

Code Text: *Where elevators are provided in buildings four or more stories above, or four or more stories below, grade plane, at least one elevator shall be provided for fire department emergency access to all floors. The elevator car shall be of such a size and arrangement to accommodate an ambulance stretcher 24 inches by 84 inches (610 mm by 2134 mm) with not less than 5-inch (127 mm) radius corners, in the horizontal, open position.*

Discussion and Commentary: In those buildings over three stories in height with one or more elevators, it is necessary that a minimum of one elevator car be of sufficient size to hold an ambulance stretcher. The elevator car must access all floor levels within the building or additional complying cars must be provided to provide such access. The minimum size requirement is based on the stretchers now commonly in use by medical service personnel and emergency responders.

Elevator used for fire department emergency access

For SI: 1 inch = 25.4 mm

The elevator car sized in a manner to accommodate the required size ambulance stretcher must be identified by the international symbol for emergency medical services (star of life). The symbol is required to be a minimum of 3 inches in height and is to be placed on both sides of the hoistway door frame.

Study Session 17

Topic: General Provisions
Reference: IBC 3007

Category: Elevators and Conveying Systems
Subject: Fire Service Access Elevator

Code Text: *Where required by Section 403.6.1* (high-rise buildings with an occupied floor more than 120 feet above the lowest level of fire department vehicle access), *every floor of the building shall be served by fire service access elevators complying with Sections 3007.1 through 3007.10. The fire service access elevator shall open into a fire service access elevator lobby in accordance with Sections 3007.7.1 through 3007.7.5. The fire service access elevator lobby shall have direct access to an enclosure for an interior exit stairway.*

Discussion and Commentary: To facilitate the rapid deployment of firefighters, fire service access elevators are required in all high-rise buildings that have at least one floor level more than 120 feet above the lowest level of fire department vehicle access. This type of elevator has a number of key features that will allow firefighters to safely access an area of a building that may be involved in a fire or to facilitate the rescue of building occupants.

Another specific type of elevator, an "occupant evacuation" elevator, is also addressed in Chapter 30. Public-use passenger elevators are specifically allowed to be used for the self-evacuation of occupants in high-rise buildings. The installation of such elevators is voluntary; however, they can be installed as an alternative to the additional exit stairway mandated by Section 403.5.2.

Study Session 17
IBC Chapters 8, 12, 25 and 30

1. Where classified in accordance with ASTM E 84, Class B interior wall and ceiling finishes have a flame spread index of _____ and a smoke-developed index of 0-450.

    a. 0-25
    b. 26-75
    c. 76-200
    d. 201 and greater

    Reference_____

2. Unless a noncombustible material or qualified by tests, an interior wall or ceiling finish that is a maximum of _____ -inch thick shall be applied directly against a noncombustible backing.

    a. $^1/_{16}$
    b. $^1/_8$
    c. $^1/_4$
    d. $^1/_2$

    Reference_____

3. The wall finish in a dining room classified as Group A-2 in a sprinklered building shall have a minimum flame spread index classification of Class _____.

    a. A
    b. B
    c. C
    d. no restrictions

    Reference_____

Study Session 17

4. The wall finish in an interior exit stairway of a four-story nonsprinklered Group B office building shall have a minimum flame spread index classification of Class _____.

    a. A   b. B

    c. C   d. no restrictions

Reference_____

5. In a patient room of a fully-sprinklered Group I-2 occupancy, the interior finish materials are permitted to be a maximum of Class C when the room has a maximum capacity of _____ persons.

    a. 1   b. 2

    c. 3   d. 4

Reference_____

6. Unless in compliance with the Method B test protocol, textile wall coverings shall have a minimum Class _____ flame spread index where installed in a fully-sprinklered art gallery.

    a. A   b. B

    c. C   d. not permitted

Reference_____

7. In a fully sprinklered Group I-1 occupancy, carpet installed as an interior floor finish in an exit passageway shall be a minimum _____.

    a. Class I   b. Class II

    c. Class A   d. DOC FF-1 "pill test"

Reference_____

8. Combustible trim, excluding handrails and guardrails, shall be limited to a maximum of _____ percent of the aggregate wall or ceiling area in which it is attached.

    a. 5   b. 10

    c. 25   d. 50

Reference_____

9. In a fully sprinklered Group A auditorium, complying flame-resistant decorative material shall be limited to a maximum of _____ percent of the aggregate area of walls and ceilings.

   a. 10   b. 25
   c. 50   d. 75

   Reference_____

10. What is the minimum required flame spread and smoke-developed index for materials, other than foam plastic, used as interior trim?

    a. Class A   b. Class B
    c. Class C   d. unlimited

    Reference_____

11. Within a sleeping unit of a Group R-1 hotel, a space-heating system shall be provided that is capable of maintaining a minimum indoor temperature of _____ at a point 3 feet above the floor on the design heating day.

    a. 65 °F   b. 68 °F
    c. 70 °F   d. 72 °F

    Reference_____

12. Where natural light by means of exterior glazed openings is utilized as the required lighting for an occupied space, the minimum net glazed area shall not be less than _____ percent of the floor area of the room served.

    a. 4   b. 5
    c. 8   d. 10

    Reference_____

13. A stairway within a dwelling unit in a Group R-2 apartment building shall be provided with a minimum illumination level on tread runs of _____ foot-candle(s).

    a. 1   b. 2
    c. 5   d. 10

    Reference_____

Study Session 17

14. A floor/ceiling assembly between a dwelling unit and a public area within the same building shall have a minimum impact insulation class (IIC) rating of _____ when field tested.

    a. 40   b. 45
    c. 50   d. not regulated

    Reference_____

15. Occupiable spaces in an office building shall have a minimum ceiling height of _____.

    a. 6 feet, 8 inches   b. 7 feet, 0 inches
    c. 7 feet, 6 inches   d. 8 feet, 0 inches

    Reference_____

16. Every dwelling unit shall have at least one room with a minimum floor area of _____ square feet.

    a. 70   b. 120
    c. 150   d. 220

    Reference_____

17. The minimum size of the required access opening to a crawl space shall be _____.

    a. 18 inches by 24 inches   b. 20 inches by 24 inches
    c. 20 inches by 30 inches   d. 22 inches by 30 inches

    Reference_____

18. The minimum size of the required access opening to an attic area over 30 inches in clear height shall be _____.

    a. 18 inches by 24 inches   b. 20 inches by 24 inches
    c. 20 inches by 30 inches   d. 22 inches by 30 inches

    Reference_____

19. Where urinal partitions are required, the partitions shall begin at a maximum height of _____ inches above the finished floor surface and extend to a minimum height of _____ inches above the floor surface.

    a. 12, 54
    b. 12, 60
    c. 15, 54
    d. 15, 60

Reference_____

20. Roof soffits are considered weather-exposed surfaces except for those portions located a minimum horizontal distance of _____ feet from the outer edges of the soffit.

    a. 3
    b. 5
    c. 8
    d. 10

Reference_____

21. Water-resistant gypsum backing board that is $^5/_8$-inch-thick is permitted for use on ceilings having a maximum frame spacing of _____ inches on center.

    a. 12
    b. 16
    c. 19.2
    d. 24

Reference_____

22. For exterior plastering, the second coat shall have a maximum variation of _____ inch in any direction under a 5-foot straight edge.

    a. $^1/_8$
    b. $^3/_{16}$
    c. $^1/_4$
    d. $^3/_8$

Reference_____

23. A maximum of _____ elevator car(s) shall be located in any single hoistway enclosure.

    a. 1
    b. 2
    c. 3
    d. 4

Reference_____

24. Holes in the elevator machine room floor for the passage of cables or other moving elevator equipment shall be limited so as to provide a maximum clearance of _____ inch(es) on all sides.

    a. ½   b. 1
    c. 1½   d. 2

    Reference_____

25. Where elevators are provided in buildings with four or more stories above grade plane, at least one elevator car shall be of such a size to accommodate a minimum _____ ambulance stretcher in the horizontal, open position.

    a. 24-inch by 76-inch   b. 24-inch by 84-inch
    c. 28-inch by 78-inch   d. 30-inch by 78-inch

    Reference_____

26. A fibrous floor covering installed in the dining area of a Group A-2 sprinklered meeting room shall comply with the requirements of _____ or ASTM D 2859.

    a. DOC FF-1 "pill test"   b. Class I
    c. Class II   d. Class A

    Reference_____

27. As a general rule, the minimum net area of ventilation openings for under-floor ventilation shall be based on 1 square foot for each _____ square feet of crawl-space area.

    a. 100   b. 120
    c. 150   d. 300

    Reference_____

28. Where a court is adjacent to window openings on both sides of the court that provide for the required natural ventilation, the minimum court width shall be _____ feet if the building is five stories in height.

    a. 3   b. 6
    c. 9   d. 10

    Reference_____

29. In the construction of a gypsum board fire-resistance-rated assembly, for which of the following applications is joint and fastener treatment required on single-layer systems?

    a. walls that extend above a fire-rated ceiling

    b. where joints occur over wood framing

    c. assemblies tested without joint treatment

    d. tongue-and-groove edge gypsum board

    Reference_____

30. In a single-elevator building where standby power is provided to operate the elevator, the transfer to standby power shall occur automatically within a maximum of _____ seconds after failure of normal power.

    a. 10
    b. 15
    c. 30
    d. 60

    Reference_____

31. Combustible decorative materials are prohibited in which one of the following occupancies?

    a. Group A-1
    b. Group I-2
    c. Group I-3
    d. Group R-4

    Reference_____

32. Where ceilings are applied directly to the underside of roof framing members to form enclosed rafter spaces, such spaces shall be provided with cross ventilation with a minimum of _____ inch(es) of airspace provided between the insulation and the roof sheathing.

    a. 1
    b. 2
    c. 3
    d. 4

    Reference_____

33. In other than dwelling units, sleeping units and toilet rooms not accessible to the public with a single water closet, walls within 2 feet of urinals and water closets shall have a smooth, hard, nonabsorbent surface to a minimum height of _____ feet above the floor.

    a. 3
    b. 4
    c. 5
    d. 6

    Reference_____

34. Fasteners used to attach gypsum board to a horizontal diaphragm ceiling shall be spaced at a maximum of _____ inches on center at all supports and located a maximum of _____ inch from the edges and ends of the gypsum board.

    a. 7, $^3/_8$ b. 8, $^3/_8$
    c. 7, $^1/_2$ d. 8, $^1/_2$

    Reference _____

35. Unless special provisions are made, exterior cement plaster shall not be applied when the ambient temperature is a maximum of _____ .

    a. 32°F   b. 35°F
    c. 40°F   d. 45°F

    Reference _____

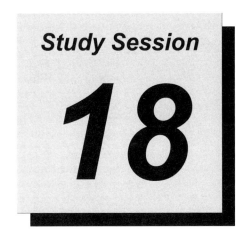

2012 IBC Chapters 24 and 26
Glazing, Skylights and Plastics

OBJECTIVE: To gain an understanding of the installation requirements for glass and glazing, glazing support and framing, safety glazing, skylights, foam plastics, light-transmitting plastics and plastic veneers.

REFERENCE: Chapters 24 and 26, 2012 *International Building Code*

KEY POINTS:
- How is the installation of replacement glass regulated?
- How must a pane of glass be identified? Tempered glass?
- How must glazing be supported?
- What are the limitations for louvered windows and jalousies?
- Sloped glazing provisions for skylights, roofs and sloped walls apply when the glazing material is installed at what minimum slope from the vertical plane?
- Which materials are permitted for sloped glazing? What limitations are placed on these materials?
- For which sloped glazing installations are screens mandated?
- Skylight frames must be constructed of noncombustible materials in which types of construction?
- When are curbs required for the mounting of skylights?
- How are unit skylights regulated?
- What is safety glazing? What are the standards that regulate safety glazing?
- How shall safety glazing be identified? What information must be included as a part of the identifying mark?
- Which types of doors are exempt from the glazing requirements for hazardous locations?
- When glazing is located adjacent to a door, how is it determined if safety glazing is required?
- Which areas of tub and shower enclosures are considered hazardous locations for glazing?
- Individual fixed or operable glazed panels exceeding nine square feet in area must be safety glazed where which three conditions exist?
- How is glazing adjacent to stairways and landings to be addressed?

485

KEY POINTS:
(Cont'd)

- What are the three types of glass permitted to be used as structural balustrade panels in rails?
- What is the maximum flame spread index for foam plastic insulation used in building construction? What is the maximum smoke-developed index?
- When is a thermal barrier necessary to separate the interior of a building from foam plastic insulation?
- Under which conditions may foam plastic insulation be incorporated as a part of a roof covering assembly?
- What are the limitations for the use of plastic veneer within a building? On the exterior wall of a building?
- How are light-transmitting plastics used as wall or roof panels regulated?
- Which specific provisions apply to plastic used as exterior wall panels? As roof panels? In skylights? In light-diffusing systems?

Topic: Allowable Glazing Materials
Reference: IBC 2405.1, 2405.2
Category: Glass and Glazing
Subject: Sloped Glazing and Skylights

Code Text: Section 2405 *applies to the installation of glass and other transparent, translucent or opaque glazing material installed at a slope more than 15 degrees (0.26 rad) from the vertical plane, including glazing materials in skylights, roofs and sloped walls. For monolithic glazing systems, the glazing material of the single light or layer shall be laminated glass with a minimum 30-mil (0.76 mm) polyvinyl butyral (or equivalent) interlayer, wired glass, light-transmitting plastic materials, heat-strengthened glass or fully tempered glass.*

Discussion and Commentary: The provisions for skylights are intended to protect such glazed openings from flying firebrands, to provide adequate strength to carry the load normally attributed to roofs, and to protect the occupants of a building from falling glazing materials.

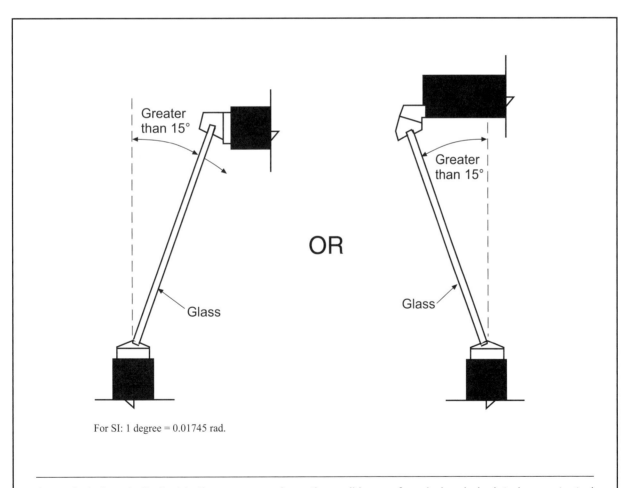

For SI: 1 degree = 0.01745 rad.

Annealed glass is limited to those areas where the walking surface below is isolated or protected from the risk of falling glass, or to specified greenhouses. Multiple-layer glazing systems must be glazed with only those materials permitted for single-layer glazing systems.

Topic: Screening
Reference: IBC 2405.3

Category: Glass and Glazing
Subject: Sloped Glazing and Skylights

Code Text: *Where used in monolithic glazing systems, heat-strengthened glass and fully tempered glass shall have screens installed below the glazing material.* See exceptions. *The screens and their fastenings shall: 1) be capable of supporting twice the weight of the glazing, 2) be firmly and substantially fastened to the framing members, and 3) be installed within 4 inches (102 mm) of the glass.*

Discussion and Commentary: Heat-strengthened glass has the undesirable characteristic of breaking into shards, whereas tempered glass has been shown to break spontaneously such that large chunks of glass may fall unexpectedly. Thus, these two types of glass require screen protection below the skylight to protect the occupants below.

For SI: 1 foot = 304.8 mm, 1 degree = 0.01745 rad.

A commonly used exception allows fully tempered glass without screening where two conditions are met: (1) the slope of the skylight is limited to 30 degrees from the vertical plane, and (2) the highest point of the glass is no more than 10 feet above the walking surface.

Topic: Impact Test
Reference: IBC 2406.2

Category: Glass and Glazing
Subject: Safety Glazing

Code Text: *Where required by other sections of the IBC, glazing shall be tested in accordance with CPSC 16 CFR 1201. Glazing shall comply with the test criteria for Category II unless otherwise indicated in Table 2406.2(1).* See exception permitting the use of Class A or B glazing material tested in accordance with ANSI Z97.1 in locations other than entrance/exit doors, storm doors, combination doors, patio doors, closet doors, and shower and tub doors and enclosures.

Discussion and Commentary: The only test standard recognized for the acceptance of safety glazing materials in all hazardous locations identified by the IBC is CPSC 16 CFR 1201. Developed by the Consumer Product Safety Commission in 1977, in cooperation with building officials and the glass industry, the standard sets forth the criteria for glazing that is required in areas subject to human impact.

TABLE 2406.2(1)
MINIMUM CATEGORY CLASSIFICATION OF GLAZING USING CPSC 16 CFR 1201

EXPOSED SURFACE AREA OF ONE SIDE OF ONE LITE	GLAZING IN STORM OR COMBINATION DOORS (Category class)	GLAZING IN DOORS (Category class)	GLAZED PANELS REGULATED BY ITEM 7 OF SECTION 2406.4 (Category class)	GLAZED PANELS REGULATED BY ITEM 6 OF SECTION 2406.4 (Category class)	DOORS AND ENCLOSURES REGULATED BY ITEM 5 OF SECTION 2406.4 (Category class)	SLIDING GLASS DOORS PATIO TYPE (Category class)
9 square feet or less	I	I	No requirement	I	II	II
More than 9 square feet	II	II	II	II	II	II

For SI: 1 square foot = 0.0929 m².

For SI: 1 inch = 25.4 mm mm, 1 foot = 304.8 mm

The tests established in CPSC 16 CFR Part 1201 vary with the category classification. As a part of the test, Category I glazing is impacted from a drop height of 18 inches. Limitations are placed on the damage that can occur due to the impact. Category II glazing is impacted from a drop height of 48 inches, resulting in a more severe test of compliance. Category II glazing is permitted in all safety glazing locations, whereas glazing only recognized as Category I materials are limited to specific locations.

Topic: Identification
Reference: IBC 2406.3
Category: Glass and Glazing
Subject: Safety Glazing

Code Text: *Except as indicated in Section 2406.3.1, each pane of safety glazing installed in hazardous locations shall be identified by a manufacturer's designation specifying who applied the designation, the manufacturer or installer and the safety glazing standard with which it complies, as well as the information specified in Section 2403.1* (general glazing identification). *The designation shall be acid etched, sand blasted, ceramic fired, laser etched, embossed or of a type that once applied, cannot be removed without being destroyed.* See exceptions for certifications of compliance and tempered spandrel glass.

Discussion and Commentary: Improper glazing installed in areas subject to human impact can create a serious hazard. Accordingly, it is critical that glazing in such locations be appropriately identified to ensure that the proper glazing is in place.

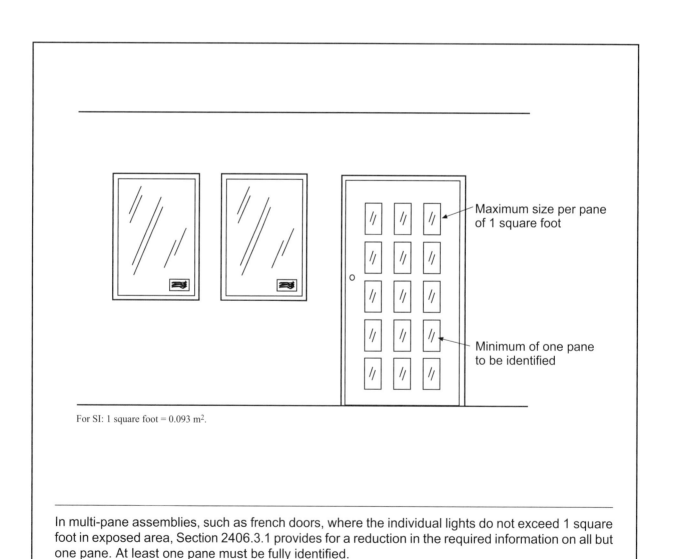

For SI: 1 square foot = 0.093 m².

In multi-pane assemblies, such as french doors, where the individual lights do not exceed 1 square foot in exposed area, Section 2406.3.1 provides for a reduction in the required information on all but one pane. At least one pane must be fully identified.

Topic: Glazing in Doors
Reference: IBC 2406.4.1

Category: Glass and Glazing
Subject: Hazardous Locations

Code Text: *The following locations shall be considered specific hazardous locations requiring safety glazing materials: Glazing in all fixed and operable panels of swinging, sliding, and bifold doors.* See four exceptions where safety glazing is not required.

Discussion and Commentary: As a general rule, any door containing glazing must be glazed with safety glass or other safety glazing material recognized by the code for that intended purpose. Glazing in doors is of particular concern due to the increased likelihood of accidental impact by individuals operating or opening the doors. In addition, a person may push against a glazed portion of the door to gain leverage in pushing it open. Therefore, it is important that only safety glazing materials be used for glazing in doors.

There are a limited number of products and applications that are exempt from the requirements for hazardous locations, including small openings in doors through which a 3-inch-diameter sphere will not pass, and specific decorative assemblies, such as leaded, faceted or carved glass.

Study Session 18

Topic: Glazing Adjacent to Doors
Reference: IBC 2406.4.2
Category: Glass and Glazing
Subject: Hazardous Locations

Code Text: *The following locations shall be considered specific hazardous locations requiring safety glazing materials: Glazing in an individual fixed or operable panel adjacent to a door where the nearest vertical edge of the glazing is within a 24-inch (610 mm) arc of either vertical edge of the door in a closed position and where the bottom exposed edge of the glazing is less than 60 inches (1524 mm) above the walking surface.* See four exceptions where safety glazing is not mandated.

Discussion and Commentary: When an individual approaches a doorway, areas adjacent to the door pose a risk when glazing is within 60 inches vertically of the walking surface. A person may slip or mistake the glass panel adjacent to a door for a passageway and walk into the glass, or a person may push against the sidelight with one hand for support while opening the door with the other hand. Therefore, safety glazing is required for any glazed opening located within 24 inches horizontally of the vertical edge of the door.

Where there is an intervening wall or similar permanent barrier between the door and the glazing, or where access through the door is to a closet or similar storage area of limited depth, safety glazing is not required, as the potential for contact is greatly reduced.

Topic: Glazing in Windows
Reference: IBC 2406.4.3
Category: Glass and Glazing
Subject: Hazardous Locations

Code Text: *The following locations shall be considered specific hazardous locations requiring safety glazing materials: Glazing in a fixed or operable panel that meets all of the following conditions: 1) exposed area of an individual pane greater than 9 square feet (0.84 m²), 2) exposed bottom edge less than 18 inches (457 mm) above the floor, 3) exposed top edge greater than 36 inches (914 mm) above the floor, and 4) one or more walking surface(s) within 36 inches (914 mm) horizontally of the plane of the glazing.* See exceptions for decorative glazing, where complying protective bar is installed or where multiple glazed units are located at least 25 feet above the surface below.

Discussion and Commentary: Large pieces of glass create a hazard where located close to a travel path because it is possible to impact glazing where no obstacle or barrier is provided as an alternative impact area. Expansive glazing may also be mistaken for a clear opening in the wall.

Protective bar alternative

Where a minimum 1¹/₂-inch protective bar is installed on the accessible sides of a large glazed panel, it is not necessary to provide the opening with safety glazing. Located between 34 inches and 38 inches above the floor, the bar must be capable of withstanding a 50 plf horizontal load.

Study Session 18

Topic: Glazing in Guards and Railings
Reference: IBC 2406.4.4
Category: Glass and Glazing
Subject: Hazardous Locations

Code Text: *The following locations shall be considered specific hazardous locations requiring safety glazing materials: Glazing in guards and railings, including structural baluster panels and nonstructural in-fill panels, regardless of area or height above a walking surface.*

Discussion and Commentary: Both intentional and unintentional contact with guards and railings are expected to occur; therefore, the IBC mandates that glazing used in such applications always be safety glazing. Safety glazing is required when the glazed infill panel is nonstructural and is supported by a structural frame system. Occasionally, glazing is used as a structural guard rail system without any other means of support. In this case, the glazing is required to be safety glazing.

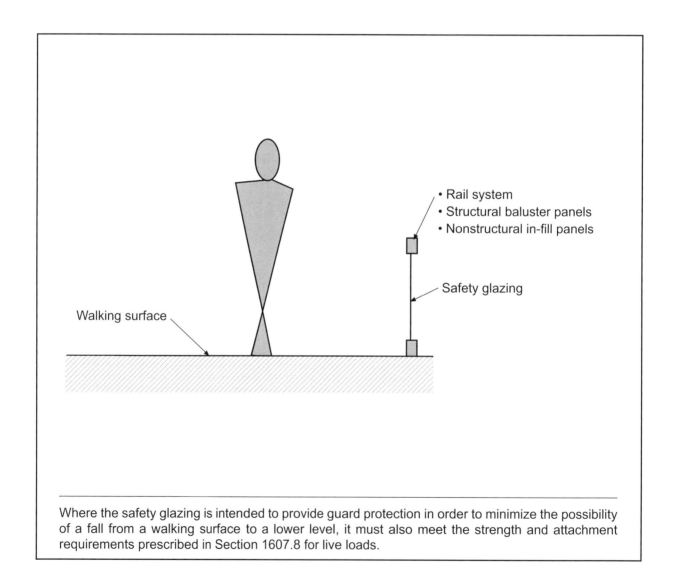

Where the safety glazing is intended to provide guard protection in order to minimize the possibility of a fall from a walking surface to a lower level, it must also meet the strength and attachment requirements prescribed in Section 1607.8 for live loads.

Topic: Glazing and Wet Surfaces
Reference: IBC 2406.4.5
Category: Glass and Glazing
Subject: Hazardous Locations

Code Text: *The following locations shall be considered specific hazardous locations requiring safety glazing materials: Glazing in walls, enclosures or fences containing or facing hot tubs, spas, whirlpools, saunas, steam rooms, bathtubs, showers and indoor or outdoor swimming pools where the bottom exposed edge of the glazing is less than 60 inches (1524 mm) measured vertically above any standing or walking surface.* See exception for glazing located at least 60 inches horizontally from the water's edge.

Discussion and Commentary: Because the standing surfaces in or adjacent to bathtubs, showers, hot tubs, swimming pools, spas and similar elements are wet and slippery, glazing adjacent to these elements must be regulated due to the potential for human impact. It is not uncommon for the user to slip while trying to enter or exit. Safety glazing is mandated where any of the glazing within the enclosed area extends to within 60 inches vertically of the standing surface.

For SI: 1 inch = 25.4 mm.

Glazing within a shower enclosure

Where a "garden tub" or similar element is installed within an alcove or similar recessed area having windows or other glazed openings, the glazing in the walls of the alcove are considered a portion of the enclosure and are thus regulated.

Study Session 18

Topic: Glazing Adjacent to Stairways
Reference: IBC 2406.4.6

Category: Glass and Glazing
Subject: Hazardous Locations

Code Text: *The following locations shall be considered specific hazardous locations requiring safety glazing materials: Glazing where the bottom exposed edge of the glazing is less than 60 inches (1524 mm) above the plane of the adjacent walking surface of stairways, landings between flights of stairs, and ramps.* See exceptions where safety glazing is not required.

Discussion and Commentary: Stairways and ramps present users with a greater risk for injury caused by falling than does a flat surface. Not only is the risk of falling greater when using a stair, but the injuries are generally more severe. Unlike falling on a flat surface where the floor will break a person's fall, there is nothing to stop someone from continuing to fall until he or she reaches the bottom of the stair.

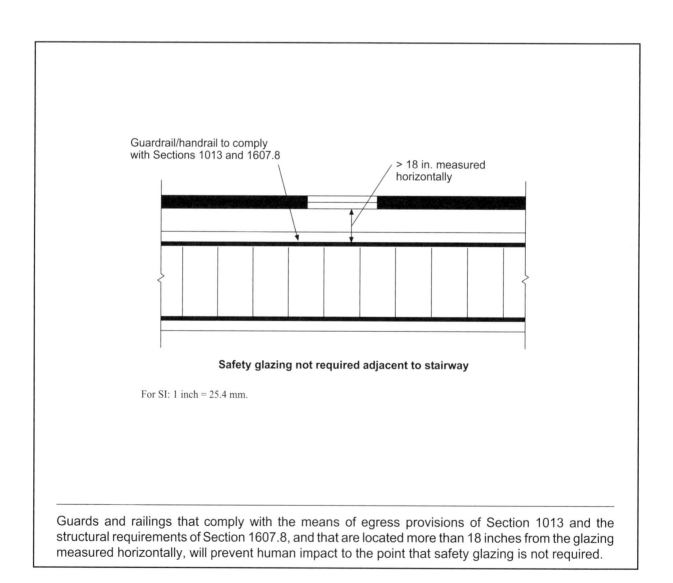

Safety glazing not required adjacent to stairway

For SI: 1 inch = 25.4 mm.

Guards and railings that comply with the means of egress provisions of Section 1013 and the structural requirements of Section 1607.8, and that are located more than 18 inches from the glazing measured horizontally, will prevent human impact to the point that safety glazing is not required.

Topic: Glazing at Bottom of Stairways
Reference: IBC 2406.4.7
Category: Glass and Glazing
Subject: Hazardous Locations

Code Text: *The following locations shall be considered specific hazardous locations requiring safety glazing materials: Glazing adjacent to the landing at the bottom of a stairway where the glazing is less than 36 inches (914 mm) above the landing and within 60 inches (1524 mm) horizontally of the bottom tread.* See exception where safety glazing is not required.

Discussion and Commentary: Historically, stairways have been considered one of the most dangerous elements of a building. Missteps and falls on stairways are quite common; therefore, it is important that any glazing that may be impacted is made of safety glazing materials. Where a complying guard rail or handrail is provided and the glazing is located a sufficient distance from the railing, safety glazing is not mandated.

Glazing adjacent to stairways

For SI: 1 inch = 25.4 mm, 1 foot = 304.8 mm.

The IBC limits the area of concern in this item to the bottom of a stair flight and its adjacent landing. Glazing that is located at least 36 inches vertically above the walking surface is not considered an impact risk.

Topic: Scope and Definition
Reference: IBC 2603, 202
Category: Plastic
Subject: Foam Plastic Insulation

Code Text: *Foam plastic insulation is a plastic that is intentionally expanded by the use of a foaming agent to produce a reduced-density plastic containing voids consisting of open or closed cells distributed throughout the plastic for thermal insulating or acoustical purposes and that has a density less than 20 pounds per cubic foot. The provisions of Section 2603 shall govern the requirements and uses of foam plastic insulation in buildings and structures.*

Discussion and Commentary: *Foam plastic* is a general term given to insulating products that have been manufactured by intentionally expanding plastic by the use of a foaming agent. Two basic concepts address the hazards created when foam plastic is exposed to fire conditions: 1) limitation of flame spread and smoke development, and 2) separation from the interior of the building by an approved thermal barrier.

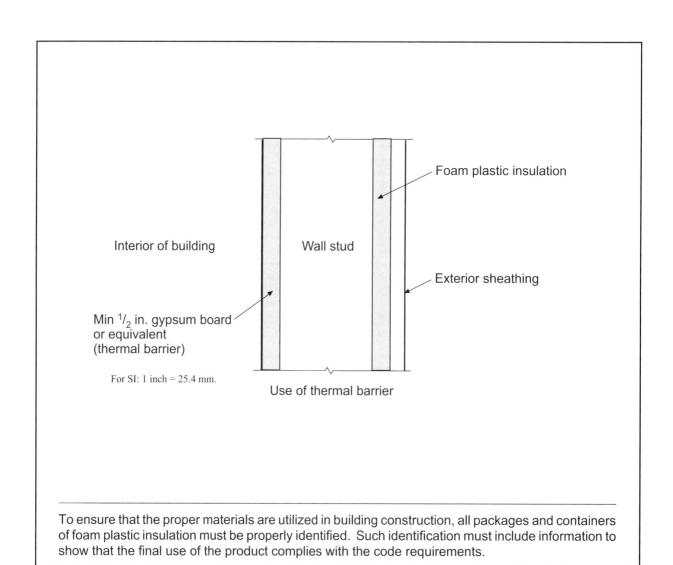

For SI: 1 inch = 25.4 mm.

Use of thermal barrier

To ensure that the proper materials are utilized in building construction, all packages and containers of foam plastic insulation must be properly identified. Such identification must include information to show that the final use of the product complies with the code requirements.

Topic: Thermal Barrier
Reference: IBC 2603.4

Category: Plastic
Subject: Foam Plastic Insulation

Code Text: *Except as provided for in Sections 2603.4.1 and 2603.10, foam plastic shall be separated from the interior of a building by an approved thermal barrier of $^1/_2$ inch (12.7 mm) gypsum wallboard or a material that is tested in accordance with and meets the acceptance criteria of both the Temperature Transmission Fire Test and the Integrity Fire Test of NFPA 275. Combustible concealed spaces shall comply with Section 718.*

Discussion and Commentary: A barrier is mandated to provide a minimum degree of protection between foam plastic materials and a building's occupants. Any type of separation equivalent to that provided by $^1/_2$-inch gypsum board is considered acceptable. Where one of the fourteen conditions established by Section 2603.4.1 is met, a thermal barrier as described in Section 2603.4 is not required.

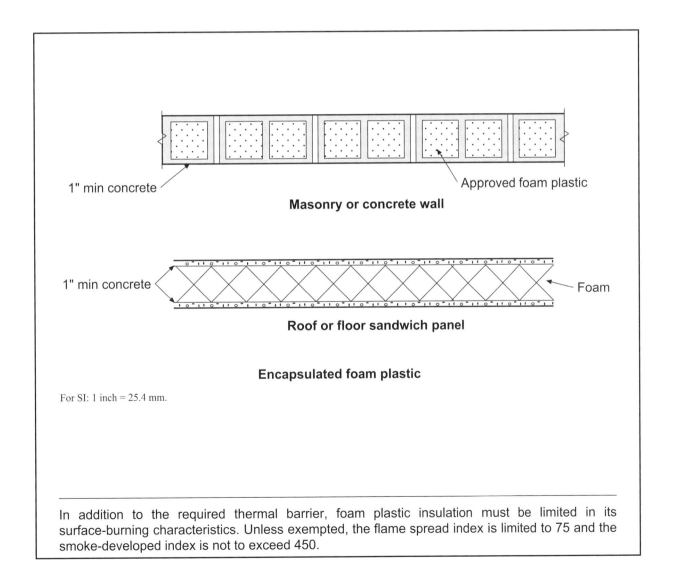

In addition to the required thermal barrier, foam plastic insulation must be limited in its surface-burning characteristics. Unless exempted, the flame spread index is limited to 75 and the smoke-developed index is not to exceed 450.

Study Session 18

Topic: Attics and Crawl Spaces
Reference: IBC 2603.4.1.6

Category: Plastic
Subject: Foam Plastic Insulation

Code Text: *Within an attic or crawl space where entry is made only for service of utilities, foam plastic insulation shall be protected against ignition by $1^1/_2$-inch-thick (38 mm) mineral fiber insulation; $^1/_4$-inch-thick (6.4 mm) wood structural panel, particleboard, or hardboard; $^3/_8$-inch (9.5 mm) gypsum wallboard, corrosion-resistant steel having a base metal thickness of 0.016 inch (0.4 mm) or other approved material installed in such a manner that the foam plastic insulation is not exposed.*

Discussion and Commentary: Although the general requirements of Section 2603.4 mandate a thermal barrier to isolate foam plastic from the interior of the building, Section 2603.4.1 reduces or eliminates the protective membrane requirements. One such reduction involves attic spaces and crawl spaces. Where access to the concealed attic or crawl space is only necessary to allow for the service of above-ceiling or under-floor utilities, such as plumbing, mechanical, electrical or communication elements, a reduced level of foam plastic protection is permitted.

If there are no utilities within the attic space or crawl space that require service, the reduced level of separation set forth in Section 2603.4.1.6 must still be provided. Where the attic or crawl space provides a usable area that exists for a purpose other than access to utilities, such as storage, a thermal barrier complying with Section 2603.4 is required.

Topic: Interior Finish and Trim
Reference: IBC 2604.2

Category: Plastic
Subject: Foam Plastic

Code Text: *For foam plastics used as interior trim: 1) the minimum density of the interior trim shall be 20 pcf (320 kg/m³), 2) the maximum thickness of the interior trim shall be ¹/₂ inch (12.7 mm) and the maximum width shall be 8 inches (204 mm), 3) the interior trim shall not constitute more than 10 percent of the specific wall or ceiling area of any room or space, and 4) the flame spread index shall not exceed 75 where tested in accordance with ASTM E 84 or UL 723. The smoke-developed index shall not be limited.* See exception to required flame spread index.

Discussion and Commentary: The general provisions regulating the use of decorations and trim in buildings are found in Section 806. They include plastics materials, other than foam plastics, used as interior trim. All foam plastic trim must comply with the provisions of Section 2604.2.

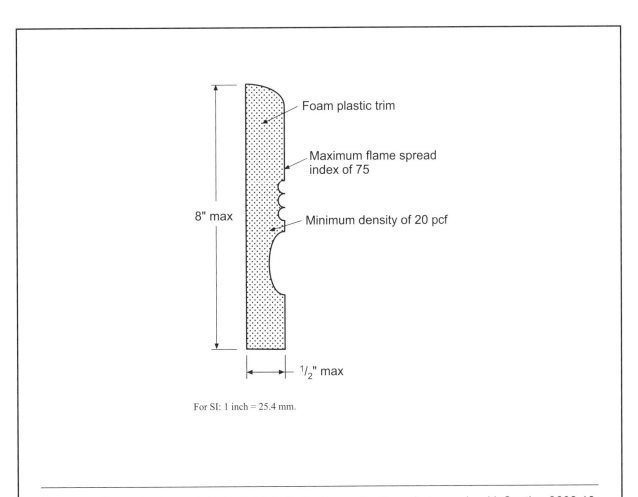

The use of foam plastics as interior finish is limited to applications that comply with Section 2603.10. This provision allows special approval based on the use of large-scale tests. Foam plastic finishes accepted by testing must also conform to Chapter 8 flame spread requirements.

Study Session 18
IBC Chapters 24 and 26

1. Wired glass in louvered windows shall have a maximum length of _____ inches.

    a. 36
    b. 42
    c. 48
    d. 60

    Reference _____

2. Where screening is required below the glazing material in a skylight, the screen shall be installed a maximum of _____ inch(es) below the glass.

    a. 1
    b. 3
    c. 4
    d. 6

    Reference _____

3. When used as the bottom glass layer in a multiple-layer glazing system installed above a walking surface, which of the following glazing materials never needs to be equipped with complying screening?

    a. fully tempered glass
    b. laminated glass
    c. heat-strengthened glass
    d. wired glass

    Reference _____

4. When set at an angle less than 45 degrees from the horizontal plane, a skylight shall be mounted on a curb a minimum of _____ inch(es) above the plane of the roof of other than a Group R-3 occupancy.

    a. 1   b. 2
    c. 3   d. 4

    Reference_____

5. Class I safety glazing with a maximum size of _____ per light is permitted in doors.

    a. 0 square inches (not permitted of any size)
    b. 100 square inches
    c. 144 square inches
    d. 9 square feet

    Reference_____

6. Which one of the following types of safety glazing materials shall not be solely identified by a certificate or affidavit?

    a. tempered glass   b. wired glass
    c. laminated glass   d. plastic glazing

    Reference_____

7. Glazing within a shower enclosure shall be safety glazing where the bottom exposed edge of the glazing is less than _____ inches above a standing surface.

    a. 60   b. 66
    c. 72   d. 78

    Reference_____

8. Where located less than 60 inches above the walking surface, glazing adjacent to a door within what maximum horizontal distance of either vertical door edge must be safety glazing?

    a. 12 inches   b. 18 inches
    c. 24 inches   d. 36 inches

    Reference_____

9. Where a protective bar is permitted as protection in lieu of safety glazing, the bar must be capable of withstanding a minimum horizontal load of _____ pounds per linear foot without contacting the glass.

   a. 15   b. 20
   c. 40   d. 50

   Reference_____

10. Glazing in walls enclosing a swimming pool solely be safety glazing where located a maximum of _____ feet horizontally from the water's edge.

    a. 5   b. 6
    c. 8   d. 10

    Reference_____

11. At other than the bottom tread, glazing adjacent to stairways does not need to be safety glazing where the plane of the glass is a minimum of _____ inches horizontally from the walking surface.

    a. 12   b. 18
    c. 24   d. 36

    Reference_____

12. Glazed openings in doors are not considered hazardous locations for safety glazing purposes, provided the openings will not allow passage of a _____ -inch sphere.

    a. 3   b. 4
    c. 6   d. 9

    Reference_____

13. Fire department glass access panels shall be of _____ glass.

    a. heat-strengthened   b. tempered
    c. laminated   d. annealed

    Reference_____

14. Where glass is used as a handrail assembly or guard section, and unless otherwise adequately supported, each handrail or guard section shall be supported by a minimum of _____ glass balusters.

    a. 2   b. 3
    c. 4   d. 6

    Reference_____

15. By definition, foam plastic insulation shall have a density less than _____ pounds per cubic foot.

    a. 2   b. 4
    c. 10   d. 20

    Reference_____

16. _____ is considered to be a plastic material that is capable of being changed into a substantially nonreformable product when cured.

    a. Approved plastic   b. Glass fiber reinforced plastic
    c. Thermoplastic material d. Thermosetting material

    Reference_____

17. Unless otherwise specified, foam plastic shall have a maximum flame spread index of _____ and a maximum smoke-developed index of _____ where tested in the maximum thickness intended for use.

    a. 25, 200   b. 25, 450
    c. 50, 200   d. 75, 450

    Reference_____

18. What material is specifically identified as an approved thermal barrier for separating foam plastic from the interior of a building?

    a. $1/2$-inch gypsum wallboard   b. $3/8$-inch Type x gypsum wallboard
    c. $3/8$-inch gypsum sheathing   d. $1/2$-inch wood structural panel

    Reference_____

19. Within a concrete floor system, all foam plastic insulation shall be covered on each face by a minimum of a _____ -inch thickness of concrete.

    a. ½   b. 1
    c. 1½   d. 2

    Reference_____

20. The maximum allowable thickness of foam plastic interior trim shall be _____ inch(es), with a maximum permitted width of _____ inches.

    a. 4, 12   b. 2, 12
    c. 1, 8   d. ½, 8

    Reference_____

21. Exterior plastic veneer, other than plastic siding, may be installed on the exterior walls of buildings of any type of construction to a maximum height of _____ feet above grade.

    a. 25   b. 35
    c. 50   d. 75

    Reference_____

22. In a nonsprinklered Group B building, Class CC2 light-transmitting wall panels used in an exterior wall with a fire separation distance of 18 feet are limited to a maximum of _____ percent of the exterior wall.

    a. 10   b. 15
    c. 25   d. 50

    Reference_____

23. In a fully sprinklered building, what is the maximum height light-transmitting plastic glazing is permitted to be installed above grade level?

    a. 35 feet   b. 55 feet
    c. 75 feet   d. unlimited

    Reference_____

24. In which one of the following occupancies may light-transmitting plastic roof panels be installed?

    a. Group A-1   b. Group H-1

    c. Group I-2   d. Group I-3

Reference_____

25. An interior wall sign of light-transmitting plastic is limited in size to _____ square feet.

    a. 16   b. 24

    c. 48   d. 100

Reference_____

26. Glass in window walls sloped a maximum of _____ degrees from the vertical shall be designed to resist the required wind loads for components and cladding.

    a. 15   b. 30

    c. 45   d. 60

Reference_____

27. For which of the following applications is the use of a removable paper label specifically permitted for identifying tempered glass for safety glazing purposes?

    a. storm doors   b. spandrel panels

    c. unframed swinging doors   d. structural balustrade panels

Reference_____

28. Only one individual pane in a multi-pane glazed door assembly requires safety glazing identification, provided each pane is a maximum of _____ in size.

    a. 64 square inches   b. 100 square inches

    c. 1 square foot   d. $1\frac{1}{2}$ square feet

Reference_____

Study Session 18

29. All glazing subject to human impact located in a(n) _____ shall comply with Category II of CPSC 16 CFR 1201 or Class A of ANSI Z97.1.

    a. school classroom   b. detention facility
    c. multipurpose gymnasium   d. manufacturing building

    Reference _____

30. Foam plastic spray applied to a sill plate of Type V construction shall have a maximum thickness of _____.

    a. 1 inch   b. 1 1/2 inches
    c. 3 inches   d. 3 1/4 inches

    Reference _____

31. Any glazing material is permitted to be installed without screens in the sloped glazing system of commercial noncombustible greenhouses, provided the maximum ridge height is _____ feet above grade.

    a. 16   b. 20
    c. 25   d. 30

    Reference _____

32. Glazing installed in sliding glass patio doors shall have a minimum category classification of Class _____.

    a. A   b. B
    c. I   d. II

    Reference _____

33. For all glazing types, the minimum required thickness for glass used as a handrail assembly or guard section shall be _____ inch.

    a. 1/4   b. 3/8
    c. 1/2   d. 9/16

    Reference _____

34. Individual panels or units of light-transmitting plastic used in a light-diffusing system shall have a maximum size of _____ square feet and be a maximum of _____ feet in length.

    a. 10, 5   b. 15, 10

    c. 30, 10   d. 36, 12

Reference _____

35. In a nonsprinklered building with no smoke and heat vents, light-transmitting plastic skylight assemblies shall each have a maximum glazed area of _____ square feet within the curb.

    a. 16   b. 36

    c. 64   d. 100

Reference _____

Answer Keys

Study Session 1
2012 *International Building Code*

1.	c	Sec. 101.2, Exc.	31.	b	Sec. 105.2, #M7
2.	c	Sec. 101.2.1	32.	a	Sec. 107.2.3
3.	b	Sec. 102.1	33.	b	Sec. 109.3
4.	a	Sec. 103.1	34.	c	Sec. 110.1
5.	d	Sec. 104.1	35.	c	Sec. 115.2
6.	a	Sec. 104.9.1			
7.	b	Sec. 104.10			
8.	b	Sec. 104.11			
9.	c	Sec. 104.11.2			
10.	b	Sec. 105.2, #B1			
11.	c	Sec. 105.2, #B13			
12.	c	Sec. 110.3.10.1			
13.	a	Sec. 105.7			
14.	d	Sec. 107.3.1			
15.	b	Sec. 107.5			
16.	c	Sec. 110.3			
17.	a	Sec. 110.5			
18.	d	Sec. 111.2			
19.	d	Sec. 111.3			
20.	b	Sec. 113.2			
21.	b	Sec. 3404.3			
22.	c	Sec. 3406.3			
23.	b	Table 3411.8.5			
24.	d	Sec. 3412.5			
25.	a	Chapter 35			
26.	b	Sec. 105.2, #B9			
27.	d	Sec. 105.2, #B12			
28.	d	Sec. 105.6			
29.	b	Sec. 108.1			
30.	c	Sec. 3412.2			

Study Session 2
2012 *International Building Code*

1.	c	Sec. 302.1, #6	31.	a	Sec. 508.3.1
2.	d	Sec. 302.1, #10	32.	c	Table 508.4
3.	a	Sec. 303.1.4	33.	c	Table 508.4
4.	c	Sec. 303.1.2, #2	34.	b	Table 508.4
5.	a	Sec. 303.1.3	35.	a	Table 508.4
6.	d	Sec. 303.5			
7.	a	Secs. 304.1, 309.1			
8.	a	Sec. 303.3			
9.	b	Sec. 307.4			
10.	a	Sec. 307.3			
11.	c	Secs. 202, 308.4			
12.	d	Table 307.1(1)			
13.	a	Secs. 202, 308.3			
14.	c	Sec. 309.1, 304.1			
15.	b	Sec. 310.4			
16.	d	Sec. 310.6. 308.4			
17.	d	Sec. 311.3, 312.1			
18.	a	Sec. 303.4			
19.	d	Sec. 307.6			
20.	c	Sec. 307.1, Exc. 11			
21.	c	Sec. 303.1.1			
22.	b	Sec. 304.1			
23.	c	Sec. 310.5			
24.	b	Sec. 202			
25.	c	Table 509, Sec. 509.4.1			
26.	d	Table 509			
27.	d	Table 509, Sec. 509.4.1			
28.	d	Sec. 509.4.2			
29.	a	Sec. 508.2.1			
30.	c	Sec. 508.2.4, Exc. 1			

Study Session 3
2012 *International Building Code*

1.	a	Sec. 602.2	31.	c	Table 601
2.	c	Sec. 602.3	32.	c	Tables 601, 602
3.	b	Sec. 602.4	33.	a	Table 602, Note d
4.	a	Sec. 602.3, 602.4	34.	d	Table 602.4
5.	d	Sec. 602.4.1	35.	a	Sec. 603.1, #2
6.	c	Sec. 602.4.2			
7.	c	Sec. 602.4.5			
8.	d	Sec. 602.5			
9.	b	Table 601			
10.	d	Sec. 202			
11.	a	Table 601			
12.	c	Table 601			
13.	b	Table 601, Note a			
14.	d	Sec. 603.1, #1.3			
15.	b	Table 601, Note b			
16.	c	Tables 601, 602			
17.	c	Tables 601, 602			
18.	b	Tables 601, 602			
19.	d	Tables 601, 602			
20.	b	Table 602			
21.	d	Sec. 603.1, #1.1			
22.	c	Table 602, Note b; Sec. 406.3.2, #2			
23.	a	Sec. 603.1, #1.2			
24.	a	Sec. 603.1, #9			
25.	d	Sec. 603.1, #4, Table 1505.1			
26.	a	Sec. 602.4.6			
27.	a	Table 601			
28.	a	Sec. 603.1, #1.3, Exc.			
29.	c	Tables 601 and 602			
30.	d	Table 601, Note c			

Study Session 4
2012 *International Building Code*

1.	b	Sec. 501.2	31.	d	Sec. 505.2.1, Exc. 1
2.	a	Sec. 202	32.	d	Sec. 505.3.1
3.	c	Table 503	33.	b	Sec. 507.12
4.	c	Table 503	34.	c	Sec. 507.10, #3
5.	b	Sec. 506.4, Exc.	35.	a	Sec. 506.3, Exc. 2
6.	d	Sec. 202			
7.	b	Sec. 506.4.1, #2			
8.	b	Sec. 504.2			
9.	c	Sec. 504.2, Exc. 1			
10.	b	Sec. 504.3			
11.	b	Sec. 505.2			
12.	b	Sec. 505.2			
13.	c	Sec. 505.2.1			
14.	a	Sec. 505.2.3, Exc. 3			
15.	a	Sec. 505.2.3, Exc. 1			
16.	b	Sec. 506.2			
17.	b	Sec. 506.2			
18.	b	Table 503			
19.	c	Sec. 504.2, Exc. 2			
20.	c	Sec. 506.3			
21.	d	Sec. 506.3			
22.	d	Sec. 507.10			
23.	d	Sec. 507.5, #2			
24.	d	Sec. 507.9			
25.	c	Sec. 510.2, #1			
26.	d	Sec. 504.3			
27.	b	Sec. 507.4			
28.	a	Sec. 507.7, #3			
29.	b	Sec. 507.8, Table 503			
30.	c	Sec. 510.6			

Study Session 5
2012 *International Building Code*

1.	a	Sec. 202	31.	b	Sec. 703.5.1
2.	b	Sec. 202	32.	b	Table 705.8, Note g
3.	b	Sec. 202	33.	d	Sec. 705.11, Exc. 3
4.	b	Sec. 202	34.	b	Sec. 705.2.1, 1406.3, Exc. 1
5.	a	Sec. 202	35.	c	Sec. 705.8.1, Exc. 1.1
6.	a	Sec. 202, 706.2			
7.	b	Sec. 202			
8.	b	Sec. 202			
9.	c	Sec. 703.2			
10.	c	Sec. 703.7			
11.	d	Sec. 703.5.2			
12.	a	Sec. 705.2			
13.	b	Table 705.2			
14.	d	Sec. 705.2.2			
15.	c	Sec. 705.3			
16.	c	Sec. 705.5			
17.	a	Table 705.8			
18.	d	Table 705.8			
19.	c	Table 705.8			
20.	c	Sec. 705.8.1, Exc. 1.2			
21.	c	Sec. 705.8.5			
22.	c	Sec. 705.8.5			
23.	b	Sec. 705.11, Exc. 2			
24.	a	Sec. 705.11.1			
25.	b	Sec. 705.11.1			
26.	d	Sec. 703.2.1			
27.	d	Sec. 705.2.3			
28.	c	Table 705.8			
29.	c	Sec. 705.11, #6; Table 705.8			
30.	a	Table 705.8, Note e			

Answer Keys

Study Session 6
2012 *International Building Code*

1.	d	Sec. 706.3, Exc.	31.	a	Sec. 706.5.2
2.	c	Table 706.4	32.	b	Sec. 709.3, Exc.
3.	b	Table 706.4, Note a	33.	b	Sec. 712.1.6
4.	a	Sec. 706.5	34.	a	Sec. 710.8
5.	b	Sec. 706.6	35.	d	Sec. 709.5, Exc. 1
6.	b	Sec. 706.6.1, Exc.			
7.	c	Sec. 706.7			
8.	b	Sec. 706.8			
9.	d	Secs. 707.3; 708.4, #4			
10.	c	Sec. 707.5.1, Exc. 3			
11.	d	Sec. 707.6			
12.	a	Sec. 707.6, Exc. 2			
13.	c	Sec. 712.1.3.1			
14.	c	Sec. 712.1.8			
15.	b	Sec. 706.2			
16.	c	Secs. 707.3.5, 1025.2			
17.	c	Sec. 708.1			
18.	c	Sec. 708.3			
19.	b	Sec. 708.4, Exc. 3			
20.	d	Sec. 709.3			
21.	b	Sec. 709.4			
22.	c	Sec. 712.1.3.2			
23.	a	Sec. 711.3.1			
24.	d	Sec. 711.3.3			
25.	c	Sec. 711.4			
26.	b	Table 707.3.10			
27.	b	Sec. 707.9			
28.	a	Sec. 710.3			
29.	a	Sec. 710.5			
30.	b	Secs. 708.3, Exc. 2; 711.3, Exc.			

Study Session 7
2012 *International Building Code*

1.	c	Sec. 713.4		31.	b	Sec. 717.3.3.4
2.	d	Sec. 714.4.2.1		32.	c	Sec. 718.3.3
3.	b	Sec. 715.1, Exc.		33.	d	Sec. 717.3.2.2
4.	d	Sec. 715.1, Exc. 9		34.	a	Sec. 718.2.6
5.	b	Sec. 716.5.5		35.	d	Sec. 719.2
6.	b	Sec. 713.13.3				
7.	d	Table 716.5				
8.	d	Table 716.5				
9.	b	Sec. 714.5				
10.	c	Sec. 713.14.1				
11.	b	Sec. 716.5.9, Exc. 1				
12.	c	Sec. 716.5.9.3				
13.	b	Sec. 716.5.9.3				
14.	a	Sec. 713.13.3				
15.	d	Table 716.6				
16.	c	Sec. 716.6.7.2				
17.	d	Table 717.3.2.1				
18.	d	Sec. 717.3.3.2, #3				
19.	c	Secs. 717.5.1, 717.5.1.1				
20.	c	Sec. 717.5.3				
21.	d	Sec. 718.2.1				
22.	b	Sec. 718.2.6				
23.	a	Sec. 718.3.1				
24.	b	Sec. 718.4.3				
25.	a	Sec. 720.2				
26.	a	Sec. 714.3.1.2				
27.	a	Sec. 714.3.2, Exc. 1				
28.	b	Table 716.6				
29.	c	Sec. 718.2.7				
30.	d	Sec. 719.4				

Study Session 8
2012 *International Building Code*

1.	b	Sec. 901.6.1, Exc. 2	31.	d	Sec. 903.3.1.2.1
2.	b	Sec. 202	32.	d	Sec. 907.2.4
3.	c	Sec. 202	33.	b	Sec. 909.20.1
4.	c	Sec. 903.2.1.2, #2	34.	c	Sec. 910.3.2.1
5.	b	Sec. 903.2.1.5	35.	d	Sec. 911.1.3
6.	c	Sec. 903.2.3, #1			
7.	b	Sec. 903.2.4.1			
8.	a	Sec. 903.2.2, #1			
9.	c	Sec. 903.2.7, #2			
10.	d	Sec. 903.2.8			
11.	d	Sec. 903.2.9, #3			
12.	c	Sec. 903.2.9.1, #2			
13.	b	Sec. 903.2.10.1			
14.	b	Sec. 903.2.11.1.3			
15.	b	Sec. 903.3.1.3			
16.	d	Sec. 903.3.2			
17.	c	Sec. 903.3.3			
18.	b	Sec. 903.3.5.2			
19.	d	Sec. 904.11.1			
20.	b	Sec. 905.5			
21.	a	Sec. 907.2.2, #2			
22.	a	Sec. 907.2.3, Exc. 1			
23.	a	Sec. 907.4.2.1			
24.	c	Table 907.5.2.3.3			
25.	d	Sec. 907.5.2.1.2			
26.	d	Sec. 904.11.1			
27.	d	Sec. 905.3.4			
28.	b	Sec. 909.20.3.3			
29.	d	Sec. 910.2.1			
30.	d	Table 910.3			

Study Session 9
2012 *International Building Code*

1.	b	Sec. 202	31.	d	Sec. 1004.4
2.	d	Sec. 202	32.	a	Sec. 1005.7.1
3.	c	Sec. 202	33.	b	Sec. 1006.2
4.	b	Sec. 202	34.	b	Sec. 1013.4, Exc. 1
5.	c	Sec. 202	35.	b	Sec. 1013.4, Exc. 6
6.	b	Table 1004.1.2			
7.	b	Table 1004.1.2			
8.	b	Sec. 1004.4			
9.	c	Sec. 1005.3.2, Exc.			
10.	c	Sec. 1005.3.1			
11.	d	Sec. 1005.5			
12.	d	Sec. 1005.7.1			
13.	b	Sec. 1003.3.1			
14.	b	Sec. 1003.3.1, Exc.			
15.	c	Sec. 1003.3.3			
16.	c	Sec. 1003.5, Exc. 1			
17.	c	Sec. 1011.1			
18.	b	Sec. 1011.6.2			
19.	b	Sec. 1006.3.1			
20.	b	Sec. 1013.2			
21.	c	Sec. 1013.3			
22.	d	Sec. 1013.4, Exc. 4			
23.	a	Sec. 1013.6			
24.	c	Sec. 1007.3			
25.	c	Sec. 1007.6.1			
26.	a	Sec. 1003.3.1			
27.	d	Sec. 1004.2			
28.	d	Sec. 1006.1			
29.	b	Sec. 1007.7.4			
30.	c	Sec. 1011.4			

Study Session 10
2012 *International Building Code*

1.	b	Sec. 1008.1.1	31.	c	Sec. 1008.1.10
2.	c	Sec. 1008.1.1	32.	c	Sec. 1009.4
3.	a	Sec. 1008.1.1, Exc. 3	33.	b	Sec. 1010.7.3, Exc. 2
4.	d	Sec. 1008.1.1.1	34.	c	Sec. 1012.6
5.	d	Sec. 1008.1.2, Excs. 1,9	35.	a	Sec. 1012.9
6.	c	Sec. 1008.1.2			
7.	b	Sec. 1008.1.3			
8.	b	Sec. 1008.1.4.1, #4			
9.	c	Sec. 1008.1.9.8, #3			
10.	b	Sec. 1008.1.5			
11.	c	Sec. 1008.1.7			
12.	a	Sec. 1008.1.9.7			
13.	c	Sec. 1008.1.9.2			
14.	c	Sec. 1009.3.1.2			
15.	c	Sec. 1009.5			
16.	b	Sec. 1009.7.2, Exc. 5			
17.	c	Sec. 1009.8			
18.	d	Sec. 1009.7.2			
19.	b	Sec. 1009.12			
20.	d	Sec. 1012.2			
21.	a	Sec. 1012.6			
22.	d	Sec. 1012.8			
23.	a	Sec. 1010.5			
24.	a	Sec. 1010.9			
25.	d	Sec. 1008.3.1			
26.	d	Sec. 1008.1.4.2			
27.	c	Sec. 1008.1.8			
28.	b	Sec. 1009.4, Exc. 1			
29.	a	Sec. 1012.3.1			
30.	a	Sec. 1010.10.1			

Study Session 11
2012 *International Building Code*

1.	d	Table 1015.1	31.	c	Sec. 1014.2, #4, Exc. 2.4
2.	d	Table 1015.1	32.	a	Sec. 1017.4
3.	c	Sec. 1015.1.1	33.	c	Sec. 1015.3
4.	b	Sec. 1015.2.1, Exc. 2	34.	d	Sec. 1017.4, Exc.
5.	c	Sec. 1014.2, #1	35.	d	Table 1018.2
6.	a	Sec. 1014.2, #4			
7.	a	Sec. 1014.2.1, Exc.			
8.	b	Sec. 1015.6			
9.	d	Sec.1015.1.1			
10.	b	Table 1016.2			
11.	c	Sec. 1016.2.1			
12.	a	Table 1014.3			
13.	c	Table 1014.3			
14.	b	Sec. 1017.3			
15.	b	Sec. 1017.3, Exc.			
16.	b	Sec. 1017.5			
17.	d	Table 1018.2			
18.	a	Table 1018.2			
19.	c	Table 1018.1			
20.	d	Table 1018.1			
21.	b	Sec. 1018.4			
22.	b	Sec. 1018.4, Exc. 2			
23.	b	Sec. 1018.5, Exc. 3			
24.	c	Sec. 1019.1, 1018.2			
25.	c	Sec. 1019.3			
26.	b	Table 1014.3			
27.	d	Sec. 1015.4			
28.	d	Sec. 1015.4			
29.	b	Table 1016.2			
30.	d	Sec. 1018.4, Exc. 3			

Study Session 12
2012 *International Building Code*

1.	b	Sec. 1021.2.4	31.	b	Sec. 1025.2
2.	d	Sec. 1021.2, Exc. 3	32.	a	Sec. 1028.10.2
3.	b	Table 1021.2(2)	33.	c	Sec. 1028.13.1
4.	b	Sec. 1022.2	34.	b	Sec. 1029.1, Exc. 3
5.	b	Sec. 1021.2.2	35.	a	Sec. 1029.2.1
6.	c	Sec. 1022.10.1			
7.	c	Sec. 1022.9			
8.	a	Sec. 1022.10			
9.	b	Sec. 1023.3			
10.	a	Sec. 1025.4			
11.	d	Sec. 1026.2			
12.	d	Sec. 1026.3			
13.	b	Sec. 1027.5, Exc., #1			
14.	b	Sec. 1027.4.1			
15.	b	Sec. 1028.10.1			
16.	b	Sec. 1028.2			
17.	c	Sec. 1028.5			
18.	d	Sec. 1028.6.1, #1, #2			
19.	d	Table 1028.6.2			
20.	a	Sec. 1028.6.2.1			
21.	d	Sec. 1028.7, Exc. 2			
22.	b	Sec. 1028.9.1, #2			
23.	b	Sec. 1028.11.1, Exc.			
24.	c	Sec. 1028.14.3			
25.	c	Sec. 1029.2, Exc.			
26.	d	Sec. 1025.1, Exc. 1			
27.	c	Sec. 1027.1, Exc. 2.2			
28.	b	Sec. 1027.5, Exc., #2			
29.	c	Sec. 1029.3			
30.	c	Sec. 1029.5.1			

INTERNATIONAL CODE COUNCIL®

People Helping People Build a Safer World™

Valuable Guides to Changes in the 2012 I-Codes®

NEW!

FULL COLOR! HUNDREDS OF PHOTOS AND ILLUSTRATIONS!

SIGNIFICANT CHANGES TO THE 2012 INTERNATIONAL CODES®

Practical resources that offer a comprehensive analysis of the critical changes made between the 2009 and 2012 editions of the codes. Authored by ICC code experts, these useful tools are "must-have" guides to the many important changes in the 2012 International Codes.

SIGNIFICANT CHANGES TO THE IBC, 2012 EDITION
#7024S12

SIGNIFICANT CHANGES TO THE IRC, 2012 EDITION
#7101S12

SIGNIFICANT CHANGES TO THE IFC, 2012 EDITION
#7404S12

SIGNIFICANT CHANGES TO THE IPC/IMC/IFGC, 2012 EDITION
#7202S12

Changes are identified then followed by in-depth discussion of how the change affects real world application. Photos, tables and illustrations further clarify application.

ORDER YOUR HELPFUL GUIDES TODAY!
1-800-786-4452 | **www.iccsafe.org/store**

HIRE ICC TO TEACH
Want your group to learn the Significant Changes to the I-Codes from an ICC expert instructor? Schedule a seminar today!
email: **ICCTraining@iccsafe.org** | phone: **1-888-422-7233 ext. 33818**

11-04865

People Helping People Build a Safer World™

Imagine... enjoying **membership benefits** that help you stay competitive in today's tight job market.

Imagine... increasing your code knowledge and sharpening your job skills with special member discounts on:

- World-class training and **educational programs** to help you keep your competitive edge while earning CEUs and LUs
- The latest **code-related news** to enhance your career potential
- **Member Discounts** on code books, CDs, specialized publications, and training materials to help you stay on top of the latest code changes and become more valuable to your organization.

PLUS – having exclusive access to **code opinions** from experts to answer your code-related questions

Imagine... extra benefits such as **Member-Only privileged access** to peers in online discussion groups
as well as access to the industry's leading periodical, the *Building Safety Journal Online*®.

Imagine... receiving these valuable discounts, benefits, and more by simply becoming a member of the nation's leading developer of building safety codes, the International Code Council®.

Imagine that!

Free code book(s) – New members receive a free book or set of books (depending on membership category) from the latest edition of the International Codes.

JOIN TODAY!
888-ICC-SAFE, x33804
member@iccsafe.org
www.iccsafe.org

11-04304

Most Widely Accepted and Trusted

ICC EVALUATION SERVICE

The leader in evaluating building products for code compliance & sustainable attributes

- Close to a century of experience in building product evaluation
- Most trusted and most widely accepted evaluation body
- In-house staff of licensed engineers
- Leader in innovative product evaluations
- Direct links to reports within code documents for easy access

Three signature programs:
1. ICC-ES Evaluation Report Program
2. Plumbing, Mechanical and Fuel Gas (PMG) Listing Program
3. Sustainable Attributes Verification and Evaluation (SAVE) Program

ICC-ES is a subsidiary of the International Code Council (ICC), the developer of the International Codes. Together, ICC-ES and ICC incorporate proven techniques and practices into codes and related products and services that foster safe and sustainable design and construction.

For more information, please contact us at es@icc-es.org or call 1.800.423.6587 (x42237).

www.icc-es.org

10-03910

Study Session 13
2012 *International Building Code*

1.	a	Sec. 202	31.	a	Sec. 1106.7.1
2.	a	Sec. 202	32.	c	Sec. 1108.2.4, Exc. 1
3.	d	Sec. 1103.2.5	33.	d	Sec. 1109.2, Exc. 3
4.	a	Sec. 1103.2.11	34.	c	Sec. 1109.5.2
5.	b	Sec. 1104.5	35.	b	Sec. 1109.15.4.1
6.	c	Sec. 1105.1			
7.	b	Table 1106.1			
8.	b	Sec. 1106.2			
9.	c	Sec. 1106.5			
10.	d	Sec. 1108.2.7.3			
11.	a	Sec. 1108.4.3.1			
12.	b	Sec. 1109.2.2			
13.	a	Sec. 1107.6.2.2.2			
14.	c	Table 1108.2.7.1			
15.	c	Sec. 1107.5.2.1			
16.	d	Table 1107.6.1.1			
17.	b	Table 1107.6.1.1			
18.	a	Sec. 1107.6.2.1.1			
19.	d	Table 1108.3			
20.	b	Sec. 1109.2.1			
21.	c	Sec. 1109.2.1.4			
22.	d	Sec. 1109.5			
23.	a	Sec. 1109.11			
24.	b	Table 1109.12.2			
25.	c	Sec. 1110.1, #1			
26.	c.	Sec. 1104.3.1, Exc. 1			
27.	a	Sec. 1104.3.2, Exc. 1			
28.	d	Sec. 1106.4			
29.	b	Sec. 1108.2.9.1			
30.	c	Sec. 1110.2			

Study Session 14
2012 *International Building Code*

1.	b	Sec. 402.8.5	31.	b	Sec. 406.3.2, #2
2.	c	Sec. 402.6.2, #3	32.	b	Sec. 202
3.	b	Sec. 202	33.	b	Sec. 412.4.1
4.	d	Sec. 403.4.8.1	34.	a	Sec. 420.2
5.	b	Sec. 404.8	35.	d	Sec. 421.6.1
6.	d	Sec. 404.9			
7.	b	Sec. 405.4.1			
8.	a	Sec. 406.4.3			
9.	b	Sec. 406.7.2			
10.	b	Table 406.5.4			
11.	d	Sec. 407.5			
12.	d	Sec. 407.4.1			
13.	c	Sec. 408.3.1			
14.	d	407.5.1, #1			
15.	b	Sec. 409.2			
16.	b	Sec. 410.3.4, 410.3.5			
17.	d	Sec. 410.3.7			
18.	a	Sec. 411.8			
19.	b	Table 412.4.6			
20.	c	Table 414.2.2			
21.	c	Sec. 415.5, Exc. 1			
22.	a	Sec. 422.2			
23.	a	Sec. 415.10.3.3			
24.	b	Sec. 417.2			
25.	b	Sec. 418.5			
26.	c	Sec. 410.5.1			
27.	d	Sec. 413.2			
28.	c	Sec. 414.2.4			
29.	b	Sec. 414.6.1.3			
30.	c	Sec. 416.2			

Study Session 15
2012 *International Building Code*

1.	b	Sec. 202	31.	b	Sec. 1405.3.2
2.	a	Sec. 202	32.	c	Sec. 1405.16.2
3.	a	Sec. 1404.2	33.	a	Sec. 1406.3, Exc. 1
4.	b	Table 1405.2	34.	d	Table 1504.8
5.	c	Sec. 1405.12	35.	b	Sec. 1810.3.10.2
6.	a	Sec. 1405.11			
7.	d	Sec. 1406.2.1, #2			
8.	b	Sec. 202			
9.	b	Table 1505.1			
10.	a	Sec. 1505.2, Exc. 2			
11.	c	Sec. 1507.2.8			
12.	d	Sec. 1507.2.8.1			
13.	c	Sec. 1507.3.6			
14.	b	Sec. 1507.5.2			
15.	d	Sec. 1507.9.8			
16.	d	Table 1507.8.7			
17.	b	Sec. 1507.14.1			
18.	c	Sec. 1509.2.1			
19.	c	Sec. 1809.3			
20.	b	Sec. 1809.4			
21.	d	Table 1806.2			
22.	b	Sec. 1808.7.3			
23.	d	Sec. 1808.8.4			
24.	b	Table 1809.7			
25.	b	Sec. 1807.1.6.3, #9			
26.	a	Sec. 1405.12.4			
27.	c	Sec. 1505.7			
28.	d	Sec. 1507.7.2			
29.	d	Sec. 1509.5			
30.	b	Sec. 1804.3			

Study Session 16
2012 *International Building Code*

1.	b	Sec. 1603.1.3	31.	c	Sec. 1607.5
2.	b	Table 1607.1, #24	32.	c	Sec. 1607.8.2
3.	d	Table 1604.5	33.	c	Sec. 202
4.	a	Sec. 1607.12.3.1	34.	c	Sec. 1705.2.2.2
5.	c	Sec. 1609.4.2	35.	b	Sec. 2103.7
6.	a	Sec. 1610.1			
7.	b	Sec. 1705.13.4.5			
8.	b	Sec. 1704.5.1, #3			
9.	b	Sec. 1906.1, Exc.			
10.	a	Sec. 1905.1.1 ACI 318, Sec. 2.2			
11.	c	Sec. 1910.3			
12.	a	Sec. 1907.1			
13.	b	Sec. 202			
14.	c	Sec. 2104.3			
15.	a	Sec. 2104.4			
16.	c	Sec. 2111.10			
17.	d	Sec. 2111.11, Exc. 4			
18.	c	Sec. 2113.9			
19.	a	Sec. 2304.11.2.5			
20.	d	Sec. 202			
21.	b	Table 2304.9.1, #27			
22.	c	Sec. 2308.3.1			
23.	a	Sec. 2308.8.3			
24.	a	Sec. 2308.9.11			
25.	b	Sec. 2308.10.5			
26.	d	Table 1607.1, #30, Note f			
27.	c	Table 1904.2			
28.	a	Table 2304.6			
29.	b	Sec. 2304.12, Exc.			
30.	d	Sec. 2308.9.4			

Study Session 17
2012 *International Building Code*

1.	b	Sec. 803.1.1	31.	c	Sec. 806.1
2.	c	Sec. 803.11.4	32.	a	Sec. 1203.2
3.	c	Table 803.9	33.	b	Sec. 1210.2.2
4.	a	Table 803.9	34.	a	Sec. 2508.5.4
5.	d	Table 803.9, Note i	35.	c	Sec. 2512.4
6.	a	Sec. 803.1.4			
7.	b	Sec. 804.4.2, Exc.			
8.	b	Sec. 806.5			
9.	d	Sec. 806.1.2, Exc. 1			
10.	c	Sec. 806.5			
11.	b	Sec. 1204.1			
12.	c	Sec. 1205.2			
13.	a	Sec. 1205.4			
14.	b	Sec. 1207.3			
15.	c	Sec. 1208.2			
16.	b	Sec. 1208.3			
17.	a	Sec. 1209.1			
18.	c	Sec. 1209.2			
19.	b	Sec. 1210.3.2			
20.	d	Sec. 202			
21.	b	Sec. 2509.3, #3			
22.	c	Sec. 2512.5			
23.	d	Sec. 3002.2			
24.	d	Sec. 3004.2			
25.	b	Sec. 3002.4			
26.	a	Sec. 804.4.1			
27.	c	Sec. 1203.3.1			
28.	c	Sec. 1206.3			
29.	a	Sec. 2508.4			
30.	d	Sec. 3003.1.2			

Study Session 18
2012 *International Building Code*

1.	c	Sec. 2403.5	31.	d	Sec. 2405.3, Exc. 3
2.	c	Sec. 2405.3	32.	d	Table 2406.2(1)
3.	b	Sec. 2405.3	33.	a	Sec. 2407.1
4.	d	Sec. 2405.4	34.	c	Sec. 2606.7.3
5.	d	Table 2406.2(1)	35.	d	Sec. 2610.4
6.	a	Sec. 2406.3, Exc. 1			
7.	a	Sec. 2406.4.5			
8.	c	Sec. 2406.4.2			
9.	d	Sec. 2406.4.3, Exc. 2			
10.	a	Sec. 2406.4.5, Exc.			
11.	d	Sec. 2406.4.6, Exc. 2			
12.	a	Sec. 2406.4.1, Exc. 1			
13.	b	Sec. 2406.5			
14.	b	Sec. 2407.1.2			
15.	d	Sec. 202			
16.	d	Sec. 202			
17.	d	Sec. 2603.3			
18.	a	Sec. 2603.4			
19.	b	Sec. 2603.4.1.1			
20.	d	Sec. 2604.2.2			
21.	c	Sec. 2605.2, #2			
22.	b	Table 2607.4			
23.	d	Sec. 2608.2, #3, Exc.			
24.	a	Sec. 2609.1			
25.	b	Sec. 2611.3			
26.	a	Sec. 2404.1			
27.	b	Sec. 2406.3, Exc. 2			
28.	c	Sec. 2406.3.1			
29.	c	Sec. 2408.3			
30.	d	Sec. 2603.4.1.13			